Technical Code for Monitoring of Building Structures

Yang Yang

Technical Code
for Monitoring of Building
Structures

 Springer

Yang Yang
School of Civil Engineering
Chongqing University
Chongqing, China

Translated by
Yang Yang
Chongqing University
China

Ruoyu Jin
University of Brighton
UK

Shaohong Yang
Hangzhou Tongda Group Co., Ltd.
China

Rui Sun
Chongqing University
China

Yao Zhang
Nanyang Technological University
Singapore

Yuner Huang
University of Edinburgh
UK

Tao Chen
Chongqing University
China

ISBN 978-981-15-1048-9   ISBN 978-981-15-1049-6   (eBook)
https://doi.org/10.1007/978-981-15-1049-6

Taking the local standards for construction "Technique Code of Building Structural Monitoring in Qinghai Province (DB 63/T1756-2019" as an example.

This Springer imprint is published by the registered company Springer Nature Singapore Pte Ltd.
The registered company address is: 152 Beach Road, #21-01/04 Gateway East, Singapore 189721, Singapore

Preface

According to the requirement of Document QJK[2019] No. 327—the Department of Housing and Urban-Rural Development of Qinghai Province, the code is prepared by the Chongqing University together with the relevant institutions of higher learning and scientific research, design, enterprises, etc.

Qinghai Province not only has the characteristics of highland areas and general areas but also contains almost all types of building structures. Although the code title is only for Qinghai Province, the content involved covers all aspects of building structure monitoring technology, and is also applicable to similar structural types in other regions.

The code is the first English translation standard in the field of engineering structural monitoring technology in China. It is helpful for experts and scholars at home and abroad who are engaged in engineering structure monitoring to understand the status quo and corresponding standards of monitoring technology in Qinghai and China.

It is noticed that the code is written in Chinese and English. The Chinese text shall be taken as the ruling one in the event of any inconsistency between the Chinese text and the English text.

If you have any questions, please don't hesitate to contact the authors (Email: yangyangcqu@cqu.edu.cn).

Chongqing, China   Yang Yang
September 2019

Local Standards for Construction of Qinghai Province

(DB 63/T1756-2019)

Technique code of building structural monitoring in Qinghai Province

2019-08-19 Published

2019-11-01 Implement

Department of Housing and Urban-Rural Development of Qinghai Province
and
Administration for Market Regulation of Qinghai Province
Joint Release.

Department of Housing and Urban-Rural Development of Qinghai Province and Administration for Market Regulation of Qinghai Province

Document QJK[2019] No. 327

Notice of Promulgation for the local standards of Construction "Technique Code of Building Structural Monitoring in Qinghai Province"

Urban and Rural Construction Bureau of Xining Municipality, Urban and Rural Construction Bureau of Haidong City and prefectures of Qinghai, and all relevant organizations:

The local standards for construction "Technique Code of Building Structural Monitoring in Qinghai Province (DB 63/T1756-2019)", edited by Chongqing University, etc., have been reviewed and approved by related experts, and are now approved for publication. It will be officially implemented on November 1, 2019. This standard is managed by the Department of Housing and Urban-Rural Development of Qinghai Province, and the authorized chief editor is responsible for interpretation.

August 19, 2019

Abstract

This Technical Code is established aiming to provide the standard for the life cycle monitoring of building structures during the construction and operation stages. It targets on the implementation and standardization of health monitoring in Qinghai Province. The Technical Committee has conducted extensive investigation and research, summarized the engineering practice experience, referred to the relevant national standards, and finally formulated this Technical Code.

The main contents of this Technical Code include: 1. General; 2. Nomenclature and Symbols; 3. Basic Regulations; 4. Hardware Installation; 5. Software System Setup; 6. High-Rise Structure; 7. Long-Span Structure; 8. Heritage Asset Structure; 9. Other structures; 10. Pre-warning threshold; 11. Engineering Inspection and Approval; 12. Appendix.

This Technical Code is managed by the Housing Construction Department of Qinghai Province, and the technical contents are specified by Chongqing University. During the implementation of this Code, it is advised that all enterprises should follow up the Code and send feedback to the Technical Committee of Code of Practice and Acceptance of Building Structure Monitoring Technology in Qinghai Province, with the post address at: The 20th Floor, Shengli Road, Xining City, Qinghai Province. (zip code: 810000, Email: yangyangcqu@cqu.edu.cn).

The chief editor, the co-editor, the main drafter and the review panel of this Code are listed below:

Chief editing institute:

Chongqing University
China Construction Third Bureau Group Co., Ltd.

Co-editing institute:

Hangzhou Tongda Group Co., Ltd.
Haidong City Development and Construction Investment Co., Ltd.
China Architecture Design and Research Institute Co., Ltd.
National Power Investment Group Qinghai Huanghe Power Technology Co., Ltd.
Changan University

Zhejiang Boyuan Electronic Technology Co., Ltd.
Sanxia University
Zhongguan Engineering Inspection and Testing Co., Ltd.
Tianjin Yixing Technology Co., Ltd.

Main Drafters:

Yang Yang, Zhen Wang, Shaohong Yang, Jiqiang Li, Rui Sun
Yiping Cao, Minjie Luo, Dechi Li, Zhigang Wu, Jun Wang
Fajiang Luo, Peiqi Zhang, Peng Li, Lin Zhou, Jin Zeng
Shengyuan Zhang, Yong Jia, Gangjie Wu, Zhihui Luo, Shuqiang Jia
Hongguang Piao, Jiachun Li, Xichen Yan, Lilei Wang, Qinglei Chi
Zhihui Shang, Hongyu Wang, Bowen Yan, Chenghong Shen, Jianping Wang

Review Experts:

Hongtao Pan, Shaowei Hu, Shuquan Zhao, Hua Cheng, Yan Feng
Lingguo Kang, Lianxin Liu, Gulan Lu, Hai Li, Jiye Zhang
Yanjun Zhang

Contents

Chapter 1
General Provisions

1.0.1 This code is established to standardize the life cycle monitoring during the construction and operation stages of the building structures in Qinghai Province, to specify the implementation of safety monitoring technology, and to ensure the safe and economic application of the advanced monitoring technology.

1.0.2 This code applies to the monitoring of a variety of building sectors, including new construction, renovation, and retrofitting of industrial buildings, high-rise and large-span spatial structures, protective buildings, endangered houses, separated (subtractive) buildings, and industrial buildings in Qinghai Province. The monitoring of other types of building structures may also refer to this code.

1.0.3 The implementation of the life cycle monitoring during construction and operation stages of the building structure should also comply with other relevant national standards.

© The Author(s), under exclusive license to Springer Nature Singapore Pte Ltd. 2020
Y. Yang, *Technical Code for Monitoring of Building Structures*,
https://doi.org/10.1007/978-981-15-1049-6_1

Chapter 2
Terms and Symbols

2.1 Terms

2.1.1 Construction monitoring

During the construction stage of the building project, a series of monitoring work is conducted on structural stress, strain, deformation, and surrounding environment to ensure the construction quality and safety.

2.1.2 Post construction monitoring

Continuous monitoring is performed over a long period of time after the project is delivered in order to collect information on structural safety status and trends of changes.

2.1.3 Regular monitoring

The same test is conducted to the structure after a certain period of time to collect the information on the development trend of a certain performance of the structure.

2.1.4 Online monitoring

Continuous automatic monitoring is carried out through various types of monitoring instruments installed on the structure and/or equipment, with the information uploaded to the terminal receiving end to allow a comprehensive evaluation of the development trend of a certain performance of the structure.

2.1.5 Sensor

A device, consisting of a sensitive component and a conversion component and capable of sensing a specified measurement and converting the information into a usable signal according to certain rules.

© The Author(s), under exclusive license to Springer Nature Singapore Pte Ltd. 2020
Y. Yang, *Technical Code for Monitoring of Building Structures*,
https://doi.org/10.1007/978-981-15-1049-6_2

2.1.6 Monitoring point

A monitoring point that directly or indirectly set on the monitored object to reflect its changing pattern.

2.1.7 Frequency of monitoring

The number of monitoring times per unit time.

2.1.8 Wire transmission

A technology to transmit signals from one party to another through a physical connection between two communication devices.

2.1.9 Wireless transmission

The technology of transmitting signals from one party to another without using any physical connection between two communication devices.

2.1.10 Structural performance

The state of the structure in terms of safety, suitability, and durability.

2.1.11 Early warning threshold

The warning value describing monitoring parameters of structural elements during the construction and operation stages due to the different degrees of abnormality or danger of the building structure.

2.1.12 Damage identification

The process of analyzing changes in structural modal parameters by using structural response data and/or physical parameters to identify structural damage.

2.1.13 Modal analysis

The process of obtaining the structural modal characteristics by measuring the structural dynamic response information.

2.1.14 Database

An organized and shareable collection of data that is stored in a computer system for the long term.

2.1.15 Cloud computing

The computation program that is broken down into countless smaller subroutines, and further the huge system composed of multiple servers being calculated and analyzed, until finally the processing result being transmitted back to the user.

2.1.16 Cloud storage

A storage system that uses cluster applications, grid technologies, or distributed file systems, through software tools to bring together a large number of different types of storage devices in the network to provide an external data storage and service access.

2.1.17 Cloud platform

A structural monitoring platform which relies on cloud computing technology; provides unified cloud host, cloud storage, core database, and other resources; provides computing, storage, network services, and data for various application systems, as well as offers processing, network information security, and system management services.

2.1.18 Structure group monitoring system

Based on the regional integrated management system established by the Big Data concept, the monitored information of multiple structures in a certain area being transmitted to the structural health monitoring system in real time for unified analysis and management, and with further feedback to the mobile terminals of the operation managers at all levels.

2.2 Symbols

MAC_{ij}	The (i, j)th element of the modal confidence criterion matrix
$COMAC$	Coordinate modal confidence criterion
$GNSS$	Global navigation satellite system
d	Day
ε	Strain
σ	Stress
H	Height
L	Length or span
T	Tension

Chapter 3
Basic Requirements

3.1 General Requirements

3.1.1 Structural monitoring should be divided into construction monitoring and post construction monitoring.

3.1.2 Monitoring scheme should be based on factors such as structural characteristics, site, and surrounding environmental conditions.

3.1.3 Engineering monitoring should include the following contents:

1. Collection and analysis of relevant information and site investigation;
2. Preparation and review of the monitoring scheme;
3. Arrangement, evaluation, and protection of monitoring point;
4. Specifying equipment and calibrate component, and measurement of the initial value of monitoring points;
5. Collection of monitoring data;
6. Processing, analysis, and feedback on monitoring information;
7. Submission of daily monitoring report, early warning report, and stage monitoring report;
8. Regular submission of monitoring work summary and data.

3.1.4 The content of the monitoring report shall comply with the following provisions:

1. Monitoring reports should include monitoring time, monitoring personnel, monitoring targets, monitoring content, monitoring methods, monitoring data, and monitoring conclusions;
2. The monitoring conclusion during construction should indicate whether the results of the monitoring parameters during the construction period are reasonable;
3. Timely reaction is required when the monitoring parameters meet the early warning requirements during construction;

© The Author(s), under exclusive license to Springer Nature Singapore Pte Ltd. 2020
Y. Yang, *Technical Code for Monitoring of Building Structures*,
https://doi.org/10.1007/978-981-15-1049-6_3

4. The monitoring conclusion during operation should indicate whether there is damage in the structural part monitored, and also the type of damage;
5. The structural damage indicated by the monitoring conclusion during operation shall be promptly warned with specified reactions that are needed.

3.1.5 The monitoring of building structure should be based on the monitoring and warning threshold. The monitoring and warning threshold should meet the engineering design and the control requirements of the monitored structure.

3.2 Monitoring Scheme and Requirements

3.2.1 The monitoring of building structure should adopt the comprehensive monitoring method combining instrument measurement, site inspection, modeling analysis, video monitoring, and other various means to collect monitoring information. It is also necessary to carry out real-time monitoring for surrounding structures and key engineering components that are at high safety risks.

3.2.2 The frequency and period of monitoring information collection shall be determined comprehensively by the construction progress, characteristics of monitored objects, surrounding environment, and natural conditions.

3.2.3 The monitoring information shall be processed, analyzed, and sent for feedback on time. A report shall be sent on time when an abnormal situation that jeopardizes the safety of the project is found.

3.2.4 The emergency rescue monitoring aiming to monitor the sudden risk of the structure should increase the monitoring points and the monitoring frequency based on the original monitoring, and should implement real-time monitor if necessary.

3.2.5 The engineering monitoring scheme should include the following contents:

1. Project overview;
2. Geological and surrounding environmental conditions on construction sites;
3. Aims and rationale of monitoring;
4. Contents of monitoring;
5. Reference point, monitoring point arrangement and protection requirements, monitoring point layout;
6. Monitoring method and accuracy;
7. Monitoring period and frequency;
8. Monitoring control values, warning levels, early warning standards, and reactions under abnormal conditions;
9. Monitoring information recording system and processing methods;
10. Feedback system of monitoring;

11. Monitoring equipment, components, and personnel;
12. Quality management, safety management, and other management systems.

3.2.6 The following items during construction and operation stages of the structure shall be monitored based on structural characteristics, monitoring requirements, and construction and operation environment:

1. Environmental monitoring;
2. Foundation settlement monitoring;
3. Deformation monitoring;
4. Structural vibration monitoring;
5. Stress–strain monitoring;
6. Structural crack monitoring;
7. Corrosion monitoring;
8. Cable force monitoring;
9. Fire accident monitoring;
10. Fatigue monitoring.

3.3 Monitoring System

3.3.1 The online monitoring system should include the following modules: sensor module, data acquisition and management module, data analysis and evaluation module, security early warning module, and integrating each module into a unified and coordinated system.

3.3.2 When it comes to construction and regular monitoring without online monitoring system, it is advisable to develop a monitoring plan based on actual needs and to select on-site monitoring equipment.

3.3.3 The selection of monitoring equipment should meet the following requirements:

1. The sensor should have good stability and anti-interference ability during monitoring, and the signal-to-noise ratio of the acquired signal should meet the actual engineering requirements;
2. It is advisable to select a sensor with compensation functions;
3. It is advisable to establish a mechanical model of the structure and to determine the type of monitoring equipment based on the analysis results, empirical judgment, the engineering properties of sensors, as well as the performance parameters and price of the sensor product;
4. It is advisable to determine the necessary and reasonable monitoring position, quantity, and installation method based on the results of on-site investigation and structural analysis;

5. It is advisable to have the ability to maintain normal work in high temperature, extremely cold and low temperature conditions.

3.3.4 The technical indicators of monitoring equipment shall be implemented in accordance with Appendix A.

3.4 Construction Monitoring

3.4.1 In addition to the special requirements of the design documents, structural analysis and structural monitoring should be carried out for the following building structures:

1. High-rise buildings constructed on a deep foundation with landfilling thickness greater than 5 m;
2. High-rise buildings with a composite foundation constructed on a Class IV self-weight collapsible site;
3. High-rise buildings whose height is more than 80 m and with a raft foundation and a supporting layer made of the Neogene, Paleogene, or strongly weathered mud-stone (not more than 1 m);
4. A large-spanned structure with a single span not less than 60 m, without full-foot support, and using the uniform unloading construction method;
5. The construction of a special structure falling into one of the following conditions: the cantilever and the outward inclined overhang span not less than 20 m; the structure inclined at an angle of not less than 15°; the aerial joint structure span not less than 30 m; the number of upper support floors not fewer than two; and the conversion structure span greater than 30 m;
6. During the construction process, there is a complicated loading condition in the whole or part of the structure. It is necessary to accurately understand the main or partial structure, or the large-sized temporary support structure under stress and deformation, including:

   (1) Large temporary support added during the construction process and to be removed after the main structure is firmly established;
   (2) Overall improvement of the structure with a span of not less than 20 m;
   (3) The cantilever structure with the outward inclined overhang span not less than 15 m, and when step extension construction method is adopted;
   (4) Large-spanned structure spanning over 90 m adopting an overall unloading scheme;
   (5) Two or more single structures separately constructed, but spatially connected in the later stage;
   (6) Other building structures that adopt special construction techniques and have certain risks during construction;

7. Flexible structure requiring high safe and firmed-shaped cable;
8. Other building structures where the actual overall or local fore within the structure is significantly different from the results of one-piece structural loading analysis.

3.4.2 It is advisable to perform monitoring during construction for the following important components and nodes:

1. Components with a significant stress change or a high stress level during construction;
2. Components or nodes that are significantly deformed during construction;
3. Components or nodes subject to large construction loads during construction, and the loadings on them not considered in the original design;
4. Critical structural nodes that control geometrical configuration during construction;
5. Other typical important components or nodes.

3.4.3 Construction monitoring should include the following contents:

1. Record of component installation process during construction;
2. Distribution, variation and value of construction personnel, construction machinery, or temporary loading;
3. The weight, the way of support, the timing of installation, and removal of the formwork during the construction process;
4. Variation record of connections between components;
5. Environmental records of where the building is located;
6. Strength test record of concrete under the same curing condition;
7. All these recorded data involving interior decoration, equipment installation, and curtain wall installation records should meet the relevant loading requirements;
8. Other relevant records required for structural analysis during the construction process.

3.4.4 The structural analysis and construction monitoring shall be prepared with special schemes and submitted to the relevant authority for approval. Analysis and monitoring schemes of major engineering structures shall undergo expert evaluation.

3.4.5 The structural analysis of the construction process should be conducted in accordance to the construction scheme, and a reasonable analysis model should be established to accurately reflect the changing structural stiffness caused by the construction. It is advised to apply the load and effects that are consistent with the real construction conditions, in order to calculate the internal force and deformation of the structure.

3.4.6 The structural analysis of the construction process should be carried out in conjunction with the construction scheme and construction organization scheme before the start of construction. When the construction scheme is adjusted, or when there is a deviation between the actual construction process and the construction scheme, it is advisable to adjust the structural analysis model in accordance with the construction process.

3.4.7 When conducting on-site monitoring, the analysis results of the construction process should be compared with the on-site monitoring results. When there is a significant difference between them, the potential causes should be analyzed and identified timely, and the model should be revised and recalculated if necessary.

3.4.8 Green construction of the project construction project should regularly monitor air pollution, water and soil pollution, noise pollution, light pollution and construction waste pollution during construction.

3.5   Post Construction Monitoring

3.5.1 In addition to the special requirements from relevant design documents, it is advisable to conduct post construction monitoring for the following building structures:

1. High-rise structures, and large-span structures should be determined by the seismic fortification intensity levels of different regions and local geological conditions;
2. Major engineering buildings built on permafrost in high altitude areas;
3. Buildings recognized as cultural heritages;
4. Endangered houses identified as Class C or Class D and temporarily unable to remove;
5. Industrial factories that affect production or operations;
6. High-rise seismically isolated structure with height exceeding 60 m or aspect ratio greater than 4;
7. Prefabricated structures with key nodes calculated and analyzed with large stress or deformation.

3.5.2 Post structural analysis should meet the following requirements:

1. Structural analysis models and basic assumptions should be consistent with the actual state of the structure;
2. Structural analysis should consider permanent and variable loads;
3. Consider temperature effect, foundation settlement, and wind load according to actual needs;
4. Consider the influence of structural reinforcement on model parameters;
5. The numerical model is corrected by using the monitoring parameters, and the correction accuracy is verified in subsequent periods.

3.5.3 Post construction structural damage identification shall comply with the following provisions:

1. It is advisable to use monitor data and structural analysis models for damage identification in a comprehensive approach;
2. It is advised to follow the steps consisting of damage observation, damage location identification, damage quantification, and damage assessment;
3. The damage observation should lead to a clear judgment of whether the structure is damaged, and an explanation of the corresponding judgment criteria or threshold;
4. Damage location identification should specify the location of the damaged unit or component;
5. The damage quantification should specify the degree of damage that occurs in the unit or component in which the damage occurred;
6. Damage assessment should comprehensively evaluate the performance degradation after structural damage, and predict the remaining life after structural damage.

3.5.4 The green construction of the project construction project shall be regularly monitored for environmental quality during use.

Chapter 4
Hardware Implementation

4.1 General Requirements

4.1.1 The selection of hardware devices should meet the following requirements:

1. It should have good stability to adapt different working environments;
2. Equipment life should meet the requirements during the monitoring period;
3. The signal-to-noise ratio should meet the requirements of national standards;
4. The resolution ratio of the device should not be lower than the minimum unit level of the required monitoring parameters;
5. The accuracy and measurement range of the hardware equipment should meet the requirements of the monitoring system;
6. Under the premise of meeting the monitoring requirements, it is advisable to adjust the accuracy or quantity of the equipment based on the cost budget.

4.1.2 The layout of the hardware equipment should meet the following requirements:

1. Monitoring points of sensor should be sensitive enough to external environmental changes and structural changes;
2. The overall monitoring layout should be determined based on structural vibration, deformation characteristics, and modal parameter identification.

4.1.3 Installed hardware should be protected to prevent vandalism and environmental corrosion.

4.1.4 The monitoring system should be equipped with emergency power.

© The Author(s), under exclusive license to Springer Nature Singapore Pte Ltd. 2020
Y. Yang, *Technical Code for Monitoring of Building Structures*,
https://doi.org/10.1007/978-981-15-1049-6_4

4.2 Hardware Installation

I Displacement Monitoring Equipment

4.2.1 The installation of displacement monitoring equipment should meet the following requirements:

1. Mechanical instruments, electrical instruments, or optical instruments can be used for displacement monitoring;
2. The three-dimensional displacement monitoring can be synthesized by monitoring the vertical displacement of the measuring points and the horizontal displacement of the mutually perpendicular planes;
3. The measuring instrument range should be 3–6 times of the estimated displacement of the measuring point;
4. The accuracy of the mechanical instrument should be less than 1/10 of the estimated displacement of the measuring point.

4.2.2 The installation of GNSS displacement measuring equipment should meet the following requirements:

1. When the monitoring pier is a concrete structure, the construction quality should meet the requirements of the "Code for Construction Quality Acceptance of Concrete Structures" GB 50204; when the monitoring pier is a steel structure, the construction quality should meet the "Code for Acceptance of Construction Quality of Steel Structures" GB 50205;
2. The tilting of the forced plummet at the top of the monitoring pier in any two vertical directions should be less than 1°;
3. The occlusion around the GNSS antenna should meet the requirements specified in the design documents. When the requirements are not available or specified in the design documents, there should be no obstruction within 2 m surrounding the antenna;
4. The protective cover of the GNSS antenna should be tightly installed to prevent rainwater from penetrating;
5. The GNSS antenna shall be located within the protection range of the lightning rod. If there is no external lightning rod, a specialized lightning rod shall be installed;
6. The ground wire of the GNSS antenna cannot be connected to the grounding conductor of another hardware.

4.2.3 The installation of the photoelectric measurement equipment should meet the following requirements:

1. The measurement instrument should be installed on a structure with a small foundation settlement. When the settlement amount is greater than the system measurement precision, the measured value should be corrected;
2. The target should be installed vertically and perpendicular to the structure to be tested, and the angular deviation should be less than ±1°;

3. When the axis of the measuring instrument is not perpendicular to the target surface, effective measures should be taken to eliminate the measurement deviation;
4. The target image should be clear;
5. Light should not be blocked when the target moves within the measurement range;
6. The mounting base of the gauge and the target should be absolutely stable (Fig. 4.2.3).

II Stress and Strain Monitoring Equipment

4.2.4 The installation of stress and strain monitoring equipment should meet the following requirements:

1. The distance between the installation position of the embedded stress–strain sensor and the design monitoring point should be less than 30 mm, and the angular deviation should be less than $\pm 1°$;
2. The distance between the mounting position of the surface-mounted stress–strain sensor and the design monitoring point should be less than 20 mm, and the angular deviation should be less than $\pm 0.5°$.

4.2.5 The installation of the embedded stress and strain sensor should meet the following requirements:

1. When the sensor is directly buried, it should be firmly tied with the structural steel bar and the auxiliary steel bar. During the concrete pouring process, the vibrator is not allowed to directly touch the sensor;
2. When the sensor is per-cast in the precast block, the preparation material of the precast concrete block should be the same as the concrete at the embedding point of the sensor, and the coarse aggregate larger than 80 mm should be removed. In the sensor test direction, the buried depth of the top surface of the precast block should be greater than 100 mm;
3. For sensors buried in the concrete monitoring hole, the buried depth of the sensor should be greater than 100 mm.

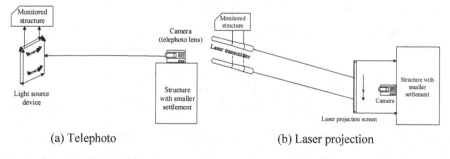

(a) Telephoto   (b) Laser projection

Fig. 4.2.3 Photoelectric target test schematic

III Acceleration and Speed Monitoring Equipment

4.2.6 Acceleration and speed monitoring equipment installation should meet the following requirements:

1. The angular deviation between the installation direction and the design direction should be less than 0.5°;
2. A rigid connection should be used between the monitoring equipment and the test member;
3. When a single bolt is used to fix the acceleration and speed sensors, effective measures should be taken to prevent the rotation;
4. The impedance between the monitoring device and the ground should be less than 1 Ω.

IV Earthquake Monitoring Equipment

4.2.7 The earthquake monitoring equipment should be installed on a special base. The surface of the base should be smooth and flat. The bolts should be pre-embedded in the middle of the base. The ground vibration sensor should be fixed by bolts and epoxy.

4.2.8 The deviation between the installation direction of the ground motion monitoring equipment and the design direction should be less than 0.5°.

V Temperature and Humidity Monitoring Equipment

4.2.9 The installation of temperature and humidity monitoring equipment should meet the following requirements:

1. The temperature sensor should not be exposed to solar radiation and should be located 1.5 m above the ground;
2. The humidity sensor should not be installed in a dead space where the air does not circulate;
3. The humidity sensor should be equipped with a breathable dust cover.

VI Wind Speed and Direction Monitoring Equipment

4.2.10 The bracket of the wind speed and direction monitoring device shall be fixed to the main structure and shall have sufficient rigidity.

4.2.11 Wind speed and direction sensor installation should meet the following requirements:

1. The angle deviation between the north standard direction of the wind direction sensor and the truth north should be less than 0.5°;
2. The installation distance of mechanical wind speed and wind direction sensor should be greater than 1 m;
3. The axis of the wind speed and direction sensor should be vertical.

4.2.12 A lightning rod should be installed at the top of the bracket, and a red copper wire should be used as the lower lead.

VII Crack Gauge

4.2.13 The installation accuracy of the crack gauge should meet the following requirements:

1. The axis of the crack gauge should be perpendicular to the crack to be tested, and the deviation of the perpendicularity should not be more than 1°;
2. The axis of the crack gauge should be parallel to the surface where the crack is located, and the parallelism deviation should be no more than 1°.

4.3 Cable Layout

4.3.1 The power cable/signal cable should be laid in accordance with the following requirements:

1. It should not be laid in areas with corrosive substances, strong magnetic or strong electrically interfered fields. When it is unavoidable, protective or shielding measures should be taken;
2. There should be a margin at the terminal wiring of the line, as well as through the expansion joint and the settlement joint;
3. Cables should not have intermediate joints. If they are unavoidable, they should be wired in the junction box or cable box. The joints should be crimped. When adopting the welding approach, non-corrosive flux should be used. The compensation wires should be crimped.

4.3.2 When entering the cabinet or box that is exposed to rain and direct sunlight, the power cord and signal cable should start from the bottom and should be waterproof.

4.3.3 Power cables and signal cables should be protected according to the design documents before entering the bridge; when the design documents are not specified, soft corrugated pipes should be used for protection.

4.4 Lightning Protection and Grounding Design

4.4.1 The installation of lightning protection and grounding devices shall comply with the following requirements:

1. The lightning protection device should be installed firmly and the wiring should be reliable. The position and sequence of installing lightning protection devices should meet the requirements of the design documents and product specifications;

2. When the monitored structure has a grounding device, the grounding of the monitoring system should be firmly welded with anti-corrosion methods adopted;
3. When the monitored structure has no grounding device, the buried position and depth of the grounding body should meet the requirements of the design documents;
4. When the grounding resistance value cannot meet the requirements of the design documents, physical or chemical resistance reduction measures should be taken.

4.4.2 The installation of the grounding wire should meet the following requirements:

1. The grounding wire must not have mechanical damage;
2. Ground terminal should be clearly marked.

4.5   Identification

4.5.1 Sensor identification layout should meet the following requirements:

1. The surface should be marked with a sign containing the type, model, quantity, and number of sensors;
2. The ID number of the leading line and channel number of the corresponding acquisition device should be marked.

4.5.2 The equipment and wiring posts inside the cabinet shall be with uniform and clear signs.

4.5.3 The signal line should be identified in the following parts:

1. Sensor connection;
2. Bridge access and terminal;
3. Access to acquisition devices.

4.5.4 The power cord should be identified in the following locations:

1. Connection to the main power supply;
2. Bridge access and terminal;
3. Access to the power equipment.

Chapter 5
System Integration

5.1 General Requirements

5.1.1 This chapter applies to structural monitoring with software systems in both construction and operation stages.

5.1.2 Software systems should have local configuration and management capabilities.

5.1.3 The software system should have an automatic recovery function to restore normal working conditions from failures.

5.1.4 Software database should be modular featured, and it should manage the structural information, monitoring system, and monitoring data in a layer-featured classified way.

5.2 System Design and Development

5.2.1 The software should be modular in design, and the functions of each module should be independent.

5.2.2 Documents that should be formed during software design and development include:

1. Software requirements specification;
2. Software interface specification;
3. Software design specification;
4. Database design specification.

© The Author(s), under exclusive license to Springer Nature Singapore Pte Ltd. 2020
Y. Yang, *Technical Code for Monitoring of Building Structures*,
https://doi.org/10.1007/978-981-15-1049-6_5

5.2.3 For the building structure with BIM model established during construction, it is advisable to combine the BIM model with the monitoring program to build a data sharing, 3D visualized, and intelligent monitoring system to achieve the full life cycle monitoring of the building structure during construction and operation.

5.3   System Requirements

5.3.1 The software system should have the function of remote control and diagnosis of hardware devices.

5.3.2 The software system should have signal filtering, compression, noise reduction, and classification of data, and data preprocessing functions.

5.3.3 The software system should have the following data management functions:

1. Supporting the classification and storage of monitoring and analysis data according to their type and time;
2. Supporting distributed data management, including distributed data storage, replication, and access;
3. Automatic generation of a variety of data reports.

5.3.4 The software system should be able to perform the following analysis functions on the monitoring data:

1. For structural modal parameter identification function, the following methods can be adopted:
   (1) Frequency domain identification method, that is, power spectrum density function, peak-picking method, frequency domain decomposition, enhanced frequency domain decomposition, least squares complex frequency domain method;
   (2) Time domain identification method, that is, ambient excitation technology (NExT), Eigen system realization algorithm (ERA), stochastic subspace identification (SSI), auto regressive moving average (ARMA);
   (3) Time–frequency domain identification method, that is, wavelet analysis, Hilbert-Huang transform.

2. As for structural damage identification function, the following methods can be used:
   (1) Static parameter method involving structural unit stiffness, strain, residual stress, elastic modulus, elementary area, and moment of inertia;
   (2) Dynamic parameter method, that is, natural frequency, natural mode, mode curvature, modal assurance criteria (MAC), z-coordinate modal assurance criteria (COMAC), strain mode shape, flexibility curvature, modal strain energy, direct stiffness method;

(3) Finite-element model updating method, that is, matrix correction method, elemental correction method, error factor correction method, design parameter correction method, and so on.
(4) Pattern recognition, that is, artificial neural networks, deep learning, fuzzy theory, genetic algorithm, particle swarm optimization algorithm.

3. As for safety assessment function, the following methods can be used:
   (1) Deterministic method, that is, analytic hierarchy process, limit analysis;
   (2) Reliability analysis method, that is, member reliability analysis method, system reliability analysis method.

5.3.5 When structural damage is detected, the software system should correctly judge the structure's security warning level (i.e., yellow warning, orange warning, or red warning), and convey the real-time warning information through voice, report, SMS, and email.

5.4 System Validation and Development

5.4.1 A detailed software testing scheme should be developed, and the hardware and network environment of the software testing should be built before the test work can be carried out.

5.4.2 Data acquisition, transmission, and storage software should be tested for software strength, including:

1. Collection task of data from more than 30% of design sensors;
2. Synchronous acquisition capability of all sensors under extreme conditions;
3. Saturation test of data transmission;
4. Data storage capacity beyond the specified amount of storage.

5.4.3 The hardware and network environment test report should be provided to the third-party software testing organization along with the software.

5.4.4 Software deployment cannot be performed before the hardware and network configuration are tested on meeting the design requirements.

Chapter 6
High-Rise Building Structure

6.1 General Requirements

6.1.1 In addition to the design documental requirements or other high-rise structures that are required to be monitored during construction, the high-rise structure shall be monitored during construction when one of the following conditions is met:

1. High-rise structure with a large temporary support structure;
2. High-rise structure with complex or integral structural forces during construction;
3. The structure of extra-long members and special sections that are affected by environmental factors such as temperature difference between day and night;
4. High-rise structure whose construction scheme has a significant influence on the internal force distribution of the structure;
5. High-rise structure with strict requirements for settlement and configuration;
6. High-rise structure affected by surrounding construction work.

6.1.2 In addition to the design documental requirements or other provisions that should be monitored during the operation stage of high-rise structures, high-rise buildings that meet the heights specified in Table 6.1.2 should be monitored during the operation stage.

6.1.3 The monitoring content during the construction of the high-rise structure can be selected by referring to Table 6.1.3.

6.1.4 During the post construction, it is advisable to monitor the parameters of deformation, wind and response, stress, temperature and humidity, and dynamic response (including ground motion). According to the actual situation of the project, the monitoring parameters may be increased or decreased appropriately.

Y. Yang, *Technical Code for Monitoring of Building Structures*,
https://doi.org/10.1007/978-981-15-1049-6_6

Table 6.1.2 Minimum height specified for high-rise building monitoring (unit: m)

Structural system	Seismic fortification intensity	
	7°	8° (0.2 g)
Steel frame-support	100	90
Frame-shear wall	110	100
Shear wall	100	80
Frame structure	50	40
Slab column-shear wall	70	60
Tube structure	110	90
Mega-structure	230	220

Note 1. The high-rise buildings in Haibei and Guoluo Tibetan Autonomous Prefecture are reduced by 10 m on the basis of Table 6.1.2; the high-rise buildings in Xining City and Yushu Tibetan Autonomous Prefecture are reduced by 5 m on the basis of Table 6.1.2
2. *g* is the acceleration of gravity

Table 6.1.3 Monitoring content during construction

Deformation monitoring			Stress and strain monitoring	Environmental monitoring		
Foundation settlement	Vertical deformation	Horizontal deformation		Wind	Temperature	Earthquake
★	★	★	★	▲	▲	○

Note ★ items ought to be monitored, ▲ items could be monitored, ○ items may be monitored

6.2 Construction Monitoring

I General Requirements

6.2.1 In the monitoring of foundation settlement, the working base point should be first tested before performing the zoning monitoring. The measuring points for settlement monitoring shall be located at a symmetrical position at the intersections of the base axes along the periphery. A minimum of four points is required for tests. The settlement point should be as consistent as possible with the horizontal displacement point. The measuring point should be set at a position that reflects the structural deformation characteristics.

6.2.2 Settlement deformation monitoring points should be arranged according to the following locations:

1. The adjacent spacing of monitoring points should be 15–30 m;
2. The base of the dome, the base of the box base and the base, and four (several) large angles;
3. Postcasting strips, new and old buildings, and sides of different load distribution junctions;

4. The intersection of the artificial foundation and the natural foundation and the boundary of the filling and excavation;
5. The sides of high- and low-rise buildings, new and old buildings, and intersection of vertical and horizontal walls;
6. The difference between the base depth and the basic form change, and the changes occur in the geological conditions;
7. Near the place where heavy objects are stacked, the parts affected by the vibration and the dark shovel (ditch) under the foundation;
8. The cantilever and large platform on the floor of the closure floor;
9. For buildings with a width greater than or equal to 15 m or less than 15 m but complex in geology or in special soil areas, monitoring points shall be provided in the middle of the load-bearing wall, and monitoring points shall be provided on the indoor ground center and the four corners.

6.2.3 For the high-rise structure, the key components of the special important structure should be under stress monitoring. For the important stressed members and nodes determined by the construction simulation, the structural stress monitoring should be conducted on site, and appropriate additional test should be carried out according to the key operation time periods.

6.2.4 Crack monitoring should integrate site measurement and monitoring.

6.2.5 For structural members with high temperature and large temperature difference between day and night, temperature monitoring should be performed; within high humidity environment, monitoring frequency should be increased.

6.2.6 Wind load monitoring includes monitoring of wind speed, wind direction, and structural surface wind pressure monitoring. The wind speed test accuracy should not be less than 0.1 m/s; the wind direction test accuracy should not be lower than 3°; and the surface wind pressure test accuracy should not be lower than 10 Pa. Wind load monitoring should be conducted in a real-time manner.

II Frame Structure System

6.2.7 In addition to the measuring points specified in 6.2.2, settlement monitoring points shall also be placed on the elevator shaft, as well as each or part of the column base or along the vertical and horizontal axes of columns.

III Shear Wall Structure System

6.2.8 In addition to the measuring points specified in 6.2.2, settlement monitoring points should be set along the horizontal direction of the shear wall spaced at 10–25 m.

IV Frame-Shear Wall Structure System

6.2.9 In addition to the measuring points specified in 6.2.2, settlement monitoring points should be set in the elevator shaft and along the horizontal direction of the shear wall spaced at 15–25 m.

V Tube Structure System

6.2.10 The vertical deformation measuring points shall be evenly distributed along the outer frame and the core tube, and the measuring points shall be added at the key inner and outer contact members (the arm truss).

VI Mega-Structure

6.2.11 The monitoring of foundation settlement, deformation, as well as stress and strain should refer to the provisions of each structural system.

6.2.12 The deformation and stress–strain monitoring should be performed on the joints of the truss and suspension members.

6.2.13 The relative deformation between the various structural systems should be monitored.

6.3   Post Construction Monitoring

6.3.1 The layout of high-rise structural deformation monitoring points should be determined according to the structural system, and each type of system should be arranged with measuring points. The arrangement of monitoring points should adopt the principle of symmetry; connection members between structural systems should be added with measurement points.

6.3.2 Sunshine deformation monitoring should be carried out on building structures which are sensitive to seasonal effects and temperature effects under uneven sunlight.

6.3.3 Stress and strain monitoring should be conducted on critical components within important structures, and certain redundancy should be guaranteed.

6.3.4 High-rise buildings, which have undergone the wind tunnel tests, should be arranged with the test points according to the prior wind tunnel tests. For high-rise buildings that have not undergone wind tunnel tests, free-field and wind-sensitive components and joint locations should be selected as arrangement points. Wind-induced response points according to the monitoring data should be coordinated with ground motion measurement points.

6.3.5 The vibration measuring points shall be set at the top of the building, above the ground floor, at the floor where the structure is abrupt (stiffness and mass abrupt change), and the center of the stiffness of the key floor with high safety requirements or near the center of the stiffness, and should be along different heights of the structure. It should be arranged at the mass center of each section of the structure and avoids the vibration type node; when the translational vibration is monitored, the measuring point should be arranged at the stiffness center of the building; when the torsional vibration is monitored, the measuring points should be placed at the point where the circumference of the structure rotates the most.

6.3.6 Earthquake monitoring of engineering structures shall be carried out in accordance with the relevant provisions of the current National Regulations on Earthquake Monitoring and Administration. The high-rise structures that have been subjected to the vibration table model test shall be arranged according to the test arrangement points. Except for the bottom and top of the building, the measuring point should also refer to the structural vibration measuring point and should select its important part.

6.3.7 It is advisable to conduct the temperature monitoring of external thermal insulation system for high-rise buildings with external thermal insulation system.

I Frame Structure System

6.3.8 It is advisable to conduct the stress and strain monitoring on the transfer girder and the transfer column.

6.3.9 Monitoring of complex structural areas with transition layer locations, joint structure joint locations (especially weak joint regions), belt reinforcement locations, body shape and cantilever locations, and staggered structure locations should be monitored. The check of the complex structure location of the monitoring project should be carried out.

6.3.10 Stress and strain monitoring should be carried out near the bottom and belt trusses.

II Shear Wall Structure System

6.3.11 Stress and strain monitoring should be carried out on the transfer girder and the transfer column.

6.3.12 Stress and strain monitoring should be carried out at the bottom.

III Frame-Shear Wall Structure System

6.3.13 Measuring points should be added to the connecting members between the structural systems.

IV Tube Structure System

6.3.14 Monitoring should be performed for vertical deformation and different vertical deformations of inner and outer frame tubes.

V Mega-Structure

6.3.15 Stress and strain monitoring should be carried out on outrigger, mega column, mega diagonal brace.

Chapter 7
Long-Span Spatial Structure

7.1 General Requirements

7.1.1 Construction monitoring of long-span spatial structures is required when one of the following conditions is met or when design is required:

1. Grid structure and multi-layer reticulated steel structure or cable membrane structure with span greater than 100 m;
2. Single-layer reticulated shell structure with span greater than 50 m;
3. Large-span composite structure with a single span of more than 30 m;
4. Steel structure with a cantilever length greater than 30 m;
5. Structures with a significant difference of structural configuration between the final version of the structure and the design target due to the construction process;
6. In areas where the annual temperature difference between day and night is greater than 20 °C for more than 100 days, the monitoring standard of the structure shall be reduced by 10 m on the basis of the above specified span.

7.1.2 It is advisable to perform post construction monitoring of long-span spatial structures when one of the following conditions is met or whenever it is required in design:

1. Grid and multi-layer reticulated steel structure with span more than 120 m;
2. Single-layer reticulated shell structure with span greater than 60 m;
3. Steel structure with a cantilever length greater than 40 m;
4. Long-span structure that is particularly complex and sensitive to structural deformation;
5. In areas where the annual temperature difference between day and night is greater than 20 °C for more than 100 days, the monitoring standard of the structure shall be reduced by 10 m on the basis of the above specified span.

7.1.3 See Table 7.1.3 for construction monitoring of large-span space structures.

Y. Yang, *Technical Code for Monitoring of Building Structures*, https://doi.org/10.1007/978-981-15-1049-6_7

Table 7.1.3 Construction monitoring content of large-span structure

	Deformation monitoring		Stress and strain monitoring	Environmental monitoring		Foundation settlement	Earthquake	Brace
	Vertical	Horizontal		Wind	Temperature			
Grid structure	★	○	★	○	▲	★	○	★
Reticulated shell structure	★	○	★	○	▲	★	○	★
Cable structure	★	○	★	○	▲	★	○	★
Membrane structure	★	○	★	○	▲	★	○	▲

Note ★ items ought to be monitored, ▲ items could be monitored, ○ items may be monitored

Table 7.1.4 Post construction monitoring content of large-span structure

	Deformation monitoring		Stress and strain monitoring	Environmental monitoring		Foundation settlement	Earthquake
	Vertical	Horizontal		Wind	Temperature		
Grid structure	★	○	▲	▲	▲	★	▲
Reticulated shell structure	★	○	▲	▲	▲	★	▲
Cable structure	★	○	★	▲	▲	★	▲
Membrane structure	★	○	★	▲	▲	★	▲

Note ★ items ought to be monitored, ▲ items could be monitored, ○ items may be monitored

7.1.4 See Table 7.1.4 for post construction monitoring of large-span space structures.

7.2 Construction Monitoring

I General Requirements

7.2.1 Settlement monitoring should be carried out during the main construction period and the initial stage of commissioning. The settlement monitoring point shall be placed at the connection part of the artificial foundation and the natural foundation, and at the symmetry part of the boundary of the structural foundation axis. There should be a minimum of four monitoring points. A minimum of five monitoring times are required during construction.

7.2.2 Selection of monitoring instruments and arrangement of monitoring positions for stress and strain monitoring should be based on the type of structure and construction method.

II Grid (Truss) Structure

7.2.3 The deflection value of the grid, the center offset and height deviation of the support points, the longitudinal and lateral length deviations of the grid, and the deformation values of the temporary support points during the removal process shall be monitored. The measurement of the deflection value of the short-span grid (truss) structure (i.e., span below 24 m) can be measured at the center of the bottom chord. The structure of the long-span grid (truss) (i.e., spans over 24 m and above) should be monitored at the center of the bottom chord and the quarter point of the chord span.

7.2.4 These strains should be monitored, including the strain of important parts of the structure, the strain of the member with larger absolute values of the stress during construction, the strain of the member with large stress changes, the strain of the main stressed member, and the strain when the temporary auxiliary support is unloaded.

7.2.5 Wind direction and wind load monitoring should be carried out. Installation should not be carried out when wind speed grade is 6 or above.

III Reticulated Shell Structure

7.2.6 The deflection of the control point, the reference axis position, the elevation, and the vertical deviation should be monitored for various construction methods and corrected on time; the deformation measurement point should be set at the center span of the reticulated shell, at the quarter span, and at the reticulated support.

7.2.7 Stress and strain monitoring should be carried out for top and bottom chords and webs near the fringe truss, the secondary fringe truss, and the support of mid-truss.

7.2.8 Wind direction and wind load monitoring should be carried out. Installation should not be carried out when wind speed grade is 6 or above.

IV Cable Structure

7.2.9 The distance between the two anchorage sections of the cable, the deformation of the key points of the monitoring structure under tension at each stage, and the deflection deformation of the structural camber should be monitored.

7.2.10 Stress and strain monitoring should be carried out on cable under tension and important parts of cable structures.

7.2.11 Wind direction and wind load monitoring should be carried out.

V Membrane Structure

7.2.12 Deformation monitoring should be carried out on the membrane face.

7.2.13 The stress and strain monitoring should be carried out by uniformly arranging the measuring points of each membrane unit, and the measuring points should be arranged at the key parts; during the tensioning process, the face membrane tension of the two orthogonal directions should be measured separately.

7.2.14 The cable tension should be monitored.

7.2.15 Wind direction and wind load monitoring should be carried out. Installation should not be carried out when wind speed grade is 3 or above, or when the temperature is lower than 4 °C.

7.3 Post Construction Monitoring

I General Requirements

7.3.1 It should be implemented with reference to Construction Code 7.2.1; the monitoring frequency should be determined according to the stability of the structure. When the settlement speed is less than 0.01–0.04 mm/day, the monitoring can be stopped.

7.3.2 Seismic monitoring should take into account the structural features. Monitoring of the dynamic response should be performed at the bottom, top and important parts of the structure. The specific implementation rules should be in accordance with Code 6.3.6.

7.3.3 Fatigue safety monitoring and evaluation shall comply with the following provisions:

1. It is advisable to use the allowable stress method or the fatigue damage index method to evaluate the fatigue state of the components at the monitoring point;
2. When the allowable stress method is used to evaluate the unusual fatigue state of the component, it is advised to re-evaluate the fatigue state according to the fatigue damage index method;
3. The fatigue cumulative damage index method should be calculated by the rain flow method and the Miner criterion. It should be executed according to the provisions provided in Table 7.3.3.

7.3.4 Continuous monitoring of wind speed, wind direction, and wind pressure on the surface of the structure should be carried out.

7.3.5 Environmental temperature change and structural temperature distribution should be monitored.

II Grid (Truss) Structure

7.3.6 The vertical displacement, horizontal displacement, bearing displacement, and geometric deformation of the grid (truss) structure should be regularly monitored

Table 7.3.3 Fatigue state classification

D value	Status of construction measuring point
0–0.05	Intact status
0.05–0.2	Fairly good status
0.2–0.45	Moderate faulted status
0.45–0.8	Severe faulted status
>0.8	Dangerous status

Note The fatigue state classification given in the table does not consider the effect of corrosion on fatigue life. When corrosion occurs, the adverse effects of corrosion on the fatigue life of steel components should be considered

Table 7.3.6 Location of grid (truss) structure deformation monitoring on post construction

	Deformation monitoring location	
	Vertical	Horizontal
Grid (truss) structure	Mid-span	Supports and ends

and checked of whether the deformation exceeds the allowable value. It can be referred to Table 7.3.6 regarding the monitoring location layout.

7.3.7 Stress and strain monitoring should be carried out on supports, main load-bearing members, and most unfavorable locations.

7.3.8 Surface wind pressure monitoring should be carried out.

VI Reticulated Shell Structure

7.3.9 The deformation monitoring shall be carried out on the supports and welding joints. The position of the monitoring points shall be implemented in accordance with Code 7.3.6.

7.3.10 The stress and strain conditions near the main stressed members and the welds of key joints should be monitored. The working state of the structural nodes should be evaluated in combination with the corrosion.

7.3.11 The wind direction and wind speed field of the reticulated shell structure should be monitored.

III Cable Structure

7.3.12 Deformation monitoring should be carried out for steel trusses, cable anchoring nodes, and structural key control points.

7.3.13 Regular monitoring of cable under tension should be carried out.

7.3.14 Wind direction and wind load monitoring should be carried out.

VII Membrane Structure

7.3.15 Regular deformation monitoring should be carried out on membrane face.

7.3.16 Stress and strain monitoring on cable of membrane structures and other key points should be carried out.

7.3.17 Wind direction and wind load monitoring should be carried out.

Chapter 8
Protective Building

8.1 General Requirements

8.1.1 This chapter applies to the safety monitoring of the following protective buildings:

1. Building structure approved as the cultural relics protection unit;
2. Building structure that has not been approved for publication as a cultural relics protection unit but has been registered as an immovable cultural relic;
3. Building structure that not listed as an immovable cultural relic, but approved as a historic building;
4. Building structure that has not been approved for publication as a historic building but has a protective value.

8.1.2 This chapter applies to the following damages to protect buildings:

1. When the building tilts, settles, deforms, affecting safety use;
2. When the component is damaged or missing, affecting the structural carrying capacity;
3. When the external environment is bad, it is not conducive to the long-term preservation and maintenance of protective building structures;
4. After suffering from a serious natural disaster;
5. When regular inspections reveal a safety hazard;
6. Others need to grasp the structural safety status.

8.1.3 The structure of the protective building should be regularly inspected and monitored, and its deformation, stress and strain, foundation settlement, temperature and humidity, and vibration project monitoring uses monitoring instruments.

8.1.4 Monitoring points should be placed at key points that reflect the characteristics of the monitoring parameters, while protecting the integrity of the structure.

© The Author(s), under exclusive license to Springer Nature Singapore Pte Ltd. 2020
Y. Yang, *Technical Code for Monitoring of Building Structures*,
https://doi.org/10.1007/978-981-15-1049-6_8

8.2 Monitoring Contents

I General Requirements

8.2.1 The basic settlement monitoring scheme shall be formulated according to actual needs, and shall be monitored regularly. The determination of the monitoring frequency shall systematically reflect the change process of the measured building deformation.

8.2.2 It is advisable to selectively monitor overall or local horizontal displacement deformation, deflection, tilt, and crack deformation based on actual structural conditions.

8.2.3 Stress and strain monitoring shall be carried out on the critical parts of the flexural members (beam, girder, purlin, rafter, grid), compression members (column foot, pillar-base), and connecting members (rabbet, mortise, bucket).

8.2.4 Online monitoring of the dynamic characteristics and response of protective building structures should be carried out.

8.2.5 Online monitoring of the environment inside and outside the structure and the temperature of key components should be carried out.

8.2.6 It is advisable to carry out on-line monitoring of the internal humidity of the building and take necessary dehumidification measures for the monitoring results.

8.2.7 Wind and wind-induced response monitoring of wind-sensitive structures should be carried out.

8.2.8 Regular monitoring of wood properties should be carried out, and pest control, anti-corrosion, and fire remedies should be developed.

8.2.9 The deformation, cracking, measuring point layout, and monitoring equipment operation status of the protective building structure should be regularly monitored, and the inspection records should be made.

II Timber Structure

8.2.10 Regular monitoring of the following items of load-bearing wood columns should be carried out:

1. Material decay and aging deterioration;
2. The degree of bending of the column;
3. Resistance and bearing condition of column foot and pillar-base;
4. Pillar-base misalignment;
5. Damage status along any part of the length of the column;
6. Status of reinforcement in the past few times.

8.2.11 Regular monitoring of the following items of load-bearing wood beam rafts should be carried out:

1. Material decay, aging, and insect conditions;
2. Bending deformation;
3. Bridge damage;
4. Status of reinforcement in the past few times.

8.2.12 Regular monitoring of the following deformations of the timber frame should be carried out:

1. Global tilt;
2. Partial incline;
3. Connection between frames;
4. Connection between beam and column;
5. Integrity of mortise and tenon.

8.2.13 Regular monitoring of materials in the following areas should be carried out:

1. Rafter system;
2. Purlin system;
3. Short column, hump of bracket;
4. Upturned roof-ridge, the edge of eaves, inverted V-shaped brace.

III Masonry Structure

8.2.14 The following items should be monitored regularly:

1. Brick weathering;
2. Wall tilt;
3. Cracking.

IV Reinforced Concrete Structure

8.2.15 The following items should be monitored regularly:

1. Foundation settlement;
2. Degree of global tilt;
3. Cracking.

V Steel Structure

8.2.16 The following items should be monitored regularly:

1. Global tilt;
2. Node horizontal displacement;
3. Node vertical displacement;
4. Material corrosion.

Chapter 9
Other Structures

9.1 Dilapidated Building

9.1.1 The C-class dilapidated buildings in the urban areas that have been identified and inhabited shall be monitored, and the D-class dilapidated buildings in the towns where the personnel have not been removed and the C- and D-class dilapidated buildings in the urban areas that are not yet inhabited but still threaten public safety shall be monitored.

9.1.2 Regular monitoring should be carried out. The frequency of monitoring should be no less than once a week for C-class dilapidated buildings and no less than twice a week for D-class dilapidated buildings. When conditions permit, online monitoring of dangerous buildings should be carried out.

9.1.3 If the dangerous houses under the following conditions are met, the frequency of regular monitoring should be increased according to the actual situation; if the conditions permit (the sensor is easy to install and the monitoring plan is feasible), it should be monitored on-line:

1. If the house is affected by other construction, the frequency of monitoring should be strengthened;
2. When the monitoring data of dangerous points in dangerous buildings is abnormal, in flood season, or in the event of extreme weather such as heavy rain or snow, earthquakes and other natural disasters, the programming unit shall encrypt the monitoring points and monitoring frequency according to the actual situation.

9.1.4 The dangerous housing monitoring program should follow the principle of "one building one scheme", according to the specific situation of the dangerous points of dangerous buildings in the dilapidated house monitoring and appraisal report and relevant information of house design, developed by the programming unit through detailed on-site investigation and combined with relevant requirements of current national and industry standards.

9.1.5 Dangerous housing monitoring methods should use monitoring tools and macro inspections to conduct dynamic monitoring of dangerous buildings.

9.1.6 The content of dangerous house monitoring shall be determined according to the specific conditions of the dangerous points of dangerous buildings in the dilapidated house monitoring and appraisal report and the relevant design of the house design, and shall be determined through detailed on-site investigation and the relevant requirements of the current national and industry standards, including the original dangerous point changes and new dangerous points condition.

9.1.7 The dangerous house monitoring record table should include the following contents:

1. Foundation monitoring;
2. Basic and superstructure component monitoring:
   (1) Basic component;
   (2) Superstructure component.

9.1.8 The layout of dangerous building monitoring points should fully reflect the deformation characteristics of houses.

9.1.9 The on-line or regular monitoring methods for dangerous houses include manual inspection, main structure tilt monitoring, and local dangerous point monitoring. The monitoring should be carried out according to the specific requirements of the national, industry and local standards, combined with the specific conditions of the dangerous houses.

9.1.10 The programming institution shall conduct a summary analysis of the monitoring data on a regular basis, clearly and specifically respond to the disposal recommendations based on the changes in the monitoring data, and promptly inform relevant institutions and individuals to take effective countermeasures for disposal as required.

9.1.11 The plan preparation institution shall determine the early warning threshold according to the specific conditions of the dangerous points of the dangerous buildings in the dangerous building monitoring and appraisal report and the relevant information of the house design, combined with the relevant requirements of the current national and industry standards, and through detailed on-site investigation.

9.2   Industrial Factory

9.2.1 The permissible vibration standards of industrial factory under industrial and environmental vibrations shall be provided by the equipment manufacturer or determined by tests; when the equipment manufacturer cannot provide information and cannot conduct tests, it may be carried out in accordance with Code 10.6.1.

9.2.2 The vibration test method of the industrial plant and the selection of the vibration control point shall be provided by the equipment manufacturer or comply with the relevant design standards; when the equipment manufacturer cannot provide or has no relevant design regulations, it shall be implemented in accordance with the following provisions:

1. The vibration test should select the appropriate test system according to the vibration direction, frequency range, amplitude, and vibration characteristics of the object to be tested. The test system should be periodically calibrated by the nationally recognized measurement department according to the current national standards;
2. The vibration test point should be set at the vibration control point. The test direction of the vibration sensor must be consistent with the vibration direction and should not be inclined;
3. The sensor and the object to be tested must be tightly fixed, no looseness should be generated during the test, and secondary vibration of the fixture should be avoided.
4. Typical test conditions should be selected for vibration testing.

9.2.3 The vibration of industrial factory structures shall meet the following requirements:

$$A \le [A] \qquad\qquad (9.2.3 - 1)$$

$$V \le [V] \qquad\qquad (9.2.3 - 2)$$

$$a \le [a] \qquad\qquad (9.2.3 - 3)$$

which: A—calculated or tested vibration displacement of industrial factory structures;

V—calculated or tested vibration speed of industrial factory structure;

a—calculated or tested acceleration of industrial factory structure;

[A]—allowable vibration displacement of industrial factory structure;

[V]—allowable vibration speed of industrial factory structure;

[a]—allowable vibration acceleration of industrial factory structure.

9.3 Mass Concrete Structure

9.3.1 The mass concrete structure should be monitored for the concrete temperature during the construction, the temperature control should comply with the relevant national standard "Code for Construction of Mass Concrete" GB50496.

9.3.2 Construction monitoring of mass concrete structures should meet the requirements of the following Table 9.3.2.

Table 9.3.2 Monitoring content during construction of mass concrete structure

Concrete					Ambient temperature	Structural crack	Settlement
Center temperature	Surface temperature	Highest internal placing temperature difference	Temperature difference in the pouring body	Pouring body cooling rate			
★	★	★	★	★	★	▲	○

Note ★ items ought to be monitored, ▲ items could be monitored, ○ items may be monitored

9.3.3 The mass concrete temperature monitoring instrument should be composed of temperature sensor, data acquisition system, and data transmission system; the system should have the function of displaying, storing, and processing temperature and time parameters, and plot the temperature curve of the measuring point in real time; the number of temperature measuring points should not be less 50; temperature monitoring instruments should be calibrated regularly; and the allowable error should not exceed 0.5 °C.

9.3.4 Construction temperature monitoring should meet the following requirements:

1. The test of temperature difference in the pouring body, cooling rate, and ambient temperature in mass concrete pouring body shall not be less than four times per day and night after concrete pouring;
2. The measurement of the mold temperature is not less than two times per shift;
3. After the concrete is poured, the temperature is measured once every 15–60 min.

9.3.5 The temperature, deformation, stress, and strain during the use of mass concrete structures shall be monitored.

9.3.6 Sensor selection and arrangement should meet the following requirements:

1. Temperature sensor error should be less than 0.3 °C;
2. Temperature sensor test range: 40–150 °C;
3. The temperature sensor should be placed on the steel bar closest to the symmetry axis of the plane view of the concrete cast unit;

4. It is advisable to arrange the sensor to measure the external temperature at 50 mm inside the cast external surface;
5. It is advisable to place the sensor at a height of 50 mm above the ground of the cast unit to measure the temperature of the bottom surface.

9.4 Prefabricated Structure

9.4.1 This section applies to post construction monitoring of prefabricated concrete structures, fabricated steel structures, and fabricated wood structures.

9.4.2 It is advisable to carry out stress and strain monitoring on the prefabricated components of the prefabricated concrete structure and to monitor the deformation, strain, and cracks at the joints of the components.

9.4.3 It is advisable to carry out stress and strain monitoring on the steel members of the fabricated steel structure and to monitor the deformation and strain at the joints of the components.

9.4.4 It is advisable to monitor the deformation, stress, and strain of the joints of the fabricated wood structures.

9.5 Isolated (Damper-Added) Structure

I General Requirements

9.5.1 In addition to the design documentation requirements or other isolated (damper-added) structures that are subject to monitoring, when the following conditions are met, isolated (damper-added) structures shall be monitored during construction or post construction:

1. High-rise isolated building structures with a structural height of more than 60 m or an aspect ratio of more than 4 should be monitored during post construction;
2. Large-span air-spaced isolated structures with a span of more than 60 m should be monitored during post construction;
3. High-rise, long-span structure with shock absorbing technology installed should be monitored during post construction;
4. When design document requirements or other special structures are required, it should be monitored during construction post construction.

9.5.2 The monitoring content during the construction of isolated (damper-added) structures shall be carried out in accordance with Table 9.5.2.

Table 9.5.2 Monitoring content during construction of isolated (damper-added) structure

	Deformation monitoring		Stress monitoring		Environmental monitoring		Foundation settlement	Vibration
	Vertical	Horizontal	Stress	Strain	Wind	Temperature and humidity		
Isolated structure	★	★	○	★	○	▲	★	▲
Damper-added structure	★	★	○	★	▲	▲	★	▲

Note ★ items ought to be monitored, ▲ items could be monitored, ○ items may be monitored

9.5.3 The monitoring content during post construction of isolated (damper-added) structures shall be carried out in accordance with Table 9.5.3.

Table 9.5.3 Monitoring content during post construction of isolated (damper-added) structure

	Deformation monitoring		Stress monitoring		Environmental monitoring		Foundation settlement	Vibration
	Vertical	Horizontal	Stress	Strain	Wind	Temperature and humidity		
Isolated structure	★	★	○	★	★	▲	★	▲
Damper-added structure	★	★	○	▲	★	▲	★	★

Note ★ items ought to be monitored, ▲ items could be monitored, ○ items may be monitored

II Construction Monitoring

9.5.4 The vertical compression deformation of the isolation bearing should be monitored.

9.5.5 The height of the horizontal isolation fracture and the width of the vertical isolation fracture should be monitored.

9.5.6 Stress, strain, and deformation on the position of the isolation beam and the damper should be monitored.

III Post Construction Monitoring

9.5.7 It is advisable to monitor the horizontal shear deformation and vertical compression deformation of isolation bearing.

9.5.8 It is advisable to monitor the isolation bearing, damper, isolation fracture, flexible connection, and damping device. The monitoring time should be 3 years, 5 years, 10 years after completion, and every 10 years in the future.

9.6   Large Public Buildings

9.6.1 The monitoring of various building structures in large public buildings should be carried out on the basis of the various chapters of this code, combined with the actual characteristics of the structure.

9.6.2 Large public buildings monitoring should meet the following requirements:

1. It should accommodate multiple structures for collaborative real-time monitoring;
2. Various structural monitoring systems are highly integrated through the cloud platform;
3. Massive data analysis and processing;
4. Monitoring data synergy analysis and performance evaluation;
5. Monitoring results are displayed uniformly.

9.6.3 The overall design of the large public buildings monitoring platform should meet the following requirements:

1. The design of the monitoring platform should be scientific, economical, and reasonable, and the on-line monitoring of the structure group can be realized;
2. The platform should be able to accommodate multiple building structural units. Through the analysis and processing of massive data in the platform, the damage of the structural group can be qualitatively positioned and quantitatively analyzed;
3. Establish a long-term monitoring database of the structure group from construction to post construction;
4. Good scalability.

9.6.4 The large public buildings monitoring platform consists of the data infrastructure layer, the data resource layer, the application support layer, and the user layer. The platform should transmit the monitoring information in the large public buildings to the cloud computing center in real time for unified analysis and management, and feedback to the mobile terminal of the operation management at all levels. Cloud platform monitoring technical requirements should be referred to Appendix B for implementation.

Chapter 10
Warning Threshold

10.1 General Requirements

10.1.1 The warning indicator should be selected based on the following principles:

1. The warning indicator should be selected according to the type, including load, key component, and whole structure;
2. The selected warning indicator should have stability, wide applicability, and strong operability;
3. When different indicators conflict with each other, the warning indicator that reflects the most unfavorable conditions of the structure should be selected;
4. For the major engineering structure, the lower limit range of the warning threshold can be selected. For the common engineering structure, the upper or average range of the warning threshold can be selected.

10.1.2 The warning threshold should include three warning levels:

1. Yellow warning (primary warning): remind construction or post construction units to start paying attention to the environment, load, overall, or partial response;
2. Orange warning (heavier warning): remind construction or post construction units to pay more attention to the environment, load, structure or partial response, and conduct follow-up observation;
3. Red warning (severe warning): warn construction or post construction units to pay close attention to the environment, load and structure response, identify the cause of the alarm, take appropriate inspections, emergency management measures to ensure safety, and timely conduct structural safety assessment.

10.1.3 The warning method should meet the following requirements:

1. The warning information should be released by a special organization or an authorized institution, and it should be released, adjusted, and lifted according to

Y. Yang, *Technical Code for Monitoring of Building Structures*,
https://doi.org/10.1007/978-981-15-1049-6_10

the development of the dangerous situation and the progress of the emergency actions.

2. The warning method should be obvious and should be diversified, including indicators, sounds, short messages, interface display, variable information board, and roadside broadcasting;
3. The warning information should form a log, including the start and end time, warning items, warning level, warning sensor number and position, early warning monitoring value, and early warning frequency.

10.1.4 During the construction process, the warning threshold should meet the following requirements:

1. When the monitored stress is greater than 0.5 times of the allowable value or the designed value, a yellow warning should be issued; when it is greater than 0.7 times of the allowable value or the designed value, an orange warning should be issued; when it is greater than 0.8 times of the allowable value or the designed value, a red warning should be issued;
2. When the monitored
   deformation is greater than 0.5 times of the allowable value or the designed value, a yellow warning should be issued; when it is greater than 0.7 times of the allowable value or the designed value, an orange warning should be issued; when it is greater than 0.8 times of the allowable value or the designed value, a red warning should be issued;
3. When the foundation settlement rate is greater than 0.1 mm/day, the yellow warning should be issued; when it is greater than 0.2 mm/day, the orange warning should be issued; when it is greater than 0.3 mm/day, the red warning should be issued;
4. If the warning threshold is set according to the analysis of the structure during construction, it should be set as 0.7, 1.1, and 1.3 times of the analysis result for yellow, orange, and red warnings, respectively.

10.1.5 During the post construction, the warning threshold should meet the following requirements:

1. When the monitored displacement is greater than 0.6 times of the allowable value or the designed value, a yellow warning should be issued; when it is greater than 0.8 times of the allowable value or the designed value, an orange warning should be issued; when it is greater than the allowable value or the designed value, or when the orange warning is issued for 10 times in a month, a red warning should be issued;
2. When structural stress and strain is greater than 0.6 times of design allowable value or specification limit value, a yellow warning should be issued; when it is greater than 0.8 times of design allowable value or specification limit value, an orange warning should be issued; when it is greater than design allowable value or specification limit value, or 10 times of orange warning happened in one month, a red warning should be issued;

3. When the monitored peak of horizontal seismic acceleration is greater than 1.0 times of the allowable value or the designed value, a yellow warning should be issued; when it is greater than 1.1 times of the allowable value or the designed value, an orange warning should be issued; when it is greater than 1.2 times of allowable value or the designed value, a red warning should be issued;
4. When the monitored maximum average wind speed is greater than 0.7 times of the allowable value or the designed value, a yellow warning should be issued; when it is greater than 0.8 times of the allowable value or the designed value, an orange warning should be issued; when it is greater than the allowable value or the designed value, a red warning should be issued;
5. When the maximum temperature, or minimum temperature, or maximum temperature difference, or maximum temperature gradient is greater than the allowable value or the designed value, an orange warning should be issued;
6. When the steel bars at the monitoring point are obviously corroded, a red warning should be issued.

10.2   High-Rise Building

I Construction Monitoring

10.2.1 The limit of global inclination of the high-rise structure shall be implemented in accordance with Table 10.2.1:

Table 10.2.1 Limit value of the global tilt of the high-rise structure in construction

Structural height	Tilt allowable value $\tan\theta$	Tilt rate
$H \leq 24$	0.003–0.006	$H/9000/d–H/4000/d$
$24 < H \leq 60$	0.0025–0.005	$H/7500/d–H/3500/d$
$60 < H \leq 100$	0.002–0.004	$H/6000/d–H/3000/d$
$H > 100$	0.002–0.004	$H/5000/d–H/2500/d$

Note H is the height of the building (meter) from the outdoor ground, and the inclination refers to the ratio of the settlement difference between the two ends of the foundation in the oblique direction and its distance

10.2.2 Under wind load or earthquake, the deformation limit at the tower is 1/380–1/200, and the deformation rate is limited to H/9000/d–H/4000/d.

10.2.3 Under the action of wind load, the horizontal displacement limit of any point of the high-rise structure is 1/150–1/50 of the height of the point from the ground.

10.2.4 Under the combination of various load standard values, the maximum crack width limit of reinforced concrete members is 0.15–0.3 mm.

10.2.5 Under the dynamic action of wind load, the amplitude of structural vibration acceleration is limited to 120–250 mm/s^2.

10.2.6 The compression deformation limit of the high-rise structure during construction is $0.3 \times H$–$0.5 \times H$ mm.

Note: H is the height of the high-rise structure (unit: m).

II Post Construction Monitoring

10.2.7 The limit of global inclination of the high-rise structure shall be implemented in accordance with Table 10.2.7.

Table 10.2.7 Limit value of the global tilt of the high-rise structure in post construction

Structural height	Tilt allowable value tanθ	Tilt rate
$H \leq 24$	0.004–0.008	H/9000/d–H/4000/d
$24 < H \leq 60$	0.0025–0.007	H/7500/d–H/3500/d
$60 < H \leq 100$	0.003–0.005	H/6000/d–H/3000/d
$H > 100$	0.002–0.004	H/5000/d–H/2500/d

10.3 Long-Span Spatial Structure

I Grid Structure During Construction

10.3.1 The deviation limit of longitudinal (lateral) length is 1/2500–1/1200 of the length and should not be greater than 30–60 mm.

10.3.2 The deviation limit of the support center is 1/3700–1/1800 of the span of the grid, and should not be greater than 30–50 mm.

10.3.3 Support height deviation limit value: between adjacent points of the grid supported by the perimeter, it should be 1/500–1/250 of the adjacent production spacing, and should not be greater than 10–30 mm; the highest and lowest points should be 30–50 mm; for multi-point support grid, the distance between adjacent supports should be 1/1000–1/500, and should not be greater than 30–50 mm.

10.3.4 The grid deflection limited is 10–25% of the designed value.

II Grid Structure During Post Construction

10.3.5 The deflection limit value of the space grid structure under the constant load and live load standard values is as specified in Table 10.3.5.

Table 10.3.5 The deflection limit value of the space grid structure during post construction

Roof structure (short span)	Floor structure (short span)	Cantilever structure (cantilever span)
1/300–1/150	1/370–1/180	1/150–1/80

Note For roof structures with suspended lifting equipment, the maximum deflection value is limited to 1/500–1/250 of the structural span

III Reticulated Shell Structure During Construction

10.3.6 The deviation limit of the distance between two control supports should be 1/2500–1/1200 of the distance, and it should not be greater than 30–50 mm.

10.3.7 When the span is less than or equal to 60 m, the height deviation limit shall not exceed the design elevation of ± 30 mm; when the span is greater than 60 m, the design elevation shall not exceed ±50 mm.

10.3.8 After the installation is completed, the vertical displacement value monitored by several control points of the reticulated shell is 1–1.8 times of the design value under the corresponding load.

IV Reticulated Shell Structure During Post Construction

10.3.9 The deflection limit of the space reticulated shell structure under the constant load and live load standard values shall be as specified in Table 10.3.9.

Table 10.3.9 The deflection limit value of the reticulated shell structure during post construction

Structural system	Roof structure (short span)	Cantilever structure (cantilever span)
Single-layer lattice shell	1/500–1/250	1/150–1/80
Double-layer reticulated shell spatial truss	1/300–1/150	1/150–1/80

V Cable Structure During Construction

10.3.10 The allowable limit of the distance between the two anchors of the cable is the smaller between $L/1800$ and 30 mm.

Note: L is the distance between the two anchor segments.

10.3.11 After tensioning at each stage, the tension is allowed to be limited to a design value of 10–15%.

10.3.12 After tensioning at each stage, the limit of camber and deflection is 4–8% of the design value.

VI Cable Structure During Post Construction

10.3.13 The high-span ratio limits of the cable dome is 1/10–1/5; the angle limits of intersection between the cable and the horizontal plane is 10°–25°.

10.3.14 The ratio of the maximum deflection to the span of the single cable roof is limited to 1/250–1/120 after the initial geometric state.

10.3.15 The maximum deflection to span ratio of the cable net, double-layer cable, and transverse stiffener roof is limited to 1/300–1/150 after the initial prestressing state.

10.3.16 The ratio of the maximum deflection to the span of the cable-stayed structure, the string structure, or the cable dome roof under load is from 1/300 to 1/150 after the initial prestressing state.

10.3.17 The ratio of maximum deflection to the span of single-layer plane cable net glass curtain wall is limited to 1/50–1/30.

10.3.18 The ratio of maximum deflection to span of curved cable net and double-layer cable system glass curtain wall is limited to 1/250–1/120 after initial prestressing state.

10.3.19 The ratio of maximum deflection to span of the lighting roof of double-layer cable system with curved surface is limited to 1/250–1/120 after the initial prestressing state.

10.3.20 The ratio of maximum deflection to span of glass roof with string structure is limited to 1/250–1/120 after initial prestressing state.

VII Membrane Structure During Construction

10.3.21 When the pre-tension is applied to the membrane, the displacement limit of the applied point is 10–15% of the design value.

10.3.22 The limit value of force for monitoring the representative application point of membrane material is 10–15% of the design value.

VIII Membrane Structure During Post Construction

10.3.23 Under the first kind of load effect combination, the maximum overall displacement limit value of membrane structure is 1/300–1/150 of span or 1/150–1/80 of cantilever length.

10.3.24 Under the second kind of load effect combination, the maximum overall displacement limit value of membrane structure is 1/250–1/120 of span or 1/100–1/60 of cantilever length.

10.3.25 Under the second kind of load effect combination, the limit value of lateral displacement of membrane structure is 1/300–1/150 of mast length.

10.3.26 The relative normal displacement of the inner surface of each membrane element in the membrane structure is limited to 1/20–1/10 of the nominal size of the element.

10.3.27 Under the second kind of load effect, the limit value of fold area caused by membrane relaxation is 10–15% of membrane area.

10.4 Protective Building

I Timber Structure During Post Construction

10.4.1 The load-bearing wood column monitoring limit value should meet the following requirements:

1. When the material of the wood is only surface decay and aging, the ratio of the area occupied by decay and aging (the total of the two) to the total cross-sectional area is 1/6–1/3; when the material of wood is only heart rot, the ratio of the area occupied by decay and aging (the total of the two) to the total cross-sectional area is 1/9–1/4; when both of the above conditions exist, the ratio of the area occupied by decay and aging (total of the two) to the total cross-sectional area is 0;
2. The bending vector height of the column is the unsupported length of the column divided by 210—the unsupported length of the column divided by 400;
3. The ratio of the actual abutment area between the bottom of the column foot and the column base to the original cross-sectional area of the column at the column foot is 1/2–1;
4. The ratio of the position of the misalignment between the column and the column base to the size of the column diameter (or column section) along the misalignment direction is 1/7–1/4.

10.4.2 The monitoring limit value of load-bearing wooden beams should meet the following requirements:

1. When the wood material is only surface decay and aging deterioration, the ratio of the area occupied by the beam body and the aging deterioration (the total of the two) to the entire cross-sectional area is 1/10–1/5; when the material of the wood is only heart rot, the ratio of the area occupied by decay and aging (total of the two) to the area of the entire cross section is 0;
2. When the high-span ratio is greater than 1/14, the vertical deflection value is $L^2/2600$–$L^2/1300$ h; when the high-span ratio is less than 1/14, the vertical deflection value is $L/180$–$L/90$;

3. For the beams and rafts that are more than 300 years, the lateral bending height
   is the sum of vertical deflection and $h/100$—vertical deflection and $h/180$;
4. The lateral bending height is $L/250$–$L/120$.

Annotation: where h is the height of the component section; L is the calculated
span.

10.4.3 The limit value of the overall deformation monitoring of timber frame should
meet the following requirements:

1. For the beam-lifting structure, the inclination along the plane of the frame is $H_0/100$–$H_0/190$ or 100–190 mm; the inclination of the vertical frame plane is $H_0/200$–$H_0/380$ or 50–90 mm; the relative displacement of the column head and the
   column foot is $H_0/70$–$H_0/140$;
2. For the column and tie construction, the amount of inclination along the plane of
   the frame is $H_0/80$–$H_0/160$ or 120–240 mm; the inclination of the vertical frame
   plane is $H_0/160$–$H_0/320$ or 60–120 mm; the relative displacement of the column
   head and the column foot is $H/60$–$H/120$;
3. The connection between the beam and the column (including the column, the
   inter-column, the column, the connection between the purline) has no tying, and
   the length of the hoe opening is greater than 3/10–3/5 of the length of the hoe
   (post and lintel construction) or 2/5–4/5 (column and tie construction);
4. The tenon and mortise have decayed, moth-eaten, split, or broken, and the
   compression is 3–6 mm.

Note: Among them, H_0 is the height of wood frame and H is the height of structure.

10.4.4 The monitoring limit value for roof structures should meet the following
requirements:

1. The rafter system deflection is 1/120–1/60 of rafter span;
2. The midspan deflection of purlin system is $L/120$–$L/60$ (when L is less than
   3 m) or $L/150$–$L/70$ (when L is more than 3 m);
3. Purlin support length is 50–90 mm (when supported on wood members) or 100–
   190 mm (when supported on masonry);

Note: Among them, L is the calculated span of purlin.

10.4.5 The floor structure monitoring limit value should meet the following
requirements:

1. The vertical deflection of grate (corrugated wood) is $L/220$–$L/100$;
2. The lateral bending height is $L/250$–$L/120$;
3. When there is no reliable anchorage at the end, the supporting length is 50–
   100 mm.

Note: L is the calculated span of the truss.

II Masonry Structure During Post Construction

10.4.6 For protective buildings with timber frame as the main load-bearing system, the monitoring limit value of brick wall should meet the following requirements:

1. When the brick is weathered for more than 1 m, the ratio of average weathering depth to wall thickness is 1/6–1/3 ($H < 10$ m) or 1/7–1/4 ($H \geq 10$ m);
2. The inclination of single-storey buildings is $H/180–H/100$ or $B/8–B/4$ (when $H < 10$ m); or $H/180–H/100$ or $B/9–B/5$ (when $H \geq 10$ m);
3. The total inclination of multi-storey buildings is $H/150–H/70$ or $B/7–B/4$ (when $H < 10$ m); or $H/150–H/70$ or $B/9–B/5$ (when $H \geq 10$ m);
4. The inter-storey inclination of multi-storey buildings is $H_i/110–H_i/60$ or 30–60 mm.

Note: H is the total height of the wall; H_i is the height of the inter-layer wall; B is the wall thickness. If the wall thickness varies from top to bottom, the average is adopted.

10.4.7 The monitoring limits of non-bearing earth or rubble walls in protective buildings should meet the following requirements:

1. Cob wall
   (1) The wall is inclined to 1/90–1/40 of the wall height;
   (2) The weathering and nitrification depth of the wall is 1/5–1/3 of the wall thickness;
   (3) The wall has obvious local sinking or bulging deformation;

2. Rubble wall
   (1) The inclination of the wall is 1/100–1/50 of the wall height;
   (2) The wall surface has been damaged seriously, which has seriously affected its use function.

10.4.8 In protective buildings with wooden roofs, the monitoring limit values of load-bearing stone pillars should meet the following requirements:

1. The ratio of weathered layer area to total section area is 1/7–1/4 on column section;
2. Horizontal or oblique cracks caused by stress have fine cracks visible to the naked eye, while longitudinal cracks (cracks only longer than 300 mm) have more than one, and the width of the cracks is greater than 0.1–0.15 mm;
3. The inclination of single-storey buildings is $H/250$ or 50 mm;
4. The total inclination of multi-layer columns is $H_i/170$ or 50 mm, and the inter-layer inclination is $H_i/120$ or 40 mm;
5. There is no reliable connection between the stigma and the upper wooden frame, or the connection has been loosened or damaged;
6. The ratio of the actual confined area between the base of column and the base of column to the base area of column is 2/3;

7. The ratio of the dislocation position between the column and the column foundation to the size of the column diameter (or section) along the dislocation direction is 1/7–1/4.`

Note: H is the total height of the wall; H_i is the height of the inter-layer wall.

10.4.9 The monitoring limit values of stone beams and loquats in protective buildings should meet the following requirements:

1. Surface weathering, the area of component section is 1/10–1/5 of the total section area;
2. There are transverse or oblique cracks;
3. At the end of the member, the depth of horizontal crack is 1/5–1/3 of the cross-section width.

III Reinforced Concrete Structure During Post Construction

10.4.10 The inclination limit values of reinforced concrete protective building structures shall be implemented in accordance with Table 10.4.10:

Table 10.4.10 Limit value of inclination of reinforced concrete protective building structures

Structural height	Inclined value
$H \leq 24$	0.004–0.008
$24 < H \leq 60$	0.003–0.006
$60 < H \leq 100$	0.0025–0.005
$H > 100$	0.0015–0.003

Note H is the height (m) of the building calculated from the outdoor surface, and the slope value is the ratio of the settlement difference between the two ends of the inclined direction of the foundation and its distance

10.4.11 The limited crack width of reinforced concrete protective building structure is 1.0–1.5 mm.

IV Steel Structure During Post Construction

10.4.12 The limit value of weekly tilt rate of steel protective buildings is $H/1200$–$H/600$, and the limit value of daily tilt rate is $H/9000$–$H/4000$.

10.4.13 The limit values of horizontal displacement and deformation rate of steel structural protective building connection joints are 30–60 mm per week and 4–8 mm per day.

10.5 Dilapidated Building

10.5.1 The limit value of foundation settlement rate of dilapidated buildings is 3–6 mm per month for two consecutive months and 0.1–0.15 mm per day for real-time monitoring.

10.5.2 Regular monitoring of masonry dilapidated buildings is limited by 2–3 mm of the difference between each two inclination monitoring, and regular monitoring of slope angle deformation rate limits are $H/1000–H/600$ per week. Real-time monitoring of slope angle deformation rate limits are $H/9000–H/4000$ per week, and deformation rate of crack width is limited to 1–1.5 mm per week.

10.5.3 Regular monitoring of concrete structures for dilapidated buildings is limited to 2–3 mm for each tilt monitoring difference, regular monitoring of the slope angle deformation rate limit value is $H/800–H/500$ per week, real-time monitoring of slope angle deformation rate limit value is $H/7000–H/4000$ per week, deformation rate of crack width is limited to 1–1.5 mm/week, deformation limit value of roof truss and main girder for regular monitoring is $L/250–L/120$ per week, and deformation limit value of real-time monitoring is $L/1800–L/900$ per day.

Note: L is the calculated span of the roof truss or main beam.

10.5.4 Regular monitoring of timber structures for dilapidated buildings is limited by 1–1.5 mm of the difference between each two inclination monitoring, regular monitoring of the slope angle deformation rate limit value is $H/150–H/80$ per week, real-time monitoring of slope angle deformation rate limit value is $H/1000–H/500$ per week, deformation limit value of roof truss and main girder for regular monitoring is $L/180–L/100$ per week, and deformation limit value of real-time monitoring is $L/1200–L/600$ per day.

Note: L is the calculated span of the roof truss or main beam.

10.5.5 Regular monitoring of steel structures for dilapidated buildings is limited by 1–1.5 mm of the difference between each two inclination monitoring, regular monitoring of the slope angle deformation rate limit value is $H/18–H/600$ per week, real-time monitoring of slope angle deformation rate limit value is $H/1000–H/500$ per week, deformation limit value of roof truss and main girder for regular monitoring is $L/300–L/150$ per week, and deformation limit value of real-time monitoring is $L/1800–L/1000$ per day.

Note: L is the calculated span of the roof truss or main beam.

10.6   Industrial Factory

10.6.1 When there is a power machine inside an industrial factory, the allowable vibration limit value should meet the following requirements:

1. The peak vibration velocity of piston compressor foundation is 4–8 and 10–15 mm/s under isolation foundation condition;
2. The peak vibration displacement of the turbine motor unit is 0.02–0.05 mm; the vibration speed is 3–6 mm/s;
3. The peak displacement of the forging hammer base is 1–1.5 mm; the vibration acceleration is 4–8 mm/s^2;
4. The peak vibration displacement of the press base is 0.4–0.8 mm;
5. The basic vibration speed of the piston engine is 10–15 mm/s;
6. The vibration test bed has a peak vibration velocity of 6–12 mm/s; the vibration acceleration peak is 0.6–1.2 mm/s^2;
7. The peak of basic vibration velocity of other general mechanical is 4–8 and 10–15 mm/s under the condition of vibration isolation.

10.7   Mass Concrete Structure

I Construction

10.7.1 The temperature rise limit of concrete pouring body is 40–80 °C on the basis of molding temperature.

10.7.2 The limit value of temperature difference between inner surface and outer surface of concrete pouring block is 20–40 °C.

10.7.3 The limited cooling rate of concrete pouring body is 2.0–3.0 °C/day.

10.7.4 The limit value of the difference between the surface of concrete pouring body and the air temperature is 15–30 °C.

10.8   Prefabricated Structure

10.8.1 The deformation limit values of fabricated concrete structural members should comply with the requirements of Table 10.8.1.

Table 10.8.1 Deformation limit values of prefabricated concrete structural members

Monitoring items	Member category		Deformation limit value
Deflection	Main bending members-main beam, bracket beam, etc.		$l_0/250$–$l_0/125$
	General flexural members	$l_0 \leq 7$ m	$l_0/150$–$l_0/75$, or 38–75 mm
		7 m $< l_0 \leq 9$ m	$l_0/200$–$l_0/100$, or 40–80 mm
		$l_0 > 9$ m	$l_0/200$–$l_0/100$
Height of lateral bending	Prefabricated roof beam or deep beam		$l_0/500$–$l_0/250$

Note l_0 is the calculation span

Table 10.8.2 Deformation limit values of prefabricated steel structural members

Monitoring items	Member category			Deformation limit value
Deflection	Main members	Grid	Roof (short direction)	$l_s/300$–$l_s/150$
			Floor (short direction)	$l_s/250$–$l_s/125$
		Main girder and joist		$l_0/250$–$l_0/125$
	General members	Other beams		$l_0/200$–$l_0/100$
		Purlin beam		$l_0/130$–$l_0/70$
Height of lateral bending	Deep beam			$l_0/500$–$l_0/250$
	Solid-web beam			$l_0/450$–$l_0/225$

Note l_0 is the calculation span; l_s is the calculation span of grid in short direction

10.8.2 The deformation limit values of fabricated steel structural members should comply with the requirements of Table 10.8.2.

10.8.3 The deformation limit values of fabricated steel structural members should comply with the requirements of Table 10.8.3.

Table 10.8.3 Deformation limit values of prefabricated timber structural members

Monitoring items		Deformation limit value
Deflection	Truss (roof truss, bracket)	$l_0/250$–$l_0/125$
	Main beam	$l_0^2/(4000\,h)$–$l_0^2/(2000\,h)$ 或$l_0/200$–$l_0/100$
	Joist or purlin	$l_0^2/(3000\,h)$ ~$l_0^2/(1500\,h)$ 或$l_0/150$–$l_0/75$
	Rafter	$l_0/125$–$l_0/65$
Height of lateral bending	Columns or other compressive members	$l_c/250$–$l_c/125$
	Beam with rectangular section	$l_0/200$–$l_0/100$

Note l_0 is the calculation span, l_c is the unsupported length of the column, and h is the height of the section

10.9   Isolated (Damper-Added) Structure

I Construction

10.9.1 Vertical isolation joints should be set around the upper structure of isolation structure, and the width of the joints should not be less than 1–2 times of the limit value of horizontal displacement of isolation bearings under rare earthquakes, and not less than 150–300 mm. For two adjacent isolation structures, the maximum horizontal displacement of the joint width should be taken as the sum, and the limit value should not be less than 300–600 mm.

10.9.2 A fully penetrated horizontal isolation joint should be set between the upper structure and the lower structure of the isolation structure, and the limited height of the joint should be 15–30 mm.

10.9.3 Isolated (damper-added) structure shall comply with the other provisions of this chapter on the limited values of structures during construction.

II Post Construction

10.9.4 Vertical compressive stress limits of rubber isolation bearings on the representative values of gravity loads shall be implemented in accordance with Table 10.9.4.

Table 10.9.4 Compressive stress limits of rubber isolation bearings

Building category	Class A building	Class B building	Class C building
Compressive stress limits (MPa)	8–10	10–12	12–15

10.9.5 The limit horizontal deformation of isolation bearings under the compressive stress listed in Table 10.9.4 should be greater than the larger values of 0.55 times the effective diameter and 3 times the total rubber thickness of the bearings.

10.9.6 The limit value of tension stress of isolation bearing is 1–1.5 MPa.

10.9.7 The limited values of the inter-storey displacement angles of structures above ground level below the isolation layer under rare earthquakes shall be implemented in accordance with Table 10.9.7.

Table 10.9.7 The limited values of the inter-storey displacement angles of structures above ground level below the isolation layer under rare earthquakes

Substructure type	$[\theta_p]$
Reinforced concrete frame structure and steel structure	1/120–1/60
Reinforced concrete frame-seismic wall	1/250–1/120
Reinforced concrete seismic wall	1/300–1/150

10.9.8 Isolated (damper-added) structure shall comply with the other provisions of this chapter on the limit values of structures during post construction.

Chapter 11
Project Acceptance

11.1 General Requirements

11.1.1 Before the handover of each project, it shall be checked and accepted, and records shall be made; the comprehensive test shall be carried out before the completion of the monitoring system, and the reasons for the unqualified items shall be analyzed and rectified in time to ensure the acceptance of the project quality and handover.

11.1.2 After the sensor installation, cable laying, collection station, and machine room construction and software deployment are completed within the scope of the design document, the acceptance of system divisional work can be carried out.

11.1.3 When the system software and hardware joint debugging is completed, in accordance with the design documents and the provisions of this specification, the monitoring system can be opened for operation and enter the trial operation phase, and the system phase acceptance can be performed at the same time.

11.1.4 The completion and acceptance of the monitoring system project shall be in accordance with the relevant provisions of the current national standard "Unified standard for constructional quality acceptance of building engineering". On the basis of checking various records, data, and inspection system construction, conclusions can be drawn on the project quality and fill in the project quality completion acceptance form.

11.1.5 After 90 days of continuous normal operation, and after passing the verification and completing the file system, the system acceptance test will be completed. The project manager shall fill in the handover, takeover record, and the relevant personnel of the construction and the supervision units shall confirm their signatures.

© The Author(s), under exclusive license to Springer Nature Singapore Pte Ltd. 2020

65

Y. Yang, *Technical Code for Monitoring of Building Structures*,

https://doi.org/10.1007/978-981-15-1049-6_11

11.1.6 The project manager shall fill in the handover record, and the relevant personnel of the construction and the supervision units shall confirm the signature.

11.1.7 The compilation and filing of the monitoring system engineering documents and the acceptance and handover of the engineering archives shall comply with the relevant provisions of the national standard "Code for Construction Project Document Filing and Arrangement" GB/T 50328.

11.1.8 Construction acceptance shall include the following contents:

1. Installation of equipment, equipment and accessories;
2. Data acquisition, transmission, conversion, and control functions of monitoring system;
3. Structural damage warning function, recording and displaying function and system fault self-checking function of monitoring system;
4. Control function, monitoring function, display function, and recording function of video and image monitoring system.

11.2   Sub-Item Project Acceptance

11.2.1 The monitoring system can submit a written application to the owner for the acceptance of divisional work after it has the conditions for acceptance of divisional work.

11.2.2 The division of divisions and sub-projects shall comply with the following provisions:

1. The sub-division project shall be divided into several sub-division projects according to the characteristics of the implementation process;
2. In the sub-project (sub-division project), it shall be divided into sub-projects according to the characteristics of the implementation process;
3. The division of the monitoring system division project and sub-project can refer to the provisions of Appendix B of "Standard for Construction Quality Acceptance" GB50300.   •
4. The sub-project of the monitoring system combined with the BIM model should include data sharing of the BIM model and visualization of the 3D model;
5. The divisional engineering and sub-project division of the cloud platform monitoring system can be performed with reference to Table 11.2.2.

Table 11.2.2 Divisional engineering and sub-project division of cloud platform monitoring system

Divisional project	Sub-division project	Sub-item project
Cloud platform monitoring system	Sensor system	Sensor installation, cable laying, electromagnetic shielding, system commissioning, commissioning
	Information network system	Computer network equipment installation, computer network software installation, network security equipment installation, network security software installation, system debugging, commissioning
	Cloud platform system	Cloud platform management software installation, large database testing, system debugging, commissioning
	Integrated wiring system	Ladder frame, tray, box and conduit installation, cable laying, cabinet, rack, distribution frame installation, information socket installation, link or channel test, software installation, system commissioning, commissioning
	Engine room	Power supply and distribution system, lightning protection and grounding system, air conditioning system, water supply and drainage system, integrated wiring system, monitoring and security system, fire protection system, interior decoration, electromagnetic shielding, system debugging, commissioning
	Lightning protection and grounding	Grounding device, grounding wire, equipotential bonding, shielding facility, surge protector, cable laying, system commissioning, commissioning

11.2.3 The hardware and software of the monitoring system should be fully tested when entering the site.

11.2.4 The number of random inspections for hardware installation quality inspection shall meet the following requirements:

1. The installation quality of the power supply equipment should be fully tested;
2. Sensors should be classified according to test principle and use. Each type of sensor should be sampled 30% and not less than one;
3. Data collection equipment should be classified according to the collection principle and use. Each type of collection equipment, data transmission, and processing equipment should be sampled at 10% and not less than one.

11.2.5 The number of inspections for system commissioning shall comply with the following requirements:

1. Sensors of the same model, specification, and production lot in the hardware should be divided into a verification batch during system verification. The number of system verification samples for the same sensor verification batch

shall not be less than 10% of the number of sensors in the batch and not less than one;

2. The data acquisition equipment and data transmission equipment in the hardware should be fully tested;
3. The monitoring software should be fully tested.

11.2.6 When the division project is accepted, the following documents should be submitted:

1. Installation and quality inspection records;
2. Concealed engineering records;
3. Grounding resistance measurement record;
4. Technical documents such as product quality certificate and manual for hardware and materials;
5. The monitoring system combined with the BIM model should submit a test record of the BIM model.

11.3 Completion Acceptance

11.3.1 After the system has the condition of completion acceptance, it should apply to the construction (supervision) unit for completion acceptance.

11.3.2 The following items shall be completed before submitting the completion acceptance report:

1. All the problems found in previous inspections and supervision have been dealt with;
2. The filing documents conform to the relevant regulations of the management of project document.

11.3.3 The list of engineering data includes:

1. Application report for acceptance of completion;
2. Completion drawing;
3. Final account sheet;
4. Drawing review record, design change consultation record;
5. Quality certificate of materials and hardware;
6. Construction records. Including necessary inspection and test record;
7. Intermediate handover record and certification;
8. Record of accident occurrence and treatment of engineering quality;
9. Other technical decisions and information on the project;
10. System specification;
11. Operating system description;

12. List of system hardware and software;
13. Internal wiring diagram of the system;
14. Systematic approval data;
15. Documents formed during the software implementation process specified in this specification;
16. System joint debugging report;
17. The monitoring system combined with BIM model should submit the debugging receiving report of data transmission, information sharing, and model visualization;
18. Third-party software testing report.

Appendix A: Technique Requirement of Monitoring Equipment

A.0.1 The types of sensors and related instruments for monitoring should be implemented in accordance with Table A.0.1.

Table A.0.1 Selection of sensor instrument type

Type of monitoring	Monitoring content	Sensor type
Load source monitoring	Wind load	Mechanical anemometer and ultrasonic anemometer
	Temperature	Thermometer
	Humidity	Hygrometer
	Earthquake	Accelerometer
Structural response monitoring	Stress	Fiber grating strain gauge and vibrating wire strain gauge
	Cable force	Accelerometer, anchor gauge, magnetic flux sensor, fiber grating cable
	Deflection	Global positioning system, inclinometer
	Spatial displacement	Global positioning system, inclinometer
	Vibration	Speedometer, accelerometer
	Geometric line	Static level, displacement gauge, inclinometer, global positioning system, automatic total station
	Expansion joint displacement	Displacement gauge

© The Author(s), under exclusive license to Springer Nature Singapore Pte Ltd. 2020
Y. Yang, *Technical Code for Monitoring of Building Structures*,
https://doi.org/10.1007/978-981-15-1049-6

A.0.2 The main technical indicators of the accelerometer should be implemented in accordance with Table A.0.2.

Table A.0.2 Main technical indicators of accelerometer

Item	Force balance accelerometer	Electric accelerometer	ICP piezoelectric accelerometer
Sensitivity (V/ (m/s^2))	0.125	0.3	0.1
Full-scale output (V)	±2.5	±6	±5
Frequency response (Hz)	0–80	0.2–80	0.3–1000
Dynamic range (dB)	≥ 120	≥ 120	≥ 110
Linearity error (%)	≤ 1	≤ 1	≤ 1
Operating temperature (°C)	−30 to +80	−30 to +80	−30 to +80
Signal adjustment	Linear amplification, integration	Linear amplification, integration	ICP conditioning amplification

A.0.3 The main technical indicators of the speedometer should be implemented in accordance with Table A.0.3.

Table A.0.3 Main technical indicators of speedometer

Item	Technical indicators	Remarks
Sensitivity (V/(m/s))	0.7–23	Adjustable
Full-scale output (V)	±5	
Frequency response (Hz)	0.14–100	Adjustable
Dynamic range (dB)	≥ 120	
Linearity error (%)	≤ 1	
Operating temperature (°C)	−30 to +80	
Signal adjustment	Linear amplification, integration, filtering	

A.0.4 The main technical specifications of the recorder used for accelerometer and speedometer should be implemented in accordance with Table A.0.4.

Table A.0.4 Main technical indicators of recorder

Item	Technical indicators	Item	Technical indicators
Number of channels	≥ 4	Sampling rate	Program control, at least 2 files, the highest sampling rate not less than 1000 SPS
Full-scale input (V)	± 10 V	Time service	Standard UTC, internal clock stability is better than 10–6, synchronization accuracy is better than 1 ms
Dynamic range (dB)	≥ 120	Data communication	RJ45 network port, 100 Mbps bandwidth communication rate
Conversion accuracy (bit)	≥ 24	Data storage	CF card flash memory, ≥ 64 GB
Trigger mode	Bandpass threshold trigger, STA/LTA ratio trigger, external trigger	Inter-channel delay	0
Ambient temperature (°C)	-20 to $+80$	Software	Includes communication programs, graphical display programs, other utilities and monitoring, diagnostic commands
Environment humidity	<80%		

A.0.5 The main technical indicators of strain gauges should be implemented in accordance with Table A.0.5.

Table A.0.5 Main technical indicators of strain gauges

Sensor	Technical indicators	Technical requirement
Strain sensor	Measuring range ($\mu\varepsilon$)	± 1500
	Resolution ($\mu\varepsilon$)	1
	Sampling frequency (Hz)	≥ 10
	Operating temperature (°C)	-30 to 85
	Measurement accuracy ($\mu\varepsilon$)	± 3
	Service life	Conforms to current national/industry equipment standards

A.0.6 The technical indicators of the fiber strain monitoring gauge should be implemented in accordance with Table A.0.6.

Table A.0.6 Main technical indicators of optical fiber strain gauge

Item	Technical indicators
Standard range ($\mu\varepsilon$)	±1500
Accuracy (FS)	0.3%
Sensitivity (FS)	0.1%
Standard distance (mm)	40
Operating temperature (°C)	−30 to +100

A.0.7 The main technical indicators of global navigation satellite system (GNSS) positioning should be implemented in accordance with Table A.0.7.

Table A.0.7 Main technical indicators of deformation sensor

Global navigation satellite positioning system (GNSS)	Static baseline accuracy	
	Horizontal	3 mm + 0.3 ppm
	Vertical	5 mm + 0.3 ppm
	Sampling frequency (Hz)	20
	Operating temperature (°C)	−30 to 80
	Service life	Conforms to current national/industry equipment standards

A.0.8 The main technical indicators of the displacement gauges should be implemented in accordance with Table A.0.8.

Table A.0.8 Main technical indicators of displacement gauges

Motion detector	Measuring range (mm)	10–1000
	Accuracy (mm)	0.1
	Resolution (mm)	0.01
	Operating temperature (°C)	−30 to 80
	Service life	Conforms to current national/industry equipment standards

A.0.9 The main technical indicators of the hygrometer should be implemented in accordance with Table A.0.9.

Table A.0.9 Main technical indicators of hygrometer

Hygrometer	Measuring range (RH)	12–99
	Accuracy	±2%
	Stability (RH/year)	<1%
	Operating temperature (°C)	−30 to 80
	Service life	Conforms to current national/industry equipment standards

A.0.10 The main technical indicators of the thermometer shall be implemented in accordance with Table A.0.10.

Table A.0.10 Main technical indicators of the thermometer

Thermometer	Measuring range (°C)	−30 to 85
	Accuracy (°C)	±0.1
	Operating temperature (°C)	−30 to 80
	Service life	Conforms to current national/industry equipment standards

A.0.11 The main technical indicators of the inclinometer shall be in accordance with Table A.0.11.

Table A.0.11 Main technical indicators of the inclinometer

Inclinometer	Range	±14.5°
	Sensitive axis misalignment	<0.15°
	Zero deviation	<0.15°
	Nonlinearity	<0.005°
	Bandwidth	∼3 dB, (Typical value): 30 Hz
	Full-scale output	±5.00 V ± 0.5%
	Sensitivity non-alignment (ppm/°C)	<100
	Zero temperature coefficient (/°C)	0.003°
	Resolution	<0.00005°
	Repeatability	<0.003°
	Noise (rms)	<0.0002°
	Operating temperature (°C)	−30 to 80
	Service life	Conforms to current national/industry equipment standards

A.0.12 The main technical indicators of the static level should be implemented in accordance with Table A.0.12.

Table A.0.12 Main technical indicators of static level

Static level	Range (mm)	≥ 10
	Accuracy (mm)	± 0.3
	Operating temperature (°C)	-30 to 80
	Service life	Conforms to current national/industry equipment standards

A.0.13 The main technical indicators of the automatic total station shall be implemented in accordance with Table A.0.13.

Table A.0.13 Main technical indicators of automatic total station

Automatic total station	Angle measurement accuracy	$\leq 1''$
	Ranging accuracy	1 mm + 2 ppm
	Range (average atmospheric conditions)	2.5 km single prism/3.5 km tri prism
	Single measurement time (s)	≤ 3.5
	Telescope magnification	30×
	ATR function	1000 m single prism/600 m 360° prism
	Operating temperature (°C)	-30 to 80
	Service life	Conforms to current national/industry equipment standards

A.0.14 The main technical indicators of the anchor gauge shall be implemented in accordance with Table A.0.14.

Table A.0.14 Main technical indicators of anchor gauge

Anchor gauge	Accuracy level	3%
	Applicable range	0.2Fnom–Fnom
	When the installation position is unchanged	0.5%
	When the installation position changes	1.5%
	Repeatability error	0.3% (0.2Fnom–Fnom)
	Linearity error	0.3%
	Temperature effect on sensitivity/10 °C	0.25% (After compensation 0.05%)
	Temperature effect on zero output/10 °C	0.2%
	Eccentricity effect	0.4%/5 mm
	Creep	0.1%/30 min
	Zero drift	0.2%/1 year
	Stored temperature (°C)	−40 to 70
	Maximum working load	120%
	Ultimate safety overload	150%
	Destructive load	250%
	Allowable stress amplitude	70%
	Operating temperature (°C)	−30 to 80
	Service life	Conforms to current national/industry equipment standards

A.0.15 The main technical indicators of the wind pressure gauge shall be implemented in accordance with Table A.0.15.

Table A.0.15 Main technical requirements of wind pressure gauge

Technical indicators	Technical requirement
Range (kPa)	±1.25
Precision	0.5%
Sampling frequency (Hz)	0.02
Service life	Conforms to current national/industry equipment standards

A.0.16 The main technical indicators of the anemometer shall be implemented in accordance with Table A.0.16.

Table A.0.16 Main technical requirements for anemometer

Sensor	Technical indicators	Technical requirement
Ultrasonic anemometer (using ultrasonic and mechanical anemometer to work together)	Wind speed measurement range (m/s)	0–60
	Wind speed resolution (m/s)	0.01
	Wind speed measurement accuracy	±3%
	Wind direction azimuth measurement range	0°–360°
	Wind direction resolution	0.1°
	Wind direction measurement accuracy	≤±1°
	Sampling frequency (Hz)	≥10
	Operating temperature (°C)	−30 to 80
	Service life	Conforms to current national/industry equipment standards
Mechanical anemometer	Wind speed measurement range (m/s)	0–60
	Wind speed measurement accuracy	±0.1 m/s
	Wind direction azimuth measurement range	0°–360°
	Wind direction measurement accuracy	≤±3°
	Sampling frequency (Hz)	≥10
	Operating temperature (°C)	−30 to 80
	Service life	Conforms to current national/industry equipment standards

A.0.17 The main technical indicators of the wind vibration monitoring system signal conditioning instrument shall be implemented in accordance with Table A.0.17.

Table A.0.17 Technical requirement of wind vibration monitoring system signal conditioning instrument

Technical indicators	Technical requirement
Magnification	1–1000
Integral constant	≥ 2
Equivalent input noise (effective value) (V)	$\leq 10^{-5}$
Input resistance (Ω)	$\geq 10^6$
Filter	Low pass or band pass
Power consumption (mA)	≤ 100
Range of operating temperature (°C)	-30 to 80
Relative humidity	80%

A.0.18 The technical indicators of the horizontal displacement monitoring sensor of the isolation bearing should be implemented in accordance with Table A.0.18.

Table A.0.18 Technical indicators of horizontal displacement gauges for isolation bearings

Item	Technical indicators
Maximum measurable displacement (cm)	± 50
Frequency range (Hz)	0–5 (when the length of pulling line equals to 5 m)
Sensitivity (V/m)	10
Linearity	$\leq 0.2\%$
Resolution (mm)	0.2

A.0.19 The signal acquisition analyzer consists of acquisition card and analysis software. The acquisition card technology of the signal acquisition analyzer should be implemented in accordance with Table A.0.19.

Table A.0.19 Main technical indicators of signal acquisition analyzer acquisition card

Item	Technical indicators
Sampling rate (SPS)	50–1000 per channel
A/D digit	Not less than 16 digits (effective digits no less than 14 digits)
Sampling method	Acquisition channel synchronization, using separate A/D per channel
Dynamic range (dB)	≥ 80
Input range (V)	± 10
Interface	USB interface, LAN interface
Data storage length	Sample data for not less than 5 h

A.0.20 The main technical indicators of the fiber grating demodulator should be implemented in accordance with Table A.0.20.

Table A.0.20 Main technical requirements of fiber grating demodulator

Acquisition equipment	Technical indicators	Technical requirement
Fiber grating demodulator	Number of channels	≥ 4
	Wavelength range (nm)	≥ 40
	Resolution (pm)	0.2
	Repeatability (pm)	2
	Dynamic range (dB)	25
	Fiber optic connector	FC/APC
	External data interface	RJ45 USB
	Operating temperature (°C)	−30 to 80
	Service life	Conforms to current national/industry equipment standards

A.0.21 The main technical indicators of data transmission equipment should be implemented in accordance with Table A.0.21.

Table A.0.21 Main technical requirements of data transmission equipment

Transmission device	Technical indicators	Technical requirement
Fiber coupler	Optical interface type	FC/APC
	Working wavelength (nm)	1310 or 1550
	Bandwidth (nm)	±20
	Accessory loss (dB)	≤ 0.1
	Uniformity (dB)	≤ 0.6
	Polarization flatness (dB)	≤ 0.1
	Directionality (dB)	≥ 55
	Operating temperature (°C)	−30 to 80
	Package form	Standard rack box
	Service life	Conforms to current national/industry equipment standards

(continued)

Table A.0.21 (continued)

Transmission device	Technical indicators	Technical requirement
Cable	Transmission performance	Exceed TIA/EIA ~ 568B.2 ~ 1 standards
	Impedance	100 Ω ± 15%, (1–600) MHz
	Transmission delay	536 ns/100 m max. @ 250 MHz
	Delay offset	45 ns max
	Conductor resistance	66.58 Ω max/km
	Capacitance	5.6 NF max/100 m
	DC resistance (Ω)	≤7.55
	Withstand voltage	300 V (AC or DC)
	Bending radius	1 in. (4 times cable diameter)
	Rated rate	70 nom%
	UL/NEC rating	CMR
	Certification	UL listed file no. E154336
	Working temperature (°C)	−30 to 80
	Storage and transportation environment temperature (°C)	−30 to 80
	Wire	—23AWG solid bare copper
	Insulator	Monomer 042 in./overall 20 in.
	Outer skin	FR PVC
	Weight (bs/mft)	261
	EIA/TIA 568/A universal line sequence	Fit
	High-speed communication of voice, data and most media	Fit
	10Base T/100Base Tx fast Ethernet, gigabit Ethernet, and 622 Mbps ATM, token ring, and many other network types	Fit
Communication cable	Loss	1300/1.0 dB/km, 1500/0.7 dB/km
	Transmission distance (km)	≥2
	Network rate (Gps)	10
	Operating temperature (°C)	−30 to 80
	Service life	Conforms to current national/industry equipment standards

(continued)

Table A.0.21 (continued)

Transmission device	Technical indicators	Technical requirement
Communication cable	20 °C core DC resistance (Ω/km)	≤ 45
	Cable inherent attenuation (800 Hz) (dB/km)	≤ 1.10
	+20 °C insulation resistance (MΩ/km)	≥ 3000
	1 min power frequency AC voltage test	1500 V not breakdown (50 Hz)
	Far-end crosstalk attenuation (dB/500)	≥ 70 (800 Hz)
	Inductance (μH/km)	≤ 800 (800 Hz)
	Cable working capacitor (μF/km)	≤ 0.06 (800 Hz)
	Working to DC resistance difference	$\leq 2\%$ loop resistance
	Single vertical burning test	MT386
	Operating temperature (°C)	-30 to 80
	Service life	Conforms to current national/industry equipment standards

Appendix B: Technique Requirement of Cloud Platform

B.0.1 The cloud platform monitoring system shall have the following functions:

1. It has real-time monitoring information, and provides two viewing modes: live viewing and catalog viewing;
2. It is able to count and compare the monitoring data, which should include statistical comparison of data, alarm data analysis, and system event statistics within a certain period of time;
3. It has the function to record historical data and analyze trend. All historical data should be saved, and a variety of historical data query methods should be provided. Trend analysis function should be provided according to historical data, and trend graphs should be given for reference by relevant personnel;
4. It has alarm and event functions. When there is problem with sensor or structure during service, it should promptly alarm and record the event.

B.0.2 The design of the cloud platform monitoring system should follow the following principles:

1. In the sensory control layer, the sensor system should collect nodes to sense and monitor surrounding objects and events, and should obtain corresponding data;
2. The network transport layer shall ensure that the information data acquired by the sensing control layer is unobstructed, efficient, reliable, and highly secure;
3. The data aggregation layer should realize the storage and management of the monitoring data of each structure;
4. The application support layer should solve the intelligent processing and efficient management of massive information, and provide human–computer interaction interface. It should support intelligent early warning, comprehensive evaluation, presenting report, and data downloading;
5. The information output layer should output information to different terminal devices.

© The Author(s), under exclusive license to Springer Nature Singapore Pte Ltd. 2020
Y. Yang, *Technical Code for Monitoring of Building Structures*,
https://doi.org/10.1007/978-981-15-1049-6

B.0.3 The monitoring database in cloud platform should fulfill the following function:

1. Cloud platforms should manage structured and non-structured data in a distributed relational database to a centralized data center manner;
2. When implementing relational database unified management system data, the massive data management and performance should be considered;
3. The security database should be responsible for recording the dynamic data and static attribute files at the structural service status, administrating the unprocessed structural basic attribute data in a standardized manner, and updating information dynamically. At the same time, it should have the functions of supervision and service. Based on the acquired various types of data, to match data information and secondary processing according to a professional assessment system, it should be able to search for information, and identify key words from the data source in need.

B.0.4 The cloud platform monitoring security information platform should have the following regulatory and service functions:

1. The function positioning and permission security constraints should be implemented hierarchically;
2. The information classification and statistical query should be provided with figures and tables;
3. The information on structural safety evaluation should be calculated with multi-period and multi-evaluation index in a programmatic and regularized manner;
4. The dynamic information evaluation results should be classified automatically. The classification rules for automatic classification results should be formulated for different needs. The management maintenance plan strategy should be developed based on the automatic classification and sorting results.

B.0.5 The design of the cloud platform monitoring system architecture should follow the following provisions:

1. A mature operating system, development language, and development framework should be used to ensure stability of the system;
2. Physical isolation, firewalls, and detailed access control should be used to ensure security of the system;
3. The design of the system should fully consider any possible future extension, and should adopt the layering design mode to ensure the interface and code.

B.0.6 The application layer shall include two modules, data display and structure access, and shall comply with the following provisions:

1. Data display module should comply with the following:

   (1) The system home page module should display the overall rating of the current structure that the user is most concerned with and reflect the real-time status of the structure;

   (2) The monitoring project data display module should provide query function of the sensor history and real-time data; after selecting the sensor type, node, and start time, the sensor data should be displayed in the form of graph and table on the web;

   (3) In the warning management module, the warning information generated by the structure under the current user can be viewed. According to the processed and unprocessed classification, the user can use the confirmation of the unprocessed warning to set the warning state to have been processed. When the system generates a new warning, the users who have installed the SMS alert functions will be notified by SMS;

   (4) In the report management, the corresponding data report should be generated according to the structure and monitoring factors. The report type has annual, monthly, and daily reports, which can be viewed and downloaded by the user.

2. Structure access module should comply with the following:

   (1) In addition to the operations of adding, modifying, and deleting structures, the structure management module should also include factors that need to be monitored for structure allocation, configure corresponding sensor groups, and set thresholds and weights for sensors;

   (2) The warning management module performs warning filtering in the background, sends the warnings to be displayed to the corresponding users, and can set rules for warning generation, and so on;

   (3) The report management module is applied to configure the report template presented to the user;

   (4) The user management template is applied to create or delete users accessing the cloud platform. After the user is created, the structure in the system should be assigned to the user. There should be a many-to-many relationship between the user and the structure.

Appendix C: Record Table of Vertical Displacement of Building

Building vertical displacement record				Number		
Project name				Measuring instrument		
Load accumulation description				Environmental conditions		
Last monitoring time				Current monitoring time		
Point number	Initial value	Last monitoring value	Current monitoring value	Current deformation value	Cumulative displacement value	Remarks
Recorder (signature)				Reviewer (signature)		

Appendix D: Record Table of Horizontal Displacement of Building

© The Author(s), under exclusive license to Springer Nature Singapore Pte Ltd. 2020

Y. Yang, *Technical Code for Monitoring of Building Structures*,

https://doi.org/10.1007/978-981-15-1049-6

Building horizontal displacement record Table				Number		
project name				Measuring instrument		
Load accumulation description				Environmental conditions		
Last monitoring time				Current monitoring time		
Point number	Initial value	Last monitoring value	Current monitoring value	Current deformation value	Cumulative deformation value	Remarks
Recorder (signature)				Reviewer (signature)		

Appendix E: Record Table of Installation of Stress and Strain Sensors

© The Author(s), under exclusive license to Springer Nature Singapore Pte Ltd. 2020
Y. Yang, *Technical Code for Monitoring of Building Structures*,
https://doi.org/10.1007/978-981-15-1049-6

Stress and strain sensor installation record table			Number		
Project name			Ambient temperature		
Structural part			Installation date		
Measuring point number	Sensor number	Sensor type	Pre-installation reading	Reading when installation is complete	Remarks

Installation diagram and site conditions:

Recorder (signature)		Tester (signature))	

Appendix F: Record Table of Stress and Strain

© The Author(s), under exclusive license to Springer Nature Singapore Pte Ltd. 2020
Y. Yang, *Technical Code for Monitoring of Building Structures*,
https://doi.org/10.1007/978-981-15-1049-6

Stress and strain observation record Table				Number						
Project name				Ambient temperature						
Structural part				Installation date						
Measuring point number	Sensor number	Sensor type	Modulus of elasticity	Initial reading	Previous reading	This reading	This increment	Cumulative increment	Stress/internal force	Remarks
Site conditions:										
Recorder (signature)			Tester (signature)							

Appendix G: Record Table of Environmental Conditions

Environmental condition record Table				Number			
Project name				Record content			
Test location				Test instrument			
Date	Time	Measuring point number	Test instrument	Temperature	Wind speed	Wind direction	Remarks
Recorder (signature)			Tester (signature)				

Appendix H: Record Table of Periodical Inspection

Project name		Record content		
Number	Structure location at inspection	Structural status description		Remarks
Recorder (signature)		Inspection time		

Y. Yang, *Technical Code for Monitoring of Building Structures*,
https://doi.org/10.1007/978-981-15-1049-6

Appendix I: Basic Information Table of Dilapidated Building

				Administrative district	Street township	Community	Street	Building number	House number
Dilapidated building location information	Dilapidated building number		Address						
Basic situation of dilapidated building	Usage of houses	□Residential □Non-residential □mixed			Residential area（m²）				
	The real function	□Residential □Non-residential □mixed			Actual area（m²）				
	Property nature	□Social private house □Self-managed public housing □Public house □Other							
	House structure	Basic type	□Strip foundation □Independent foundation □Pile foundation □Skeleton foundation □Box foundation						
		Foundation rock and soil properties	□Rock foundation □Soil foundation □Geotechnical mixed foundation						
		Main structure	□Masonry structure □Reinforced-concrete structure □Timber structure □Brick and timber structure □Simple structure □Steel structure □Other						
	Year of construction			Number of floors			Number of households		
	House belonging	Property owner		Property owner contact number		Number of people living		Relationship between domestic and property owners	
Decrepit house identification monitoring information	Decrepit house level	□Class C □Class D							
	Identification unit								
	Identification report number								
	Identification time								
	Regulatory unit				Regulatory unit telephone number				
	Monitoring person				Monitoring person telephone number				

Form filler: Reviewer: Date of filling the form: Year   Month   Day

Note:1   The information in the form must be consistent with the data of the housing safety system, and the file number should be compiled by the district;

    2   The "Houses belong" column is filled in by households, and a large number of households can be attached to the list.

Appendix J: Monitoring Point Layout Table of Dilapidated Building

Y. Yang, *Technical Code for Monitoring of Building Structures*,
https://doi.org/10.1007/978-981-15-1049-6

Dilapidated building number			House address					
Compilation unit			Technology advisor at the compilation unit area			Phone numbe r		
Monitoring person		Phone number	Township (street) in charge			Phone numb er		
Monitor Point arrangement graph								

Monitoring content	Local dangerous point measurement	Cr ac k	Homeowner (house number):	Monitoring part:□Wall □Ground □Column □Beam □Board floor ： Floor	Initial crack value	Ruler 1	Ruler 2	Ruler 3	Ruler 4
						Seam width （mm）	Seam width （mm）	Seam width （ mm ）	Seam length （cm）
			Monitoring point number:						
			Date of construction:			Early warning threshol d	Early warning threshold	Early warnin g thresh old	Early warning threshol d （cm）
	Oblique observation of the main structure	Tilt	Homeowner (house number):	Monitoring point location	Vertical height			Horizontal displacement distance	
			Monitoring point number:	Date of constructio n:					
	Macro inspection content								
Monitoring frequency		a week, inspections should be carried out after rainfall (or extreme weather). More frequent inspection should be carried out after observing abnormal results,		Monitoring item	□Macro inspection □Tilt □Crack □Other （ ）			
Monitoring report								
Primary protecting object	Household:		Houses:		Other:		Property:	

Layout person in charge：   Monitor：   Form filler：

Date of filling the form：   Year   Month   Day

Note: The contents of the table can be adjusted according to the actual situation of the danger zone.

Appendix K: Monitoring Record Table of Dilapidated Building

Local dangerous point measurement	Crack	Homeowner (house number):	Monitoring point number:	Monitoring part:□Wall □Ground □Column □Beam □Board floor: __Floor	Ruler 1	Ruler 2	Ruler 3	Ruler 4
					Initial slit width	Initial slit width	Initial slit width	Initial seam length
					Monitoring slit width	Monitoring slit width	Monitoring slit width	Monitoring slit width
					Value added	Value added	Value added	Value added
					Early warning threshold	Early warning threshold	Early warning threshold	Early warning threshold
Oblique observation of the main structure	Tilt	Homeowner (house number):	Monitoring point number:	Vertical height		Horizontal displacement distance		Displacement direction	Early warning threshold
Macro inspection									
Other changes in the monitoring point									
Additional dangerous points									

Monitor:　　　　　　　　　　　　　　　　　Monitoring date:　　Year　Month　Day

© The Author(s), under exclusive license to Springer Nature Singapore Pte Ltd. 2020
Y. Yang, *Technical Code for Monitoring of Building Structures*,
https://doi.org/10.1007/978-981-15-1049-6

Appendix L: Disaster Prevention Plan Table of Dilapidated Building

Dilapidated building number			Dilapidated building level	□Class C □Class D	
House address					
Threat object	Household		Houses		
	Property		Other		
Deformation of dilapidated building					
Stability analysis					
Priming factor					
Forecast of disaster status					
Monitor	Monitoring method	□Macro inspection □Oblique observation of the main structure □Local danger point measurement			
	Monitoring frequency	A week, inspections should be carried out after rainfall (or extreme weather). More frequent inspection should be carried out after observing abnormal results.			
	Monitoring site				
Township (street) in charge:		Phone number:	Monitor:		Phone number:
Warning signal		□Horn □Phone □Shout □Whistling □other (　)	Alerter:		Phone number:
Personnel evacuation route			Scheduled safe haven		
Prevention advice					
Dangerous zone and evacuation route Immediately after the danger occurs, evacuate to the outside of the house along the evacuation route.					

Note: The warning signal can be telephone, shouting, knocking, whistling, loud speaker, etc.

Appendix M: Single (Joint) Item Commissioning Report

© The Author(s), under exclusive license to Springer Nature Singapore Pte Ltd. 2020 107
Y. Yang, *Technical Code for Monitoring of Building Structures*,
https://doi.org/10.1007/978-981-15-1049-6

Project name		Debugging project	
Debug content		Construction unit	

National norms and technical standards (or design requirements):			
Debugging record			
Problems and comments to solution			
conclusion			

Construction unit	Design unit	Supervisory unit	Contractor
Site person in charge:	Person in-charge:	Supervising engineer: Chief Engineer:	Quality inspection engineer: Technical director: project manager:
Year Month Day	Year Month Day	Year Month Day	Year Month Day

Appendix N: Divisional Project Completion Application

Y. Yang, *Technical Code for Monitoring of Building Structures*,
https://doi.org/10.1007/978-981-15-1049-6

Project name			Division project name	
Contract start date		Year Month Day	Contract completion date	Year Month Day
Actual start date		Year Month Day	Actual completion date	Year Month Day
Project scope and content				
Early or delay notice				
Reporting requirements	The project items that are included in the project contract have been completed on year month day . After self-evaluation on the project quality, it has demonstrated that the project has reached relevant requirements. The construction unit is now applying to organize the completion application on year month day.			
Owner unit		Supervisory unit		Construction unit
Project manager:

(Official seal)
Year   Month   Day | | Chief Engineer:

(Official seal)
Year   Month   Day | | Project manager:

The person in charge:
(Official seal)
Year   Month   Day |

Appendix O: Divisional Project Quality Acceptance Record

Project name			Division project name		
Construction unit			Project manager		
Serial number	Sub-project name		Inspection lot	Construction unit inspection result	Comments
1					
2					
3					
4					
5					
6					
7					
8					
9					
10					
Quality control information					
Single and joint commissioning report					
Function acceptance					
Comprehensive acceptance conclusion					

Construction unit	Supervisory unit	Design unit	Contractor
Project manager:	Chief Engineer:	Project manager:	Project manager:
（Official seal） Year Month Day	（Official seal） Year Month Day	（Official seal） Year Month Day	（Official seal） Year Month Day

1 Explanation of Wording in This Code

1. In order to implement this code in different situations or conditions, words denoting the different degrees of strictness are explained as follows:

    (1) Word denoting a very strict or mandatory requirement:
    "must" is used for affirmation;
    "must not" is used for negation.
    (2) Words denoting a strict requirement under normal conditions:
    "shall" is used for affirmation;
    "shall not" is used for negation.
    (3) Words denoting a permission of slight choice or an indication of the most suitable choice when conditions allow:
    "should" or "may" is used for affirmation;
    "should not" is used for negation.

2. "be in compliance with" is used to indicate that it is compulsory to implement items in relevant codes and standards. "Refer to" is used to indicate that it is not compulsory to implement items in relevant codes and standards.

© The Author(s), under exclusive license to Springer Nature Singapore Pte Ltd. 2020   113
Y. Yang, *Technical Code for Monitoring of Building Structures*,
https://doi.org/10.1007/978-981-15-1049-6

2 Quoted Standard List of This Code

1. Technical code for monitoring of building and bridge structures GB 50982
2. Technical code for acceptance of dam safety monitoring system GB/T 22358
3. Standard for appraisal of reliability of civil buildings GB 50292
4. Standard for appraisal of reliability of industrial buildings and structures GB 50144
5. Code for construction of mass concrete GB 50496
6. Standard for allowable vibration of building engineering GB 50868
7. Code for engineering surveying GB 50026
8. Code for acceptance of construction quality of building electrical engineering GB 50303
9. Code for construction and quality acceptance of automation instrumentation engineering GB 50093
10. Technical code for construction and acceptance of machine room of electronic information system GB 50462
11. Standard for construction and acceptance of cable line electric equipment installation engineering GB 50168
12. Information technology-software life cycle processes GB/T 8566
13. Specification for computer software documentation GB/T 8567
14. Information security technology-common security techniques requirement for information system GB/T 20271
15. Specification of computer software testing GB/T 15532
16. Code for quality acceptance of concrete structures construction GB 50204
17. Code for acceptance of construction quality of steel structures GB 50205
18. Code for acceptance of construction quality of timber structures GB 50206
19. Technical code for monitoring of building excavation engineering GB 50497
20. Technical code for safety of temporary electrification on construction site JGJ 46
21. Technical code for computer science engineering in construction site JGJ/T 90
22. Code for deformation measurement of building and structure JGJ 8
23. Standard for dangerous building appraisal JGJ 125

© The Author(s), under exclusive license to Springer Nature Singapore Pte Ltd. 2020   115
Y. Yang, *Technical Code for Monitoring of Building Structures*,
https://doi.org/10.1007/978-981-15-1049-6

24. Technical code for construction process analyzing and monitoring of building engineering JGJ/T 302
25. Code for construction and acceptance of building isolation engineering JGJ 360
26. Specifications for verification and test of dam safety monitoring instruments SL 530
27. Lightning technical specifications for building electronic information systems GB 50343
28. Building structure load specification GB50009
29. Unified standard for construction quality acceptance of construction engineering GB50300
30. Construction engineering document filing specification GB/T50328

Technique Code of Building Structural Monitoring in Qinghai Province

Explanation

3 Basic Requirements

3.2 Monitoring Scheme and Requirements

3.2.6 Fire monitoring shall be carried out in accordance with relevant fire prevention standards. The code for building fire protection design GB50016 shall provide an electrical fire monitoring system for non-firefighting electrical loads of the following buildings or places:

1. Class B, C, and D warehouses with a building height greater than 50 m, and outdoor fire fighting water consumption of more than 30 L/s (warehouse);
2. Class I high-rise residential buildings;
3. Cinemas and theaters with more than 1500 seats, gymnasiums with more than 3000 seats, shops and exhibition buildings with a floor area of more than 3000 m^2, radio, television, telecommunications and financial, and financial buildings at the provincial (city) level and above. Other public buildings with outdoor fire water consumption greater than 25 L/s;
4. National-level cultural relics protection units focus on ancient buildings of brick or wood structure.

Except for the buildings or places specified in the Code for Building Fire Protection Design GB50016, the electrical fire monitoring system shall be provided for the distribution circuit of non-firefighting loads of the following public buildings:

1. Distribution mains in the store building business hall with a building area of over 5000 m^2;
2. Public buildings with a single building area of 40,000 m^2;
3. Civil airport terminal building, primary and secondary passenger transport stations, first- and second-level port passenger stations;
4. Hotel buildings with a total construction area of more than 3000 m^2.

Y. Yang, *Technical Code for Monitoring of Building Structures*,
https://doi.org/10.1007/978-981-15-1049-6

Relevant monitoring projects such as environment and corrosion all include partial durability tests. Some areas of the Qinghai–Tibet Plateau are likely to cause concrete damage under the influence of long-term freeze-thaw environment. It is recommended to regularly monitor structural durability. The durability evaluation of concrete in freeze-thaw environment should meet the following requirements:

1. As the limit of concrete durability due to obvious freezing and thawing damage (the surface cement mortar is peeled off and the coarse aggregate is exposed);
2. Consider the impact of freezing and thawing damage on the evaluation of steel corrosion durability.

3.3 Monitoring System

3.3.1–3.3.2 Real-time monitoring which is continuous and uninterrupted mainly focuses on the timeliness of monitoring. Online monitoring means that monitoring is conducted online, and it is a convenient way for real-time monitoring to monitor the target objects timely and accurately. Real-time monitoring contains more applications than online monitoring: online monitoring generally focuses on certain locations while real-time monitoring is more flexible. Online monitoring is usually conducted by specific equipment, including various sensors, and real-time monitoring can be conducted by technicians with or without equipment (e.g. inspection by naked eyes). When online system is under maintenance, the data which is collected offline can also be considered as real-time monitoring data.

3.3.3 Sensor system consisting of loading and environmental sensors, global and local structure response sensors is used to measure loading and environmental parameters and structure response, and record image and video information. In order to work effectively in extreme environmental conditions, the sensors like accelerometer, speed transmitter, displacement transducer, and strain gauge should have a working temperature range of −30 to 80 °C.

3.4 Construction Monitoring

3.4.3 In practice, it is recommended to record the installation information of a group of components in a certain period rather than each individual component; the length of period can be selected based on the accuracy requirement of structural analysis during construction. The component installation record should contain special procedures like delayed installation of components, postcasting strip connection, the timing of mutual transformation of pin and fixed connection. The environment record should contain temperature, wind, humidity, and so on.

3.4.4 Structural analysis and monitoring during construction is a multi-party collaborative work involving design, construction, operation, monitoring, and supervision, based on the rationality of design documents, the accuracy of structural analysis during construction, the reliability of monitoring data, the integrity of supervision, and the timeliness of treatment to abnormal conditions. Contractor and/or supervisor can liaise with all collaborators and be responsible for the whole process of construction, so that the responsibility is clear, and the process is controllable.

3.4.7 The on-site monitoring can be influenced by many factors, which may have certain uncertainties, such as variable load during construction, foundation settlement, temperature effect, sensor drift, creep, shrinkage, and so on. Therefore, when the monitoring result does not match the analysis results of the construction process, the model should be rechecked and revised.

3.4.8 Green construction monitoring items shall comply with the provisions of DB63/T1307 "Green Construction Regulations for Construction Projects".

3.5 Post Construction Monitoring

3.5.1 The Qinghai–Tibet Plateau is the highest and largest geomorphic step on the Eurasian continent. It spans 10–11 latitudes in the north and south, and passes through more than 30 longitudes in east and west. Its high and large features not only determine the distribution temperature and thickness of permafrost in the plateau. It has vertical zoning, and at the same time, the above-mentioned frozen soil characteristics have obvious zonal variation: the average latitude increases from south to north by 1°N, and the lower boundary of frozen soil decreases by 80–100 m. The permafrost located on the Qinghai–Tibet Plateau is the highest altitude, the lowest latitude, and the youngest plateau frozen soil.

Permafrost in the Qinghai–Tibet Plateau has the characteristics of high ground temperature, thin thickness, and tectonic-geothermal melting zone. Its stability is poor and it is sensitive to climate change. Affected by rising climate, it may lead to the melting of permafrost soils. Therefore, it is necessary to monitor the impact of major engineering structures on permafrost soils. The so-called major engineering buildings refer to important or typical engineering structures identified by local government agencies.

3.5.3 Wind load should be considered if the structure is subject to significant wind load during operation. The value of basic wind pressure can be selected based on the monitoring results during construction with suitable reduction factor considering recurrence interval; area load—wind pressure coefficient, wind load factor, and gust factor can be selected and adjusted based on the surrounding conditions and stiffness during operation.

Seismic load should be considered since the life span of structures is usually long. It can be selected based on the design and the monitoring results during operation with suitable reduction factor considering recurrence interval.

Wave load should be considered if the structure is subject to significant wave load in special cases.

Uniform temperature change of the structure should be considered if the stress and deformation of the structure are greatly affected by the ambient temperature; in particular, the un-uniform temperature change of the structure due to sunlight should also be considered. The parameters of temperature effect should be selected based on historical meteorological records, theoretical analysis of ambient temperature, and on-site temperature measurement results.

3.5.4 The monitoring items during the use of green buildings shall comply with the provisions of DB63/T1340 "Qinghai Province Green Building Design Standards".

4 Hardware Implementation

4.1 General Requirements

4.1.2 The layout of the sensors should consider crucial components and sections based on structural analysis and vulnerability without damaging the normal operation and the appearance of the structure. It should also consider the safety of equipment during construction, and the convenience of sensor installation, maintenance, and upgradation. Moreover, the length of cables and the transmission range of a single sensor should be optimized to meet the monitoring requirements.

Vulnerability analysis is to investigate the weak components of the structure by analyzing the combination of structural members and the possible failure modes of the structure. When a structure is damaged by accident or earthquake causing failure of a component, the load it suffers is redistributed to the neighboring components which may be damaged accordingly, and the whole structure fails. This is a typical failure path of a structure. The first component that fails and eventually causes the failure of the whole structure is the weakest, most important, and vulnerable component of the structure. Vulnerability analysis is used to find the failure path of structures, and then determine the key components, which is very effective for structural health monitoring.

4.1.3 Sensors should be placed in special protection boxes; cables should be placed in special cable or pipe; data collection equipment and lighting arresters should be protected by special cabinets.

4.1.4 The monitoring management center shall have a set of AC uninterruptible power supply (UPS) with a capacity of 3 kVA. The storage battery should be able to maintain the normal operation of equipment for at least 3 days. The input AC 220 V loop is taken from the power distribution equipment of the management center, and the output 220 V supplies power to the equipment in the station.

Each monitoring management station shall have a set of AC uninterruptible power supply (UPS) with a capacity of 3 kVA. The storage battery should be able to maintain the normal operation of equipment for at least 60 min. The input AC 220 V loop is taken from the power distribution equipment of the management center, and the output 220 V supplies power to the equipment in the station.

4.2 Hardware Installation

4.2.1 The mechanical test instrument includes dial gauges, horizontal displacement gauge of tensional wire, extensometers, goniometers, and inclinometers. It plays an irreplaceable role in engineering monitoring because it has the advantages of high accuracy, strong adaptability, and ease of use; however, it also has disadvantages of less sensitivity and weaker amplification.

The electrical test instrument converts the displacement into an electrical signal for measurement, including electronic inclinometer, connected liquid level deflection meter, static level, electronic dial indicator, electronic dial gauge, resistive displacement sensor, strain beam displacement sensor, differential transformer displacement sensor, crack width gauge, and so on. It is widely used in engineering applications since it can be digitalized and integrated easily and has high sensitivity.

The optical instrument includes level, total station, theodolite, range finder, reading microscope, and so on. It has the advantages of well-established technology and high accuracy, but it requires many reference points which can be hardly achieved in urban area where buildings are densely constructed.

4.2.2 The ground wire of the GNSS antenna may introduce external interference into the antenna system; therefore it should be separated from other systems.

4.2.3 The photoelectric target measurement subsystem includes two types: one is to use the light source device and camera with telephoto lens as the target and the measuring instrument, respectively; the second is to use the laser projection screen instead. When the target and the measurement instrument are not in the same place, the displacement of the structure can be obtained by dividing the measured value by cosine of elevation angle. It should also be noted that the measurement precision after adjustment must meet the design requirements.

4.2.4 The requirement of the distance between the installation position of the embedded stress–strain sensor and the design monitoring point and the angular deviation in this code is higher, compared to "Technical code for monitoring of building and bridge structures" (GB50982).

4.2.6 Experimental investigation shows that better spectrum and frequency response can be obtained by using sensors rigidly fixed on the structure than using sensors attached to the structure by glue with or without block. Moreover, the sensors attached to the structure by glue may fail during the long-term monitoring. Therefore, it is required that the accelerometer and speed transmitter should be rigidly fixed on the structure. Dynamic testing is vulnerable to interference from external electromagnetic and radio frequency signals; hence ground wire is required for accelerometer and speed transmitter.

4.2.10 The bracket of the wind speed and direction monitoring device shall have enough rigidity because unstable bracket may introduce the unexpected vibration of the device itself and reduce the measurement accuracy.

4.4 Lightning Protection and Grounding Design

4.4.1 Besides the above requirements, the requirements in "Technical code for protection of building electronic information system against lightning" (GB50343) should also be met.

4.5 Identification

4.5.1 Detailed information of all sensors on the structure should be identified for easy management and maintenance during operation.

5 System Integration

5.1 General Requirements

5.1.2 The software system mainly includes data acquisition software, data transmission software, database software, security evaluation software, and interface software. Parameters can be configured, and the software can be managed remotely in the monitoring center.

5.1.3 Self-start function which can automatically recover the system from the fault should be included because there may be a crash due to interface or hardware failure during the operation of the software system. However, the recovery process cannot affect normal data collection, storage, and evaluation.

5.1.4 The database should include the structure information sub-database, the monitoring system information sub-database, the finite element model sub-database, the real-time raw data sub-database, the statistical analysis sub-database, the structural safety assessment sub-database, the construction monitoring sub-database, and the load test sub-database.

The structural information sub-database should contain structural design, as-built drawings, and scientific research data and the database tables should be classified accordingly.

The monitoring system information sub-database should contain information of sensors, data acquisition and transmission equipment, and data processing and administration software, for example, equipment installation locations, technical parameters, brands, and specifications.

The finite element model sub-database should contain finite element models of the structure at different stages, which should be generated by finite element analysis software in standard format.

The real-time raw data sub-database should contain the time history of all variables.

© The Author(s), under exclusive license to Springer Nature Singapore Pte Ltd. 2020   127
Y. Yang, *Technical Code for Monitoring of Building Structures*,
https://doi.org/10.1007/978-981-15-1049-6

The statistical analysis sub-database should contain the statistical analysis results by using different methods.

The structural safety assessment sub-database should contain warning threshold, safety assessment methods and conclusions, and warning history.

The construction monitoring sub-database should contain the construction process of the structure, the construction monitoring information of all previous maintenance and construction control process, and the reports of construction monitoring at all stages.

The load test sub-database should contain information of previous load tests including static and dynamic loads and boundary conditions, and so on, and reports of all measured data from various sensors and analysis results.

Data storage and administration can be conducted on local computers and cloud storage and cloud management technologies should be adopted.

5.2 System Design and Development

5.2.1 The structural health monitoring system should have several functions, including data acquisition and control, data storage and transmission, structural safety evaluation and user interface, and so on, which can be generated as independent modules for easy debugging and maintenance.

5.2.2 The documents which should be handed over with other related materials during acceptance are mainly to facilitate the maintenance of software during operation.

5.2.3 The communication interface between BIM model and structure monitoring module can be developed for sharing real-time monitoring data. The structure design, construction, and management with whole life cycle structural health monitoring can also be improved due to the highly integrated BIM platform.

The data and information from the sensors can be transmitted to the database of BIM model in real time and the information sharing and 3-D visualization can be conducted in the monitoring system. The location and other related information of sensors in the BIM model can be visualized so that the contractor can reserve space for sensors in advance during construction and improve smooth and effective communication with inspector. The information sharing of BIM model during construction and operation can combine the monitoring works in these two stages and further integrate access control, elevator system, and fire protection system.

The structural monitoring information can also be visualized so that the operation status of the monitoring system, the safety evaluation and warning, and the condition of each component can be visualized online, which makes the staff understand the massive monitoring data easily.

The integration of monitoring system into BIM platform makes the upgradation and function expansion of the monitoring system more convenient, enables the integration of monitoring, communication, ventilation, and lightening systems, and fully improve the utilization of BIM model.

5.3 System Requirements

5.3.4 Peak peaking method. Peaks related to natural frequencies of the structure can be observed by plotting the frequency response function, and the natural frequencies can be evaluated by the peaks.

Power spectral density. Peak picking method can also be applied on power spectral density to evaluate the nature frequencies. Power spectral density is the power of the signal in unit frequency band. It represents how the signal power changes with frequency and the distribution of signal power in frequency domain.

Frequency domain decomposition. Singular value decomposition (SVD) can be applied to the power spectral density of the signal of a multiple degree of freedom system to obtain a series of power spectral density of the signals of several single degree of freedom systems, then the frequencies can be extracted by peak picking method.

Enhanced frequency domain decomposition. Inverse Fourier transform is applied in this method to the de-composited power spectral density of the signals of several single degree of freedom systems; and the frequencies and damping can be obtained by analyzing the corresponding signals in time domain based on logarithmic decrement.

Least squares complex frequency domain method. Self-regression model and Prony polynomial can be generated by impulse response function and the modal parameters can be identified by evaluating the residue and the poles.

Natural excitation technique. The cross-correlation function of two points in the structure subject to environmental white noise is similar to impulse response function, which can be used to identify the modal parameters in time domain.

Eigen system realization algorithm. General Hankel matrix can be generated based on impulse response function matrix from multiple inputs and outputs, and the discrete time system can be realized by using a few lower modes by the SVD; therefore, the modal parameters can be obtained accordingly.

Stochastic subspace identification (SSI). The compressed input format can be obtained by assuming the input and noise can be replaced by white noise, the stabilization diagram can be generated, and the modal parameters can be estimated accordingly.

Auto-regressive moving average model (ARMA). The modal parameters can be obtained by applying parametric model directly to the time history of the response signal from the structure.

Wavelet analysis. Wavelet analysis is an approach of time–frequency analysis. It can meet the requirement of resolution of time and frequency and focus on any details in the signal, by controlling the scaling and shifting factors of wavelet functions.

Hilbert-Huang transform. Hilbert-Huang transform consists of empirical mode decomposition (EMD) and Hilbert transform. It uses the EMP method to decompose a signal into a series intrinsic mode functions and a residue and then applies Hilbert spectral analysis to evaluate the instantaneous frequencies.

Static parameter-based method. The structural damage can be identified by the change of static parameters (element stiffness, displacement, strain, residual stress, Young's modules, element area, moment of inertia, etc.).

Dynamic parameter-based method. The structural damage can be identified by the change of dynamic parameters (natural frequency, mode shape, mode shape curvature, modal assurance criteria, coordinate modal assurance criterion, strain mode, flexibility stiffness, modal stain energy, direct stiffness method, etc.)

Finite element model updating method. The structural parameters can be identified by several techniques like matrix correction, element correction, error factor correction, and design parameter correction, in which the mass matrix, stiffness matrix, and damping matrix are updated by the measured static and dynamic responses of the structure.

Pattern recognition method. The structural damage can be identified by applying pattern recognition algorithms (artificial neural network, deep learning, fuzzy theory, genetic algorithm, particle swarm optimization algorithm, etc.) to measured static and dynamic responses of the structure.

Analytic hierarchy process. The structural safety can be analyzed and evaluated by the analytic hierarchy process in which all related factors are pair-wise compared to determine the hierarchy of goals, criteria, and alternatives.

Limit state analysis. The structural safety can be evaluated by the limit state analysis in which the stress and deformation of structure or component in the limit state are analyzed to check if they are still acceptable.

Reliability analysis. The structural safety can be assessed by calculating the reliability of component (component reliability analysis) and system (system reliability analysis) which represents the probability of the structure meeting the designed requirements.

5.4 System Validation and Development

5.4.1 The module testing, integration testing, configuration testing, and system testing of the software should meet the requirements in "Specification of computer software testing" (GB/T15532).

5.4.3 The third-party test organization only performs functional testing, so one single sensor for each type should be provided.

6 High-Rise Building Structure

6.1 General Requirements

6.1.2 Through the analysis and monitoring of the construction of the Haidong Sports Center in Qinghai Province, it is shown that the structural deformation caused by the temperature difference between day and night is 20 °C, which is equivalent to the deformation caused by its own gravity load. In combination with other engineering monitoring projects and local characteristics of Qinghai, the influence of temperature factors should be considered when monitoring the structure during use.

Based on the meteorological data of day and night temperature difference in Qinghai Province in recent years, the following adjustments should be made to the height of the structure to be monitored:

1. The temperature difference between day and night in Haibei Tibetan Autonomous Prefecture and Guoluo Tibetan Autonomous Prefecture is more frequent, and the height of the high-rise building structure should be reduced by 10 m on the basis of Table 6.1.2;
2. The temperature difference between day and night in Xining City and Yushu Tibetan Autonomous Prefecture is more varied, and the height of the high-rise building structure should be reduced by 5 m on the basis of Table 6.1.2.

In Qinghai Province, the seismic fortification intensity is 7° (0.15 g), and the corresponding minimum height can be averaged 7° (0.1 g) and 8° (0.2 g).

6.1.3–6.1.4 Deformation monitoring should be carried out during both construction and post construction stages, including horizontal displacement monitoring, settlement monitoring, compression deformation monitoring, main body inclination monitoring, and vertical monitoring of ground layer.

Y. Yang, *Technical Code for Monitoring of Building Structures*,
https://doi.org/10.1007/978-981-15-1049-6

Building excavation monitoring is to monitor the surroundings of the high-rise building, including the inclination, the vertical and horizontal displacements of the surrounding buildings, the building cracks and ground surface cracks, and the deformation of nearby pipelines. The important buildings in the area which is 1–3 times of the foundation pit depth outside the edge of the foundation pit should be monitored and the area can be expanded, if necessary. The building excavation monitoring should meet the requirements in "Technical code for monitoring of building excavation engineering" (GB50497).

The monitoring scheme should consider the objectives of monitoring:

1. Monitoring of structural dynamic properties. The structural dynamic properties are directly related to the structure performance; therefore, the condition of the structure can be evaluated by analyzing the monitoring data.
2. Monitoring of structure deformation. The displacement and settlement should be monitored during construction; the vertical and horizontal displacements, inclination, crack and sunlight-induced deformation, and so on should be monitored during operation.
3. Local monitoring of structure. The stress, deformation (crack), durability, strength, temperature, and humidity of the components and joints which may induce the whole high-rise building unstable and unsafe should be monitored.
4. Monitoring of load. The objective of load monitoring is to record the time history of the various time-dependent loads, which can be used for diagnosis of structure condition. Generally, the main loads for high-rise buildings are wind and seismic load.

6.2 Construction Monitoring

6.2.1–6.2.2 The vertical and horizontal displacements of surrounding buildings and pipelines in the affected area should be monitored from the beginning of foundation pit excavation to the end of foundation pit backfill. The monitoring duration should be extended by one month in the soft soil area.

Generally, settlement monitoring adopts the method of leveling measurement, in which the settlement observation points should be measured by the vertical datum and they should be in a level loop. The staff, equipment, and the loop should be fixed in the monitoring. The monitoring of the elevation reference point is monitored by round-trip even station. The settlement monitoring point is monitored by the first round-trip even station. Afterwards, the single-pass even-number station monitoring is adopted. The monitoring of the elevation reference point and the settlement monitoring point should be monitored by closed-level route. The monitoring indicators should conform to the relevant national standards.

The settlement monitoring of high-rise buildings should be accurate enough to faithfully record the settlement of high-rise buildings with increasing load. In "Code for deformation measurement of building and structure" (JGJ8), the second-level leveling measurement method should be adopted in general settlement monitoring.

The settlement observation point should be fixed into the characteristic location of the high-rise building, following the requirements in "Code for deformation measurement of building and structure" (JGJ8). Based on the experience of settlement monitoring of high-rise buildings, the settlement observation points should be uniformly distributed on surrounding buildings and structures, and those on the buildings should be vertically and horizontally symmetrical.

The settlement monitoring frequency can be selected based on basic loading, project schedule, and properties. For high-rise building, the settlement monitoring during construction should be conducted before and during load increasing. In addition, the settlement monitoring during construction should be conducted once every 1–5 floors are constructed. The settlement monitoring frequency should be adjusted according to the actual condition during the construction, for example, the monitoring can be more frequent if there is a rain in a long time or the loading is increasing suddenly. Once the settlement speed changes in the monitoring, the settlement monitoring frequency should also be adjusted.

6.2.3 The construction of ultra-high-rise buildings is complex, and it influences the stress of the key components significantly. Therefore, the stress of key components should be monitored to ensure safe construction.

Generally, direct measurement should be used for strain monitoring in which resistive strain gauges, vibrating wire strain gauges, and fiber optical sensors are frequently used. If the sensors cannot be installed on the surface or inside the structure, indirect measurement can be used in which the strain can be calculated by the measured displacement. The stress can be obtained by the measured strain directly, and it can also be obtained indirectly by measuring the force, displacement, natural frequency, magnetic flux, and so on.

The strain and stress of the structure and the equipment introducing large temporary loading to the structure during construction should be monitored. The number and arrangement of monitoring points should meet the following requirements:

1. For the component subject to uniaxial bending moment, the monitoring points should be located along the height direction on the cross-section carrying largest bending moment. Generally, the number of monitoring points should be at least two for each section. If the strain distribution along the height direction is investigated, the number of monitoring points should be at least five for one section. For the component, the number of monitoring points should be at least four for one section.
2. For the component subject to axial compression, the monitoring points should be placed along the axial direction, and they should be uniformly distributed on one section. The number of monitoring points should be at least two for each section.
3. For the component subject to torsion, the monitoring points should be placed along the direction of 45° with respect to the torsion axis on the lateral surface.

4. For the component subject to multiple loadings, the monitoring points should be arranged carefully so that the principal stress (value and direction) on the measured section can be calculated based on the measured strain.

The strain–stress monitoring frequency should meet the following requirements:

1. The strain–stress monitoring should be conducted at least once every month during construction.
2. The strain–stress monitoring should be conducted at least once when every 3–6 floors are constructed.
3. The strain–stress monitoring should be conducted at least once at important milestones during construction.
4. The strain–stress monitoring should be conducted more frequently when the load applied on the structure changes significantly or special construction processes are being carried out.

6.2.4 Crack monitoring should include measurement of crack length and width, and the following requirements should be met:

1. The crack length or the width of large crack can be measured by a steel ruler or other mechanical instruments. Crack scale card and electrical crack meter can be used for direct measurement and each crack should be measured for at least three times. The resolution of crack width card should be no more than 0.05 mm and the measurement accuracy of electrical crack meter should be at least 0.02 mm.
2. For surface crack with width less than 1 mm and crack inside concrete, electrical instrument can be used and the measurement accuracy should be 0.1 mm.
3. The structural crack monitoring sensor, including the vibrating wire strain gauge, the fiber optic sensor, and so on, can be used to monitor the displacement of the two points on both sides of the crack. The sensor should be perpendicular to the crack and the measurement range should be larger than the threshold of crack width.
4. The crack monitoring sensor should be used if the crack propagation in a long term is under investigation.
5. The crack monitoring frequency should be determined by the rate of crack change.
6. The specification of sensors should be configured based on either crack width or strain, depending on the presence of crack.

6.2.5 Temperature monitoring includes ambient temperature and structural temperature monitoring. For ambient temperature monitoring, the temperature sensor should be placed in a 1.5 m high, air-circulated thermometer screen; for structural temperature monitoring, the temperature sensors can be placed inside the structure independently or on the same position as the strain gauge. When it is necessary to monitor the temperature difference of the structural surface induced by sunlight, it is preferred to place the sensors on the surface subject to direct sunlight and the opposite surface, and inside the structure along the height direction.

The monitoring point for ambient temperature measurement can be placed in a representative space, 1.5 m high from the floor inside the structure to represent the average structural temperature. For most structures, only the structural temperature without sunlight should be measured; for the important component affected by non-uniformly distributed temperature, the non-uniformly distributed temperature can be monitored; for the special component affected by non-uniformly distributed temperature, the non-uniformly distributed temperature should be monitored to assist further structural analysis.

The temperature monitoring can be conducted by mercury thermometer, contact temperature sensor, thermistor temperature sensor, or infrared thermometer. The measurement accuracy should not be lower than 0.5 °C.

Automatic continuous monitoring system should be used to monitor the daily temperature change; when reading is conducted manually, the frequency should be at least once per hour.

Humidity monitoring can be conducted simultaneously with temperature monitoring and the monitoring points can be placed in the space where the humidity changes greatly and has greater influence on the durability of the structure. Humidity monitoring focuses on relative humidity measurement: the measurement range should be 12–99% and the measurement results should contain daily average humidity, daily maximum and minimum humidity.

6.2.6 The wind load has more adverse effect on the structure during construction than operation. If the load combination which contains the wind load is dominant in the design, the wind load can be monitored.

The monitoring points to measure the load wind and the wind-induced structural response should follow the requirements: for the high-rise building the scaled model of which has been tested in the wind tunnel, the layout of monitoring points should be the same as those in the wind tunnel test; for others, the layout of the monitoring points should focus on components and joints sensitive to the wind effect and it should also be compatible to the layout of monitoring points for seismic load, and so on.

The measurement accuracy of wind speed and pressure should not be less than 0.5 m/s and 10 Pa.

The anemometer should be installed on the dedicated mast on top of the structure during construction. When it is necessary to monitor the distribution of wind pressure on the surface of the structure, the wind pressure box is monitored on the surface of the structure. The anemometer should be placed on an open site about 100–200 m away from the building, at a height of 10 m from the ground.

6.2.10 Considering the simulation results of the whole process of construction, the vertical displacement monitoring plays an important role in the ultra-high-rise building monitoring during construction. The monitoring points should be uniformly distributed along the vertical direction so that the global vertical displacement field of the structure can be represented. In particular, more monitoring points are required on key connection components for tube structures.

6.3 Post Construction Monitoring

6.3.2 The sunlight on the sunny and shady surfaces of the ultra-high-rise building is different due to the self-occlusion, resulting in non-uniform distribution of temperature and great temperature gradient along the height direction. It is an important environmental load affecting the ultra-high-rise building; therefore it should be monitored during the operation of the structure.

6.3.3 The monitoring points should not be distributed sparsely to facilitate data transmission. The location of "hot-spot stress" is the location of maximum stress in the structure or the location most vulnerable to stress change. The layout of the monitoring points should follow the sub-zone criterion: the structure is divided into several sub-zones based on strain and stress, and the monitoring points in each sub-zone should be distributed intensively.

6.3.4 The wind load should be determined by test in the wind tunnel for the structure which is higher than 200 m and the field test data can be compared to the wind tunnel test results. For the high-rise building which is prone to wind-induced vibration or the structure, the aerodynamic characteristics are unknown. The wind load and wind-induced response should be monitored.

When the high-rise building or the ultra-high-rise building is subject to strong wind, the wind monitoring should be conducted simultaneously with the measurement of wind speed at top of the structure, wind direction, pressure, and displacement of the top of the structure.

6.3.5 The monitoring points to measure vibration should be placed on various important floors, including first floor, top floor, standard floor, strengthened floor, and the floor with variable section. For each area, the distribution of internal and external points at each obliquely symmetric edge column should be selected, and the monitoring points on beam column joints are optional.

6.3.6 In the earthquake monitoring, the sensors should be installed in three directions on the free field to measure the input of the seismic load on the structure. Three sensors should also be placed in three directions on the geometric center of the selected floor: two of them are in the horizontal directions and one of them is in the vertical direction.

The accelerometer must be included in the monitoring scheme, but the speed transmitter and displacement transducer can be optional based on actual condition.

The dynamic response is very important for high-rise buildings, and the monitoring points should be optimized. The distance between the measuring points should not exceed 1/2–1/3 seismic wavelengths, at least 1.5–2 times the height of the structure.

6.3.7 External wall insulation system has been applied in China for many years, but its weather resistance in severe cold areas has yet to be tested. In recent years, the detachment of the external thermal insulation system due to strong wind or severe weather can be observed frequently, and periodic or online monitoring of the internal and external temperature should be conducted to avoid hollowness, cracking, and detachment.

6.3.10 The stress of the element is important to evaluate the structure safety during construction; hence the strain–stress monitoring should focus on the "hot-spot" zone where the stress is large and complex. For the frame structure system, the "hot-spot" zone is usually near the bottom and the belt truss.

6.3.11 Transmission discontinuity of horizontal force and change of stiffness can be observed in the transfer structure; therefore, the strain and the stress should be monitored.

6.3.12 The "hot-spot" zone of shear wall where the stress is maximum is usually at the bottom.

6.3.14 The vertical deformation is dominant for the high-rise tube structure, so the vertical deformation, the vertical deformation difference, and the preset construction deformation value of the core and outer tubes should be focused. For the external frame-core tube system displacement monitoring, the vertical maximum total displacement floor of the outer frame should be selected; the vertical maximum displacement floor of the core tube; the vertical maximum relative deformation floor of the core tube and the outer frame.

7 Long-Span Spatial Structure

7.1 General Requirements

7.1.1 This section is to be developed for the following structure types:

1. Grid (truss) structure

Grid (truss) structure is an indeterminate grid-like structure consisting of two-force members only, including plane truss, triangular pyramid space truss, and quadrangular pyramid space truss.

2. Reticulated shell structure

Reticulated shell structure is a spatial structure in which the truss elements are regularly placed along a certain curved surface. It is like the thin shell structure because the main load is suffered by the truss elements which can only resist axial tension or compression. The reticulated shell structure can be classified by the shape of the curved surface: spherical reticulated shell, cylindrical reticulated shell, hyperbolic paraboloid reticulated shell, and elliptical paraboloid reticulated shell.

3. Cable structure

Cable structure is a structure which consists of a series of regularly assembled tensile cables suspended to the supports and it resists the load by the stretching of the cables. The cable structure can be classified by the composition method or mechanical characteristics: single-layer suspension system, prestressed double-layer suspension system, prestressed saddle cable net, suspension structure with stiff components, suspended thin shell, suspended steel membrane, and combined suspension structure.

Y. Yang, *Technical Code for Monitoring of Building Structures*,
https://doi.org/10.1007/978-981-15-1049-6

4. Membrane structure

Membrane structure can be classified as pneumatic structure and tensile membrane structure. The pneumatic structure uses the pressurized air to inflate the envelope so that it can maintain its shape; the tensile membrane structure uses pretensioned membrane to resist external load.

The foundation settlement, deformation, strain and stress, and wind load should be monitored according to the characteristics of the long-span spatial structure during construction. In the stage of operation, the ambient temperature, humidity, foundation settlement, strain and stress, dynamic response (wind and earthquake induced vibration) should be monitored.

7.2 Construction Monitoring

7.2.1 Foundation settlement monitoring should focus on vertical datum, working point, and settlement observation point. The vertical datum should be placed outside the deformation affected zone; the working point should be placed at a location where the measurement can be conducted conveniently; the settlement observation point should be placed at a location where the foundation settlement can be observed. The monitoring reference network consists of the vertical datum and some working points and it should be rechecked every six months; the settlement monitoring network consists of some vertical data, working points, and settlement observation points.

7.2.2 Strain and stress are important to judge if the long-span spatial structure is safe and the stress of the key components is significantly influenced by the construction process; therefore, the stress of key components should be monitored to ensure safe construction. The key components and monitoring position can be determined by simulation results of the whole construction process.

7.2.4 The strain monitoring points should be placed at the area where the stress is large or the components are intensive (e.g. main force rod, open position of curved grid, lower support ribbon), so that the stress of the components which suffers large and complex load and are important to the capacity and stability of the whole structure can be obtained. Considering the data transmission, the monitoring points should not be distributed sparsely, and the layout of the monitoring points should follow the sub-zone criterion.

7.2.9 Before the cable is tensioned, it should be determined that whether the tensile force or the structural displacement is dominant. For the key component, both the tensile force and the structural displacement should be monitored, and the permitted deviation should be determined. When the cable is being tensioned, the permitted deviation of the camber and the deflection should not be more than 5% of the designed value.

If the anchorage connected to the cable is not stable, the cable should not be installed; otherwise, temporary support should be used to maintain the anchorage stable, and the cable can be installed under the condition of relaxation. In addition, the tensile force suffered by the two end joints should be used to check the displacement of the two joints, so that the deviation between the joint location after construction and the designed location can meet the acceptance requirements.

7.2.10 The cable structure cannot be stable before the prestress is applied; therefore, the tensile force should be monitored during construction to ensure correct tensile force and safe construction. The common measurement methods include pressure gauge/sensor test, hydraulic jack test, cable frequency test, and three-point flexural bending test, and so on. The tensile force and the elongation should be monitored to meet the requirement of initial tensile force; when the cable is being tensioned, the permitted deviation of the tensile force should not be more than 10% of the designed value.

7.2.11 When the wind force is above level IV, or in the case of rain and snow, the cable should not be installed; the wind speed and direction should be monitored at the same time to avoid excessive swing of the cable. When there is lightening, the work should be stopped. When the wind force is above 5, the anchorage should be inspected in time and measures should be taken to stabilize the cable.

7.2.12 The relaxation should be paid attention because it may reduce the stiffness of the membrane structure, bring significant wind induced vibration and even lead to tearing of the membrane. In addition, it may also adversely affect the structural appearance and the drainage performance. The area of the wrinkles caused by the relaxation should not exceed 10% of the area of the membrane.

7.2.13 For the membrane structure, it is necessary to apply the pretension to the membrane to ensure its initial equilibrium state and stable geometric shape. The pretension which can meet the design requirements during construction is important to the construction quality and the safety and stability of the structure.

The predetermined tension should be accurately applied to the membrane during the tensioning to obtain the designed equilibrium state and geometric shape, which requires the construction engineers to monitor and adjust the membrane tension in real time to avoid stress relaxation and intension and membrane tearing due to great tension.

For the membrane structure to which the pretension is applied by concentrated force, the applied force should be controlled by the displacement of the application point of force, the deviation of which should not be more than 10%.

7.3 Post Construction Monitoring

7.3.3 Refer to Chapter 8 of JT/T1037-2016 Technical Regulations for Highway Bridge Structure Safety Monitoring System:

$$D = \sum_{i=1}^{k} D_i = \sum \frac{n_i}{N_i} \quad (i = 1, 2, \ldots, k)$$

where n_i is the frequency corresponding to the equivalent stress, and N_i is the life of the S–N curve stress S.

7.3.6 Because the structural deformation is sensitive to the condition changing of the grid (truss) structure, it can be monitored and used to determine the reaction force, detect the local damage, and adjust the numerical model of the structure.

The monitoring points should be placed on the joints of the grid (truss) structure and the reference points should be placed at the location which is stable and around the structure.

7.3.7 During the operation of the grid (truss) structure, the stress intensity may be observed in some components due to several factors like component dimension, material properties, installation misalignment, joint, welding residue stress, changing temperature, corrosion and construction quality, and so on. Therefore, the strain and stress should be monitored on the key parts of the structure (e.g. the welding between the bar and the spherical joint). The layout of the monitoring points should focus on the side span, interior span, and other key locations, and 2–3 components or bars should be selected for each key location.

7.3.9 The load capacity of the reticulated shell structure which is initially determined by the design is influenced by the structural deformation and the amplitude and rate of the support settlement. Therefore, the structural deformation and the support settlement should be monitored to adjust the original model of the structure and further improve the structural analysis so that the structure can maintain its designed function during its service life.

7.3.10 The load and deformation capacity, the durability, and the serviceability of the reticulated shell structure may be affected by manufacture, transmission and assembly, onsite welding, and defects due to environmental and human factors in the service life. The stress exceeding the limitation at the key components and joints or the stress redistribution may danger the structural safety. Therefore, the strain and stress at the key components and joints should be monitored in accompany with the environmental factors, vibration, deformation, and corrosion to evaluate the structural condition.

7.3.11 The wind speed and direction should be monitored to obtain the wind load and its effect to the reticulated shell structure, and then to evaluate the structural safety and comfortability accordingly.

7.3.12 The tension force is important to make the cable structure safe and stable. The structural deformation due to creep and relaxation of the cable is inevitable during the service life of the cable structure; therefore, the displacement of key points and the structural deformation should be monitored, and warning should be issued in time.

7.3.13 The tension force may be reduced due to creep and relaxation of the cable, affecting the normal service of the whole cable structure; therefore, the strain of the cable should be monitored periodically to evaluate the cable condition in a long term.

7.3.15 The deformation monitoring should be conducted by using the total station or the laser scanner. The geometric model of the actual membrane structure can be constructed by measuring the exact location of every control point and it should be compared to the original design periodically.

For the integral tension structure and cable-supported membrane structure, the maximum global displacement of the structure subject to the first load combination should not exceed 1/300–1/150 of the span or 1/150–1/80 of the overhang length; the maximum global displacement of the structure subject to the second load combination should not exceed 1/250–1/120 of the span or 1/100–1/60 of the overhang length.

For the overall tensioned membrane structure like stadium stand canopy, the global displacement can be considered as the maximum displacement of the inner ring; for the cable-supported membrane structure, the global displacement can be considered as the maximum displacement of the midspan. The lateral displacement of the mast top subject to the second load combination should not be more than 1/300–1/150 of the mast length. For the skeleton-supported membrane structure, the maximum displacement should meet the requirements of the relevant design codes.

The normal displacement of the inner surface of each membrane element should not be more than 1/20–1/10 of the nominal size of the element, where the membrane element is the membrane surrounded by either flexible boundary or rigid boundary. The nominal size of a membrane element can be defined as 2/3 of the length of the shortest side for the triangular element. For the quadrilateral element, the nominal element size can be defined as minimum span between the boundaries of the maximum displacement point.

Under the second load combination, the area of the wrinkles due to relaxation should not exceed 10–15% of the area of the membrane (Table 7.3.15).

Table 7.3.15 The load combination of membrane structures

Type	Load
The first load combination	G, Q, P(p)
The second load combination	G, W, P(p)
	G, W, Q, P(p)
	Others

Note 1 In this table, G is the permanent load, W is the wind load, Q is the imposed load or snow load whichever is larger, P is the initial tension force, and p is the air pressure in the pneumatic structure.

2 The characteristic value and the combination value of a variable load should be selected in accordance with "Load code for the design of building structures" (GB50009). The characteristic value and the combination value of P or p can be set as 1.0.

3 "Other load combination" indicates the load combination considering the specific situation of the structure including temperature effect, uneven support settlement, and construction load, and so on.

7.3.16 The surface tension should be monitored at the location where the surface tension is large or small or complex. At least one measurement point should be placed every 100 m^2 or every membrane element. If the surface tension decreases due to relaxation, it should be increased according to the monitoring results.

7.3.17 The membrane structure should be inspected after strong wind and repair actions should be conducted if necessary.

8 Protective Building

8.1 General Requirements

8.1.3 Daily safety monitoring including daily inspection, monthly inspection, annual inspection, and special inspection shall be carried out based on actual conditions. The structural deformation, cracking, monitoring equipment and monitoring points, and others in the monitoring scheme should be inspected. The monitoring method which is detrimental to the protected site and its heritage components is prohibited, for example sudden changing of temperature, strong light and vibration, and so on. In case of any danger or relic with inscription, the site shall be protected, and the case should be reported immediately. The staff on site should not take any action without authority.

The process of building restoration should be inspected regularly, including the structural deformation, cracking, monitoring equipment and monitoring points, construction process, and others in the monitoring scheme considering the design requirements and local experience. The inspection should be more frequent in case of key conditions and extreme weathers.

In order to ensure the safety of the relics and the quality of monitoring, the inspection should follow the requirements in the relevant codes seriously, and the monitoring points should be selected carefully.

The duration of the monitoring should not be less than 3 years. The measurement should be conducted for three times continuously after the equipment installation and the initial value of the equipment can be determined accordingly.

8.1.4 1 Monitoring points should be in a reversible manner including protective building features or pasting signs. The monitoring points should not be placed on the important parts of the relic itself, and they should not affect the building characteristics and appearance.

2 The points to monitor the horizontal vibration of the masonry structure should be placed at or near the center of stiffness of each floor.

Y. Yang, *Technical Code for Monitoring of Building Structures*,
https://doi.org/10.1007/978-981-15-1049-6

3 The points to monitor the horizontal vibration of the timber structure should be placed at the top and bottom of columns supporting the interior span for each floor.

4 The monitoring points to install anemometers should be placed at a certain distance away from the building so that the measurement of wind speed is not affected by the surrounding buildings.

5 The points for video monitoring should be placed at the location where the lightening equipment is available. However, if the lightening equipment is not available, the infrared equipment should be used instead.

8.2 Monitoring Contents

8.2.4 The long-term slight vibration induced by the external environmental factors (visit, traffic, and industrial operation, etc.) has adverse influence on the protected building. Therefore, the dynamic characteristics and responses of the protected building should be monitored continuously in real time.

The dynamic characteristics of the protected building should be tested before the dynamic monitoring. When the building is symmetrical, the test can be conducted along any principal axis; while if the building is asymmetrical, the test should be conducted along every principal axis. The layout of measurement points should meet the following requirements:

1. The measurement points should be placed at the necessary location of the modes which are to be identified. The layout of the measurement points should cover the whole structure and the number of the measurement points can be increased so that the local information can be obtained.
2. If the key point of the mode shape cannot be identified, the layout of measurement points should meet the following requirements: the points to monitor the horizontal vibration of the masonry structure should be placed at or near the center of stiffness of each floor; the points to monitor the horizontal vibration of the timber structure should be placed at the top and bottom of columns supporting the interior span for each floor.
3. The layout of measurement points should be optimized if the number of measurement points is large.

According to "Technical specification for protection of historic buildings against man-made vibration" (GB/T50542), the acceptable vibration velocity of the column top for protected timber structure should meet the following requirements: for those at the national level, it should be 0.18–0.22 mm/s; for those at the provincial level, it should be 0.25–0.30 mm/s; for those at the city level, it should be 0.29–0.35 mm/s.

8.2.5 Temperature is an important environmental factor affecting the performance of the protected heritage building. High temperature promotes chemical reactions such as oxidative decomposition, accelerates structural aging, and increases the reproduction and activity of molds and insects. Expansion due to large temperature difference makes the structure (especially for timber structure) fatigue and the mosaic decoration may fall accordingly.

8.2.6 Humidity has a great influence on the protected heritage building (especially for timber structure). The moist environment is suitable for the reproduction of termites. The high moisture content of wood makes the possibility of deterioration, insect attack and fungal erosion increase, greatly reducing the load capacity and the earthquake resistant performance.

8.2.8 Many protected heritage buildings are the timber structure and the main load bearing components are made of wood which is vulnerable to corrosion and insert attack. The regular monitoring of the properties of wood should include:

1. General properties: mechanical properties, moisture content, density, and shrinkage rate;
2. Defects: knot, slope of grain, twist, crack, and pith, and so on;
3. Deformation and damage: effect of deterioration, insert attack, split, and disaster;
4. Monitoring of the decay of wood and the monitoring of insects: visual observation, non-destructive monitoring of decay, and construction of wormwood on protective buildings using a stress wave analyzer.

8.2.9 The regular monitoring of protected heritage building should include:

1. The building: local and global incremental deformation, crack initiation and propagation at the beam, column and wall, roof leakage, crack condition under extreme temperature at the bottom of column and wall, deformation and crack under extreme snow load, settlement of foundation;
2. The surrounding environment: bad geological condition, great vibration source, human and traffic flow;
3. The monitoring equipment: integrity of equipment and protection, obstacles affecting the monitoring.

9 Other Structures

9.1 Dilapidated Building

9.1.1 The classification of dilapidated building should be in accordance with "Standard for appraisal of reliability of civil buildings" (GB50292), or "Standard for appraisal of reliability of industrial buildings and structures" (GB50144), or "Standard for dangerous building appraisal" (JGJ125).

9.1.3 The Grade C or D dilapidated building which cannot be demolished temporarily may be further damaged if there are tunnels or foundation pits nearby. Therefore, the monitoring frequency for the Grade C and D dilapidated building should be adjusted to at least twice per week and three times per week, and it should be further adjusted according to the construction progress.

9.1.7 The foundation can be monitored by measuring the inclination, crack, and displacement of the superstructure induced by non-uniform settlement. The foundation sliding, ground cracking, and soil mass deformation should also be monitored at the same time.

The monitoring of foundation components should focus on the vertical, horizontal, and diagonal stair-step crack at the joint of foundation and load-bearing wall, the horizontal crack at the joint of foundation and column, and the inclination.

The monitoring of components of the superstructure should include:

(a) Masonry structural component

The monitoring of the masonry structural component should focus on the joints to other components with different materials. The vertical and diagonal crack at the joints of walls with vertical and horizontal reinforcement, the deformation of the load-bearing wall, crack and displacement of the arch, as well as the damage of the ring beam and the structural column should be monitored. In particular, the width, length, orientation, distribution, and propagation of the crack should be monitored.

(b) Concrete structural component

The monitoring of the concrete structural component should focus on the structural force crack at the wall, column, beam, slab and roof, the concrete cover spalling induced by the rebar corrosion, the crack at the top and bottom of the column, the roof inclination, the stability of supporting system, and so on.

(c) Timber structural component

The monitoring of the timber structural component should focus on the deterioration, insert attack, joint failure, deflection, eccentricity, and cracks in the end of the wooden truss, the out-of-plane deformation of the truss, the stability of the roof support system, and so on.

(d) Load-bearing component of retaining structure

The monitoring of the load-bearing component of retaining structure should focus on masonry self-supporting walls, infill walls for horizontal loads, door and window openings, beams, awnings, parapets, and so on. It should be classified based on the type of component.

9.1.8 The layout of the monitoring points should follow the requirements of relevant national and industry standards and the sampling principle, considering the mechanical characteristics of the damaged and deformed components of the dilapidated building, the initial design, and the on-site investigation and inspection.

9.1.9 The monitoring should meet the following requirements:

1. Global inspection

Global inspection is to inspect the dilapidated building by naked eyes, and it should focus on the displacement, inclination, cracking, damaged components (deterioration, insert attack, flexural deformation, brick weathering, etc.), spalling, and initial dangerous situations. All controlled deformation of the dilapidated structure should be inspected during the global inspection.

2. Main structure inclination monitoring

When the overall inclination of the top of the main structure to the bottom is measured, the monitoring points at the top and the corresponding points at the bottom should be placed along the same vertical line; when the local inclination is measured, the upper monitoring points and the lower points should be placed along the same vertical line.

Commonly used simple monitoring methods include the hanging ball method: at the position of the monitoring point at the top of the house or at the required height, directly or at the position of the fixed strut (fixed at each position), the appropriate weight of the pointed ball is suspended, in the vertical line. The bottom establishes a fixed millimeter grid reading board (the direction should be set on the reading board) and other reading devices, directly read or measure the horizontal displacement and displacement direction of the upper fixed point relative to the bottom monitoring point on the reading board and take a photo record. It is suggested to avoid the time period when the wind load is greatly affected in inclination monitoring.

3. Local dangerous point monitoring

Local dangerous point monitoring focus on the crack monitoring and the method should be simple and easy to operate. According to the requirements of the monitoring scheme, the crack should be monitored by using the simple measurement tools like steel tape measure, ruler, Vernier scale, and so on and the information of location, area, severity, and propagation trend should be recorded and labeled by text, photograph, and video. The requirements of the monitoring are as follows:

(1) The location, distribution, orientation, length, width, and propagation should be measured in the monitoring and each crack to be monitored should be numbered consistently.
(2) For each crack to be monitored, there should be at least three sets of monitoring label. One should be at the widest position and the other two should be at two ends of the crack. For each set, there should be two labels which are located on both sides of the crack.
(3) For short-term monitoring, the painting of parallel line or the metal sheet label attached by constructional glue can be used; for long-term monitoring, the metal sheet (rod or wedge plate) label mosaiced or embedded into the wall can be used.
(4) The change of the crack can be obtained by measuring the distance between the labels by simple measurement tools like steel tape measure, steel ruler, and Vernier scale.
(5) The accuracy of the crack width should not be less than 1.0 mm, and the accuracy of the crack length should not be less than 10.0 mm.
(6) Each measurement should be recorded carefully, and the crack should be photographed with date of measurement.
(7) An alarm can be used to alert the changing of crack width. The residents can be informed by the alarm sound to ensure the safety of resident's life and property.

If the monitoring data of the dilapidated building reaches the warning threshold during the monitoring or any abnormality is observed during the global inspection, the situation should be reported in time and immediate action should be taken to ensure safety.

9.1.10 The monitoring scheme should be modified in time according to the monitoring data to ensure safety.

The monitoring data should be collected and analyzed regularly and then reported to the administrative department and local government in the form of "monthly report", "quarterly report", "semi-annual report", and "annual report" as required; and the accident should be reported as required by the "special report".

The online monitoring data should be reported to the administrative department in time in the form of "daily report".

9.1.11 When the monitoring data of the dilapidated building reaches the warning threshold during the monitoring or any abnormality is observed during the global inspection, the situation should be reported in time and immediate action should be taken, and the local government and administrative department should organize an emergency meeting and launch emergency plan to ensure safety.

9.3 Mass Concrete Structure

9.3.1 The mass concrete structure has the following characteristics: great solidity, large volume of concrete, complicated engineering conditions (generally underground cast-in-place reinforced concrete structure), high construction technology requirements, large cement hydration heat (expected to exceed 25 °C), and easily generated temperature-induced deformation. For the mass concrete structure, in addition to the requirements to minimum cross section and internal and external temperature, there are some requirements to the dimension.

9.3.3 If there are 3–5 measurement points at each monitoring location, there can be 10–17 monitoring locations if data acquisition equipment with 50 temperature measurement channels is adopted, which can be enough for the temperature monitoring of a general mass concrete structure. For other requirements like display, storage, and analysis, they are very general to the temperature measurement device, and those provided by most domestic suppliers can meet the requirements.

9.3.5 The monitoring method during the use of mass concrete structures can be arranged with reference to monitoring during construction.

9.4 Prefabricated Structure

9.4.1 The component and the connection of the prefabricated structure should be monitored periodically. They may be repaired or strengthened according to the analysis based on monitoring data. The load capacity of the repaired component and connection should not be less than before.

9.4.2 The prefabricated concrete structure generally consists of precast column, precast beam, precast slab, precast stair, and so on. The strain and stress at the key component should be monitored because stress intensity may be observed due to the complicated conditions such as component dimension, material property, installation deviation, connection, welding residual stress, temperature change, joint corrosion and construction quality, and so on. The strain and stress at the key connection should also be monitored because the superimposed stress may occur if the connection quality is not good. The monitoring points should be placed in side span, interior span, and special key positions. For each key position, 2–3 components should be selected. The deflection of the control point, the position of the datum axis, the elevation and the vertical deviation should be monitored and corrected for various construction methods. The monitoring points to measure deformation should be placed at the midspan, quarter-span, and support of the bending component. The displacement of the component after installation should be controlled to ensure the deviation of the joint position after construction from the designed value meets the requirements of acceptance. The reaction force at the connection area is significantly affected by the construction process; therefore, more monitoring points should be placed at the connection area and the strain and stress should be monitored online.

9.4.3 The monitoring of prefabricated steel structure should follow the requirements in the explanation of 9.4.2.

9.4.4 The monitoring should focus on the safety of the prefabricated timber structure, including load capacity, configuration, displacement (deformation) not suitable for load bearing, crack, dangerous deterioration, and insert attack. For the prefabricated timber structure, the joint and the connection of component are important and the construction quality of bolted connection, toothed plate connection, mortise and tenon connection, and general connection by metal connector should be monitored. The connection structure shall comply with the requirements of the current national design code and the monitoring should focus on the defects at the connection such as relaxation and deformation, sliding, and cracking along the shear plane.

The bolted connection should meet the requirements in "Code for acceptance of construction quality of timber structures" (GB 50206), "Standard for design of timber structures" (GB 50005), and "Technical code of glued laminated timber structures" (GBT 50708).

The integrity of the mortise and tenon connection should be inspected, including:

1. Deterioration and insert attack;
2. Crack, break, and defect at the visible part of the tenon;
3. Break at the mortise;
4. Connection loose.

The monitoring of tenon pull-out should meet the following requirements:

1. When the tenon surface pull-out of the component is not consistent, the maximum value should be used;
2. For the mortise and tenon connection between column and beam, the percentage of the tenon pulled out from the mortise should meet the requirements in the following Table 9.4.4-1

Table 9.4.4-1 Critical value of the percentage of the tenon pulled out from the mortise

Type	Tailiang system	Chuandou system	Fortification intensity is 8° or 9°
The percentage of the tenon pulled out from the mortise	0.4	0.5	0.25

The monitoring of the gap between mortise and tenon should meet the following requirements:

1. The gap between mortise and tenon at every edge should be measured by wedge feeler gauge. For the dougong bracket, the tolerance is 1 mm; for other types, the tolerance can be found in Table 10.4.4.2 (Table 9.4.4-2).

Table 9.4.4-2 The tolerance of the gap between mortise and tenon

Column diameter	<200	200–300	300–500	>500
Tolerance (mm)	3	4	6	8

2. If there is no gap between mortise and tenon, it is necessary to check whether local compression damage exists (local indentation, wrinkle, shearing of local wood fiber, etc.).
3. The rotation of the mortise tenon commotion should be monitored; if it is not consistent with the component rotation, it is necessary to check whether there is break at the tenon.
4. The compression deformation at the tenon and the mortise should be monitored, and the transverse compression deformation should not be more than 4 mm.

The monitoring of general connection by metal connector should meet the following requirements:

1. The installation method, installation error, deformation, loose, and pull-out of tooth should be monitored, in which visual inspection and caliper can be used. The monitoring should comply with "Code for acceptance of construction quality of timber structures" (GB 50206).

2. The gap between the components at the connection, and the local gap between the compression surfaces of the component should be monitored in which caliper or wedge feeler gauge can be used. The monitoring should comply with "Code for acceptance of construction quality of timber structures" (GB 50206).
3. For the metal teethed plate connection, the area of wood surface defect, the area of the plate tooth lodging, and the breaks should be monitored, in which caliper or wedge feeler gauge can be used. The monitoring should comply with "Code for acceptance of construction quality of timber structures" (GB 50206).

9.5 Isolated (Damper-Added) Structure

9.5.1 The isolated (damper-added) structure is a structure in which the level of response due to earthquake can be significantly reduced from the level occurring in general non-isolated building (generally by 80%) by adopting a safe and reliable isolation layer between the superstructure and the foundation. Because of this isolation system, the superstructure can be decoupled from the ground motion and therefore it has large shear deformation capacity.

The isolated (damper-added) structure should be monitored during construction and service life, including the structural displacement and deformation, to ensure the performance of the isolation (damping) rubber bearing in both construction stage and service life. The abnormal deformation should be monitored if the structure is subject to long-term loading, and once it is observed, necessary actions should be taken. The dynamic change of the isolation system should also be monitored in the event of earthquake.

The isolation system consisting of isolation (damping) rubber bearing (including connectors) and viscous dampers (including connectors) should be installed between the foundation and the superstructure to consume seismic energy, eliminate or reduce the transmission of seismic energy to the superstructure, and protect the superstructure more effectively.

9.5.2 Based on the characteristics of load bearing and deformation of the isolation system, the monitoring of the isolation system should include the load monitoring and response monitoring.

Load monitoring: the load including seismic load, wind load (for high-rise building), temperature and foundation settlement, and so on applied on the superstructure should be monitored, because the mechanical properties of the damping rubber bearing which is frequently used in the isolation system may change with deformation and temperature.

Response monitoring during construction: the vertical and horizontal displacement of the damping rubber bearing, the strain of the key parts of the main beam in the isolation layer should be monitored during construction. Because the damping rubber bearing installed between the foundation and the superstructure has low lateral stiffness, it reduces the constraint of the foundation to the superstructure, resulting in the obvious vertical and horizontal displacement of the damping rubber bearing due to uneven constructional load and concrete shrinkage in the super-structure, which however cannot be completely recovered after construction. Therefore, the vertical load bearing area decreases, limiting the allowable horizontal displacement and reducing the seismic energy consumption capacity, and therefore the dynamic response of the superstructure subject to seismic loading is enlarged. Moreover, the constraint of the main beam in the isolation layer decreases due to the damping rubber bearing, and transverse crack may initiate and propagate in the main beam due to multiple factors like concrete shrinkage in the superstructure, and so on.

9.5.3 Due to the particularity of the isolation system, the monitoring during the service life should include the static vertical and horizontal displacement of the damping rubber bearing (for investigation of the relationship between the dis-placement and the loading like temperature), the dynamic displacement of the damping rubber bearing subject to seismic and wind loading (for investigation of the mechanical property changing when the damping rubber bearing is subject to dynamic loading, and the low cycle fatigue and high cycle fatigue), and the acceleration of the superstructure.

9.5.4 For one building, the monitoring points should be placed at four corners and every 2–3 supports under the columns (or every 10–15 m on the exterior wall), and there should be at least three points for each side. The single point settlement gages should be installed at important locations such as the boundary between different foundations, the boundary between different structures, both sides of expansion joint, seismic joint and severe cracking, both sides of the interface of new and old buildings (or tall and low buildings), and the isolation layer.

9.5.6 During the construction process, the crack initiates in the main beam of the isolation layer due to constructional load, temperature changing, and concrete shrinkage. The isolation layer is in the open-air environment where the temperature difference is large, and the concrete shrinkage usually occurs in the early stage after pouring; therefore, the strain and stress of the main components in the isolation layer should be monitored.

9.5.7 The initial condition of the damping rubber bearing can be considered as its normal working condition after the construction of the main structure, the maxi-mum horizontal sear deformation should not be more than 50 mm, and the maxi-mum vertical compression deformation should not be more than 5 mm.

9.5.8 The horizontal deformation and vertical compression deformation should be measured by using instrument, and others can be monitored by visual inspection. Emergency monitoring should be conducted in time after a disaster like earthquake and fire which may damage the isolation system.

9.6 Large Public Buildings

9.6.1 This section of the large public buildings for the construction of structural clusters, the monitoring content can be selected with reference to the high-rise structure, large-span space structure, industrial plant and other chapters, which can be monitored with reference to the monitoring of dangerous housing.

9.6.2 The safety condition can be evaluated by the real-time monitoring data on the platform.

The long-term monitoring database for structure group should be built and the data should be screened effectively for subsequent data usage and performance assessment of complex structure group.

9.6.31. Infrastructure layer

(1) Monitoring hardware (sensors, data acquisition device, wireless data transition equipment and others, to monitor strain and stress, deformation, environment factors, vibration, etc.);
(2) Data communication and computer network (data transmission between various subsystem and data acquisition system and data communication between monitoring center and management department);

2. Data resource layer

   (1) Environmental factors monitoring database;
   (2) Strain and stress monitoring database;
   (3) Deformation monitoring database;
   (4) Loading and vibration monitoring database;
   (5) Cloud data center;

3. Application support layer

The information sharing between the user layer and the data resource layer, and the common functions of the application system are realized, providing technical support for the business collaborative work.

4. User layer

   (1) Units with different levels and authorities;
   (2) Display interface (desktop client and mobile client).

10 Warning Threshold

10.1 General Requirements

10.1.1 The early warning threshold shall be formulated in accordance with the following principles:

1. It should be given in a quantitative manner and regularly tested, supplemented, corrected, and optimized;
2. It can be set according to design tolerance, theoretical calculation value, numerical analysis value, monitoring data value, and mature experience.

Major engineering structure refers to the construction of a local or central financial investment, including key planning plans and typical engineering structures with demonstration effects, and the rest can be classified as ordinary engineering structures.

10.1.3 Refer to T/CECS 529-2018, Chapter 3 of the Standard for Establishing Early Warning Thresholds for Long-Span Bridge Structural Health Monitoring Systems.

10.1.5–6 Refer to JT/T1037-2016 Chapter 8 of "Technical Regulations for Highway Bridge Structure Safety Monitoring System".

10.2 High-Rise Building

10.2.1 Refer to Chapter 5 of GB50007-2011 "Code for Design of Building Foundations".

10.2.2–6 Refer to Chapter 3 of JGJ3-2010 "Technical Regulations for Concrete Structures of High-rise Buildings".

The main factors inducing the compression deformation of high-rise buildings includes elastic deformation of vertical structural components, the creep and shrinkage of vertical components in composite structures, and temperature-induced expansion. The monitoring points should be placed at necessary locations on the same floor and they should be placed once every 8–10 floors during construction. The monitoring data should be recorded in time.

10.2.7 Refer to Chapter 5 of GB50007-2011 "Code for Design of Building Foundations".

10.3 Large-Span Spatial Structure

10.3.9 Refer to Chapter 3 of JGJ7-2010 "Technical Regulations for Space Grid Structure" and Chapter 10 of GB50011-2010 "Code for Seismic Design of Buildings".

10.3.10–20 Refer to Chapter 3 of JGJ257-2012 "Technical Regulations for Cable Structures".

10.3.21–27 Refer to Chapters 3 and 5 of CECS-158:2004 Technical Specifications for Membrane Structures.

10.4 Protective Building

10.4.1–9 Refer to Chapter 4 of GB50165-92 "Technical Specifications for Maintenance and Reinforcement of Ancient Building Wood Structures".

10.4.10 Refer to Chapter 5 of GB50007-2011 "Code for Design of Building Foundations".

10.5 Dilapidated Building

10.5.1–5 Refer to Chapters 4–5 of JGJ125-2016 "Dangerous Housing Appraisal Standards".

10.6 Industrial Factory

10.6.1 Refer to Chapter 5 of GB50868-2013 "Permissible Vibration Standards for Construction Projects".

10.7 Mass Concrete Structure

10.7.1–4 Refer to Chapter 3 of GB50496-2009 "Code for Construction of Mass Concrete".

10.8 Prefabricated Structure

10.8.1 Refer to Chapter 5 of GBT51231-2016 "Technical Standards for As-built Concrete Buildings" and Chapters 5–6 of GB50011-2010 "Code for Seismic Design of Buildings".

10.8.2 Refer to Chapter 5 of GBT51232-2016 "Technical Standards for Assembled Steel Structure Construction" and Chapter 8 of GB50011-2010 "Code for Seismic Design of Buildings".

10.8.3 Refer to Chapter 4 of GB50165-92 "Technical Specifications for Maintenance and Reinforcement of Ancient Building Wood Structures" and Chapter 11 of GB50011-2010 "Code for Seismic Design of Buildings".

10.9 Isolated (Damper-Added) Structure

10.9.1–2 Refer to Chapter 5 of JGJ360-2015 "Code for Construction and Acceptance of Building Seismic Engineering".

10.9.4–7 Refer to Chapter 6 of GB20688.3-2006 "Rubber Bearing Part 3: Building Isolation Rubber Bearing" and Chapter 12 of GB50011-2010 "Code for Seismic Design of Buildings".

The Graph Theoretical Approach in Brain Functional Networks

Theory and Applications

The Graph Theoretical Approach in Brain Functional Networks: Theory and Applications
Fabrizio De Vico Fallani and Fabio Babiloni

ISBN-13: 978-3-031-00516-9   paperback
ISBN-13: 978-3-031-01644-8   ebook

DOI 10.1007/978-3-031-01644-8

A Publication in the Springer series
SYNTHESIS LECTURES ON BIOMEDICAL ENGINEERING

Lecture #36
Series Editor: John D. Enderle, *University of Connecticut*
Series ISSN
Synthesis Lectures on Biomedical Engineering
Print 1930-0328   Electronic 1930-0336

Synthesis Lectures on Biomedical Engineering

Editor
John D. Enderle, *University of Connecticut*

Lectures in Biomedical Engineering will be comprised of 75- to 150-page publications on advanced and state-of-the-art topics that spans the field of biomedical engineering, from the atom and molecule to large diagnostic equipment. Each lecture covers, for that topic, the fundamental principles in a unified manner, develops underlying concepts needed for sequential material, and progresses to more advanced topics. Computer software and multimedia, when appropriate and available, is included for simulation, computation, visualization and design. The authors selected to write the lectures are leading experts on the subject who have extensive background in theory, application and design.

The series is designed to meet the demands of the 21st century technology and the rapid advancements in the all-encompassing field of biomedical engineering that includes biochemical, biomaterials, biomechanics, bioinstrumentation, physiological modeling, biosignal processing, bioinformatics, biocomplexity, medical and molecular imaging, rehabilitation engineering, biomimetic nano-electrokinetics, biosensors, biotechnology, clinical engineering, biomedical devices, drug discovery and delivery systems, tissue engineering, proteomics, functional genomics, molecular and cellular engineering.

The Graph Theoretical Approach in Brain Functional Networks: Theory and Applications
Fabrizio De Vico Fallani and Fabio Babiloni
2010

Models of Horizontal Eye Movements, Part II: A 3rd Order Linear Saccade Model
John D. Enderle and Wei Zhou
2010

Models of Horizontal Eye Movements, Part I: Early Models of Saccades and Smooth Pursuit
John D. Enderle
2010

Biomedical Technology Assessment: The 3Q Method
Phillip Weinfurt
2010

Fundamentals of Respiratory Sounds and Analysis
Zahra Moussavi
2006

Advanced Probability Theory for Biomedical Engineers
John D. Enderle, David C. Farden, and Daniel J. Krause
2006

Intermediate Probability Theory for Biomedical Engineers
John D. Enderle, David C. Farden, and Daniel J. Krause
2006

Basic Probability Theory for Biomedical Engineers
John D. Enderle, David C. Farden, and Daniel J. Krause
2006

Sensory Organ Replacement and Repair
Gerald E. Miller
2006

Artificial Organs
Gerald E. Miller
2006

Signal Processing of Random Physiological Signals
Charles S. Lessard
2006

Image and Signal Processing for Networked E-Health Applications
Ilias G. Maglogiannis, Kostas Karpouzis, and Manolis Wallace
2006

The Graph Theoretical Approach in Brain Functional Networks

Theory and Applications

Fabrizio De Vico Fallani and Fabio Babiloni
University of Rome – "Sapienza"

SYNTHESIS LECTURES ON BIOMEDICAL ENGINEERING #36

ABSTRACT

The present book illustrates the theoretical aspects of several methodologies related to the possibility of i) enhancing the poor spatial information of the electroencephalographic (EEG) activity on the scalp and giving a measure of the electrical activity on the cortical surface. ii) estimating the directional influences between any given pair of channels in a multivariate dataset. iii) modeling the brain networks as graphs. The possible applications are discussed in three different experimental designs regarding i) the study of pathological conditions during a motor task, ii) the study of memory processes during a cognitive task iii) the study of the instantaneous dynamics throughout the evolution of a motor task in physiological conditions. The main outcome from all those studies indicates clearly that the performance of cognitive and motor tasks as well as the presence of neural diseases can affect the brain network topology. This evidence gives the power of reflecting cerebral "states" or "traits" to the mathematical indexes derived from the graph theory. In particular, the observed structural changes could critically depend on patterns of synchronization and desynchronization - i.e. the dynamic binding of neural assemblies - as also suggested by a wide range of previous electrophysiological studies. Moreover, the fact that these patterns occur at multiple frequencies support the evidence that brain functional networks contain multiple frequency channels along which information is transmitted. The graph theoretical approach represents an effective means to evaluate the functional connectivity patterns obtained from scalp EEG signals. The possibility to describe the complex brain networks sub-serving different functions in humans by means of "numbers" is a promising tool toward the generation of a better understanding of the brain functions.

KEYWORDS

brain networks, graph theory, small-world, EEG, spinal cord injury, memory

Contents

Acknowledgments

We wish to thank Donatella Mattia, Febo Cincotti, and Laura Astolfi for their kind support and for being a constant source of challenging ideas.

Fabrizio De Vico Fallani and Fabio Babiloni
Rome, April 2010

Preface

This book describes some advanced mathematical signal processing techniques applied to the estimation of the cortical connectivity in humans from non-invasive electroencephalographic (EEG) recordings.

Although it can be thought that mathematics could not be the proper tool for a full comprehension of the brain functions, often this is not the case. In the last ten years, many different brain-imaging devices have conveyed a lot of information about the brain functioning in different experimental conditions. In every case, the biomedical engineers, together with mathematicians, physicists and physicians are called to elaborate the signals related to the brain activity in order to extract meaningful and robust information to correlate with the external behaviour of people. In such attempt, different signal processing tools used in telecommunications and other field of engineering or even social sciences have been adapted and re-used in the neuroscience field.

In particular, the science of complex networks has produced an increasingly interest in the study of complex systems where interaction networks are crucial. Recently, the analysis of real networks led to a series of important results in various fields and to the identification of the basic principles common to all the networks that are being considered. Scientists have found that several systems can be represented as networks and that the study of the whole web of links connecting different parts rather than the analysis single elements, could give a better comprehension of the system itself.

In this sense, the analysis of the brain functional connectivity through a network-based approach is representing one of the most promising procedures to study the brain functioning during motor or cognitive tasks.

The present book intends to offer a concise presentation of the theoretical aspects concerning i) the possibility to achieve the cortical functional connectivity of the human brain from standard EEG signals and ii) the advantages of a graph approach to extract the topological features of the estimated brain networks.

The aim is to have generated a practical book for guiding the reader in the application of the proposed technologies in the fascinating field of neuroscience.

Fabrizio De Vico Fallani and Fabio Babiloni
Rome, April 2010

CHAPTER 1

Introduction

Over the last decade, there has been a growing interest in the detection of the functional connectivity in the brain from different neuroelectromagnetic and hemodynamic signals recorded by several neuro-imaging devices such as the functional Magnetic Resonance Imaging (fMRI) scanner, electroencephalography (EEG), and magnetoencephalography (MEG) apparatus. Many methods have been proposed and discussed in the literature with the aim of estimating the functional relationships among different cerebral structures (Clifford, C., 1987; Gevins et al., 1989; Tononi et al., 1994; Inouye et al., 1995; Nunez, P., 1995; Urbano et al., 1998; Stam and van Dijk, 2002; Stam et al., 2007; David et al., 2004; Babiloni et al., 2005; Astolfi et al., 2006).

However, the necessity of an objective comprehension of the network composed by the functional links of different brain regions is assuming an essential role in the Neuroscience (Horwitz, B., 2003). Consequently, there is a wide interest in the development and validation of mathematical tools that are appropriate to spot significant features that could describe concisely the structure of the estimated cerebral networks. The extraction of the salient characteristics from brain connectivity patterns is an open challenging topic, since often the estimated cerebral networks have a relative large size and complex structure.

Recently, it was realized that the functional connectivity networks estimated from actual brain-imaging technologies (MEG, fMRI, and EEG) can be analyzed by means of graph theory (Stam and Reijneveld, 2007). Since a graph is a mathematical representation of a network, which is essentially reduced to nodes and connections between them, the use of a graph theoretical approach seems relevant and useful as firstly demonstrated on a set of anatomical brain networks (Stephan et al., 2000; Strogatz, S., 2001). In those studies, the authors have employed two characteristic measures, the *average shortest path PL* and the *clustering index C*, to extract, respectively, the global and local properties of the network structure. They have found that anatomical brain networks exhibit many local connections (i.e., a high C) and few random long distance connections (i.e., a low PL). These values identify a particular model that interpolate between a regular lattice and a random structure. Such a model has been designated as "small-world" network in analogy with the concept of the small-world phenomenon observed more than 30 years ago in social systems (Milgram, S., 1967).

In the brain connectivity context, these properties have been demonstrated to reflect an optimal architecture for the information processing and communication among the involved cerebral structures (Lago-Fernandez et al., 2000; Sporns et al., 2000). Many types of brain functional networks have been analyzed according to this mathematical approach. In particular, fMRI studies in healthy subjects have been addressed during simple finger-tapping tasks (Eguiluz et al., 2005), resting state (Salvador et al., 2005) and no task (Achard and Bullmore, 2007). Networks from MEG recordings in healthy subjects have been also analyzed during no task states (Stam, C., 2004), simple motor acts (Bassett et al., 2006) and in patients with brain tumors (Bartolomei et al., 2006). The use of a graph approach for the networks estimated from EEG signals was recently addressed because the usual number of elements is rather small (from 19 to 128 scalp electrodes) when compared with the number of nodes available from the other brain-imaging techniques. However, interesting results have been achieved in patients with Alzheimer during no-task state (Stam et al., 2007) and healthy subjects during working memory tests (Micheloyannis et al., 2006). The general outcome of all these studies, which are based on different imaging techniques and connectivity estimates, indicates that the structure of the brain network holds an optimal scheme of communication during its normal functioning. However, the performance of cognitive and motor tasks as well as the presence of neural diseases can affect such a small-world topology, as revealed by the significant changes of PL and C.

This evidence gives the power of reflecting cerebral "states" or "traits" to the indexes able to evaluate the average shortest distances and the clustering properties of the functional network. In particular, the structural changes could critically depend on patterns of synchronization and desynchronization - i.e., the dynamic binding of neural assemblies (Singer, W., 1999) - as supported from a range of electrophysiological studies (Varela et al., 2001). Moreover, the fact that these synchronization patterns occur at multiple frequencies might mean that brain functional networks contain multiple frequency channels along which information is transmitted (Sporns and Honey, 2006).

Moreover, some functional brain networks have been mostly found to be very unlike the random graphs in their *degree distribution*, which gives information about the allocation of the functional links within the connectivity pattern. It was demonstrated that the *degree distributions* of these networks follow a power-law trend. For this reason, those networks are called "scale-free" (Barabási and Albert, 1999). They still exhibit the small-world phenomenon but tend to contain few nodes that act as highly connected "hubs." Scale-free networks are known to show resistance to failure, facility of synchronization and fast signal processing. Hence, it would be interesting to see whether the scaling properties of the functional brain networks are altered under various pathologies or experimental tasks.

Beyond the investigation of the network main global features, such as the presence of a small-world behavior (Watts and Strogatz, 1998; Strogatz, S., 2001; Sporns et al., 2004; Sporns and Zwi, 2004), and of scale-free degree distributions (Barabási and Albert, 1999; Jeong et al., 2000; Amaral et al., 2000), the understanding of the basic structural elements composing brain networks would reveal important information about the basic type of communication among cerebral areas. In particular, a study of *link reciprocity*, i.e., the degree of mutual interaction between couples of nodes (Garlaschelli and Loffredo, 2004), and the investigation of recurring patterns of interconnections, known in the literature as *motifs* (Milo et al., 2002), represent two adequate candidates for the structural analysis of cerebral networks.

The present book proposes a graph theoretical approach in order to evaluate the functional connectivity patterns obtained from high-resolution EEG signals. In this way, the "Brain Network Analysis" (in analogy with the Social Network Analysis that has emerged as a key technique in modern sociology) represents an effective methodology improving the comprehension of the complex interactions in the brain

The present book is then divided into six chapters. The first three chapters describe the used mathematical methods, while the last three just present the results from the application of those methods on three different EEG data sets.

In the first chapter, the most advanced algorithms for the estimation of functional connectivity from brain signals are described. These algorithms are capable to give functional connectivity patterns between different cerebral areas during the execution of any task.

In the second chapter, the methodology for the analysis of the brain functional networks through a theoretical graph approach is proposed. In particular, all the mathematical measures employed in the present book are described carefully.

In the third chapter, the high-resolution EEG technique is illustrated. This procedure allows estimating the cortical activity from the original scalp signals. All the functional networks analyzed in the present study are estimated from cortical signals.

In the fourth chapter, the study of the functional cortical network in a group of spinal cord injured patients during the foot-movement attempt is presented. The principal aim of the study is to evaluate how the "architecture" of the cortical network in patients losing the sensori-motor feedback might change with respect to healthy people.

In the fifth chapter, the cortical connectivity changes during the visualization of television commercials are investigated. The main purpose is to assess the presence of significant differences in the cortical network structure during the visualization of the TV spots that will be remembered and those that will be forgotten

In the sixth chapter, the aim of the study is to evaluate the functional time-varying dynamics of the cortical network during the preparation and the execution of the foot movement. This study is particularly attractive since it reveals the capability to track the structural properties of the changing cortical network during each single time sample.

All the experimental data here employed to present practical results of the brain network analysis in humans during motor and cognitive tasks were gathered at the laboratories of Neuroelectric Imaging and Brain Computer Interface of the S. Lucia Foundation.

CHAPTER 2

Brain Functional Connectivity

The concept of brain connectivity is viewed as central for the understanding of the organized behavior of cortical regions, beyond the simple mapping of their activity (Lee et al., 2003; Horwitz, B., 2003). This organization is thought to be based on the interaction between different and differently specialized cortical sites. Cortical connectivity estimation aims at describing these interactions as connectivity patterns which hold the direction and strength of the information flow between cortical areas.

To this purpose, several methods have been developed and applied to data gathered from hemodynamic and electromagnetic techniques (Buchel and Friston, 1997; Urbano et al., 1998; Gevins et al., 1989; Taniguchi et al., 2000; Brovelli et al., 2004). In general, functional connectivity is defined as the temporal correlation between spatially remote neurophysiologic events. The methods proposed in literature typically involve the estimation of some covariance properties between the different time series measured from the different spatial sites, during motor and cognitive tasks, by EEG and fMRI techniques (Urbano et al., 1998; Gevins et al., 1989; Jancke et al., 2000; Brovelli et al., 2004).

Due to the evidence that important information in the EEG signals are coded in frequency rather than in time domain (reviewed in Pfurtscheller and Lopes da Silva, 1999), attention was focused on detecting frequency-specific interactions in EEG or MEG signals, for instance, by means of the spectral coherence between the activity of pairs of channels (Bressler, S., 1995; Gross et al., 2001, 2003).

However, coherence analysis has not a directional nature (i.e., it just examines whether a link exists between two neural structures, by describing instances when they are in synchronous activity), and it does not provide directly the direction of the information flow. In this respect, multivariate spectral techniques called Directed Transfer Function (DTF) or Partial Directed Coherence (PDC) were proposed (Kaminski and Blinowska, 1991; Baccalà and Sameshima, 2001) to determine the directional influences between any given pair of channels in a multivariate data set. Both DTF and PDC can be demonstrated (Kaminski et al., 2001; Baccalà, 2001) to rely on the key concept of Granger causality between time series (Granger, C., 1969).

Granger theory mathematically defines what a "causal" relation between two signals is. According to this theory, an observed time series $x(n)$ is said to cause another series $y(n)$ if the knowledge of $x(n)$'s past significantly improves prediction of $y(n)$; this relation between time series is not necessarily reciprocal, i.e., $x(n)$ may cause $y(n)$ without $y(n)$ causing $x(n)$. This lack of reciprocity allows the evaluation of the direction of information flow between structures.

Kaminski and Blinowska (1991) proposed a multivariate spectral measure, called the Directed Transfer Function (DTF), which can be used to determine the directional influences between any given pair of channels in a multivariate dataset. DTF is an estimator that simultaneously characterizes the direction and spectral properties of the interaction between brain signals and requires only one multivariate autoregressive (MVAR) model to be estimated simultaneously from all the time series. The advantages of MVAR modeling of multichannel EEG signals in order to compute efficient connectivity estimates have recently been stressed. Kus et al. (2004) demonstrated the superiority of MVAR multichannel modeling with respect to the pair-wise autoregressive approach.

Another popular estimator, the Partial Directed Coherence or PDC (Baccalà and Sameshima, 2001), based on MVAR coefficients transformed into the frequency domain was recently proposed, as a factorization of the Partial Coherence. The PDC is of particular interest because of its ability to distinguish direct and indirect causality flows in the estimated connectivity pattern. If another "true" flow exists from region x_2 to region x_3, the PDC estimator does not add an "erroneous" causality flow between the signals recorded from region x_1 to region x_3. This property is particularly interesting in its application to brain signals, where the interpretation of a direct connection between two cortical regions is straightforward.

2.1 MULTIVARIATE AUTOREGRESSIVE MODELS

In statistics and signal processing, an autoregressive (AR) model is a type of random process which is often used to model and predict various types of natural phenomena. An autoregressive model is essentially an all-pole infinite impulse response filter with some additional interpretation placed on it.

The approach based on multivariate autoregressive models (MVAR) can simultaneously model a whole set of signals.

Let X be a set of estimated cortical time series:

$$X = [x_1(t), x_2(t), \ldots, x_N(t)] \tag{2.1}$$

where t refers to time and N is the number of cortical areas considered. Given an MVAR process which is an adequate description of the data set X:

$$\sum_{k=0}^{p} \Lambda(k) X(t-k) = E(t) \tag{2.2}$$

where $X(t)$ is the data vector in time; $E(t) = [e_1(t), \ldots, e_N]$ is a vector of multivariate zero-mean uncorrelated white noise processes; $\Lambda(1), \Lambda(2), \ldots \Lambda(p)$ are the $N \times N$ matrices of model coefficients ($\Lambda(0) = I$); and p is the model order. The p order is chosen by means of the Akaike Information Criteria (AIC) for MVAR processes.

In order to investigate the spectral properties of the examined process, the Eq. (2.2) is transformed into the frequency domain:

$$\Lambda(f) X(f) = E(f) \tag{2.3}$$

where:

$$\Lambda(f) = \sum_{k=0}^{p} \Lambda(k) e^{-j2\pi f \Delta t k} \tag{2.4}$$

and Δt is the temporal interval between two samples. Eq. (2.3) can then be rewritten as:

$$X(f) = \Lambda^{-1} E(f) = H(f) E(f) \tag{2.5}$$

$H(f)$ is the transfer matrix of the system, whose element H_{ij} represents the important information about the connection between the jth input and the ith output of the modeled system.

2.1.1 DIRECTED TRANSFER FUNCTION

The Directed Transfer Function $\theta_{ij}^2(f)$, representing the causal influence between the cortical waveform estimated in the jth ROI and the cortical waveform estimated in the ith ROI at the frequency f, is defined in terms of elements of the transfer matrix $H(f)$:

$$\theta_{ij}^2(f) = |H_{ij}(f)|^2 . \tag{2.6}$$

In order to compare the results obtained from cortical waveforms with different power spectra, an index normalization can be performed by dividing each estimated DTF by the squared sums of all elements of the relevant row, thus obtaining the so-called normalized DTF $\gamma_{ij}^2(f)$:

$$\gamma_{ij}^2(f) = \frac{|H_{ij}(f)|^2}{\sum_{m=1}^{N} |H_{im}(f)|^2} \tag{2.7}$$

$\gamma_{ij}^2(f)$ expresses the ratio of influence of the cortical waveform estimated in the jth ROI on the cortical waveform estimated in the ith ROI, with respect to the influence of all the estimated cortical waveforms. Normalized DTF values are in the interval [0 1], and the normalization condition:

$$\sum_{n=1}^{N} \gamma_{in}^2(f) = 1 \tag{2.8}$$

is applied.

2.1.2 PARTIAL DIRECTED COHERENCE

In order to distinguish between direct and cascade flows, a more effective estimator describing the direct causal relations between signals, (i.e. the Partial Directed Coherence PDC), was proposed in 2001. Like DTF, it is defined in terms of MVAR coefficients transformed to the frequency domain. However, it refers to the model coefficients matrix $\Lambda(f)$ rather than the transfer matrix $H(f)$.

The formal definition of the normalized Partial Directed Coherence (PDC) is the following:

$$\pi_{ij}(f) = \frac{\Lambda_{ij}(f)}{\sqrt{\sum_{k=1}^{N} \Lambda_{ki}(f)\Lambda_{kj}^{*}(f)}} . \tag{2.9}$$

The PDC from j to i, $\pi_{ij}(f)$, describes the directional flow of information from the activity in the ROI $x_j(t)$ to the activity in $x_i(t)$, *whereupon common effects produced by other ROIs $x_k(t)$ on the latter are subtracted leaving only a description that is exclusive from $x_j(t)$ to $x_i(t)$.*

Similarly, to the normalized DTF $\gamma^2(f)$, PDC values are in the interval [0 1] and the normalization condition:

$$\sum_{n=1}^{N} |\pi_{ni}(f)|^2 = 1 \tag{2.10}$$

is verified.

According to this condition, $\pi_{ij}(f)$ just represents the fraction of the time evolution of ROI j directed to ROI i, as compared to all of j's interactions with other ROIs.

Figure 2.1 shows the basic scheme for the estimate of the functional connectivity through MVAR models from a set of high-resolution EEG signals.

2.2   ADAPTIVE MVAR MODELS

Recent studies have stressed the limits of conventional pairwise methods with respect to the multivariate spectral measures based on the autoregressive modeling of multichannel EEG, in order to compute efficient connectivity estimates (Astolfi et al., 2006; Kus et al., 2004). Among the multivariate methods, the directed transfer function (Kaminski et al., 2001; Kaminski and Blinowska, 1991) and the partial directed coherence (Baccalà and Sameshima, 2001) are estimators characterizing, at the same time, direction and spectral properties of the interaction between brain signals and require only one multivariate autoregressive modeling (MVAR) model to be estimated from all the time series.

However, the classical estimation of these methods requires the stationarity of the signals; moreover, with the estimation of a unique MVAR model on an entire time interval, transient pathways of information transfer remains hidden. This limitation could bias the physiologic interpretation of the results obtained with the connectivity technique employed.

To overcome this limitation, different algorithms for the estimation of MVAR with time dependent coefficients were recently developed (Ding, L., 2000). The aim of these methods is to capture the transient functional relationships within a fixed time interval (see Figure 2.2).

Moeller et al. (2001) proposed an application to MVAR estimation of the extension of the recursive least squares (RLS) algorithm with a forgetting factor. This estimation procedure allows for the simultaneous fit of one mean MVAR model to a set of single trials, each one representing a measurement of the same task. In contrast to short-window techniques, the multi-trial RLS

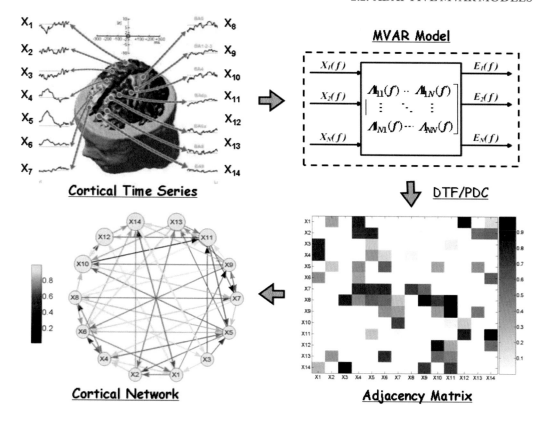

Figure 2.1: From a set of cortical time series the MVAR method estimates the spectral index DFT or PDC for each pair of regions. The estimated values can be stored into an adjacency matrix that can be modeled subsequently by means of a connectivity pattern.

algorithm does not require the stationarity of the signals, and involves the information of the actual past of the signal, whose influence decreases exponentially with the time distance to the actual samples.

The advantages of such estimation technique are an effective computation algorithm and a high adaptation capability. It was demonstrated in (Moeller et al., 2001) that the adaptation capability of the estimation (measured by its adaptation speed and variance) does not depend on the model dimension. Simulations on the efficacy of time-variant Granger causality based on AMVAR computed by the RLS algorithm were also provided (Hesse et al., 2003).

The usefulness of such time-varying approach has been achieved from a set of high-resolution EEG data in a group of healthy subjects during the preparation of a simple motor act (Astolfi et al.,

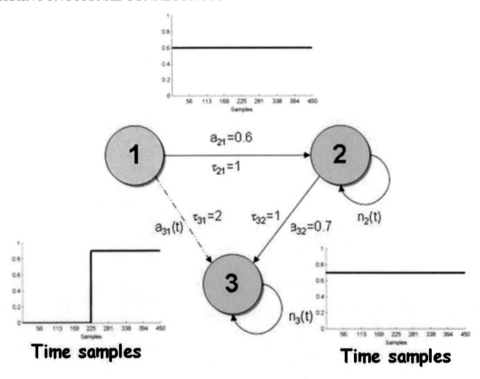

Figure 2.2: Connectivity pattern imposed between different regions of the cortex during simulation. The values of the connection strengths $a_{i,j}$ are time dependent and are represented by the blue plots near each connectivity arrow. τ is the constant delay in the propagation from area j to area i, expressed in samples. $\tau = 1$, $\tau = 2$, and $\tau = 1$ (Corresponding to 4, 8, and 4 ms, respectively).

2008). In Figure 2.3, it is possible to observe the changing connectivity pattern throughout three equidistant moments characterizing the movement preparation.

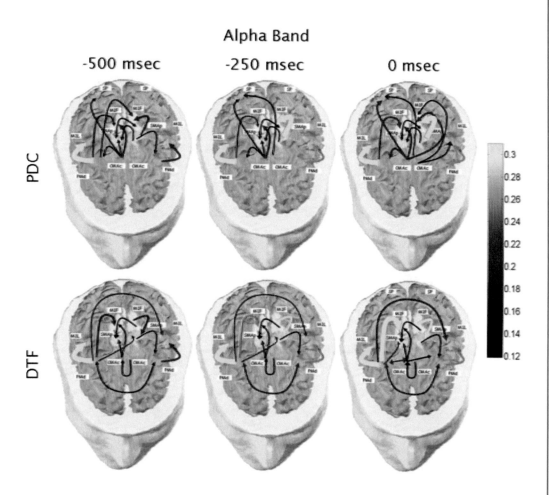

Figure 2.3: Time-varying connectivity patterns in the Alpha band, extracted at 500, 250, and 0 ms before the movement onset. First row: results obtained with time-varying DTF. Second row: results of time-varying PDC. Reconstruction of the head and cortex of the subject, obtained from sequential MRIs. The different ROIs selected are depicted in different colors and described by the labels. The color and size of the arrows code for the interaction strength (see color bar at the right side). Similarities can be seen between the results obtained with the two methods, as well as an evolution in time of the connectivity during the movement preparation.

CHAPTER 3

Graph Theory

Networks can be tangible objects in the Euclidean space, such as electric power grids, the Internet, highways or subway systems, and neural networks. Or they can be entities defined in an abstract space, such as networks of acquaintances or collaborations between individuals.

Historically, the study of networks has been mainly the domain of a branch of discrete mathematics known as graph theory. Since its birth in 1736, when the Swiss mathematician Leonhard Euler published the solution to the Königsberg bridge problem (consisting in finding a round trip that traversed each of the bridges of the Prussian city of Königsberg exactly once), graph theory has witnessed many exciting developments and has provided answers to a series of practical questions such as: what is the maximum flow per unit time from source to sink in a network of pipes, how to color the regions of a map using the minimum number of colors so that neighboring regions receive different colors, or how to fill n jobs by n people with maximum total utility. In addition to the developments in mathematical graph theory, the study of networks has seen important achievements in some specialized contexts, as for instance in the social sciences.

The last decade has witnessed the birth of a new movement of interest and research in the study of complex networks, i.e., networks whose structure is irregular, complex, and dynamically evolving in time, with the main focus moving from the analysis of small networks to that of systems with thousands or millions of nodes, and with a renewed attention to the properties of networks of dynamical units. This flurry of activity, triggered by two seminal papers, that by Watts and Strogatz on small-world networks, appeared in Nature in 1998, and that by Barabási and Albert on scale-free networks appeared one year later in Science, has seen the physics' community among the principal actors, and it has been certainly induced by the increased computing powers and by the possibility to study the properties of a plenty of large databases of real networks. These include transportation networks, phone call networks, the Internet and the World Wide Web, the actors' collaboration network in movie databases, scientific coauthorship and citation networks from the Science Citation Index, but also systems of interest in biology and medicine, as neural networks or genetic, metabolic and protein networks.

The massive and comparative analysis of networks from different fields has produced a series of unexpected and dramatic results. The first issue that has been faced is certainly structural. The research on complex networks begun with the effort of defining new concepts and measures to characterize the topology of real networks. The main result has been the identification of a series of unifying principles and statistical properties common to most of the real networks considered.

In mathematics and computer science, graph theory is the study of graphs: mathematical structures used to model pairwise relations between objects from a certain collection. A graph is an abstract representation of a network. It consists of a set of N vertices (or nodes) and a set of L edges (or connections) indicating the presence of some of interaction between the vertices. The adjacency matrix A contains the information about the connectivity structure of the graph. When a weighted and directed edge exists from the node i to j, the corresponding entry of the adjacency matrix is $A_{ij} \neq 0$; otherwise $A_{ij} = 0$.

3.1 NETWORK DENSITY

The simplest attribute for a graph is its density k, defined as the actual number of connections within the model divided by its maximal capacity; density ranges from 0 to 1, the sparser is a graph, the lower is its value. The mathematical formulation of the network density is given by the following:

$$k(A) = \frac{1}{N(N-1)} \sum_{i \neq j \in V} a_{i,j} \ . \tag{3.1}$$

Where A is the adjacency matrix and a_{ij} is the value of the respective arc from the point j to the point i. $V = 1 \ldots N$ is the set of nodes within the graph.

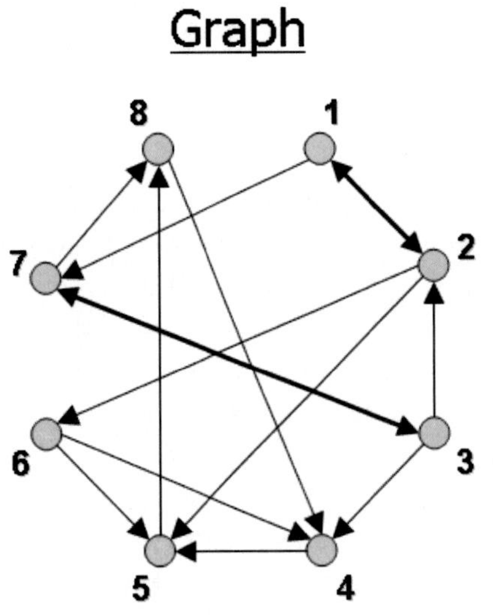

Adjacency Matrix

0	1	0	0	0	0	1	0
1	0	0	0	1	1	0	0
0	1	0	1	0	0	1	0
0	0	0	0	1	0	0	0
0	0	0	0	0	0	0	1
0	0	0	1	1	0	0	0
0	0	1	0	0	0	0	1
0	0	0	1	0	0	0	0

$N = 8$

$L = 15$

$$k(A) = \frac{15}{56} = 0{,}268$$

Figure 3.1: Example of connection density computation. At the left of the figure, the graph consists of $N = 8$ nodes and $L = 15$ edges. At the right of the figure, the respective adjacency matrix is shown.

Average connection densities can vary widely, depending on the particular neural structure, on the level of analysis (i.e., populations or single cells), and on the spatial extent of the neural network (e.g., entire brain versus local circuit). While the average connection density between cells across the entire cerebral cortex is approximately 10^6-10^7, local connection densities are significantly higher within single cortical columns (10^1-10^3). Matrices of connection pathways linking cortical areas tend to have $k \sim 0.2$-0.4. Connection matrices of patches of local cortical circuits comprising multiple columns would likely have very low $k \sim 0.01 - 0.001$ (Sporns, O., 2002).

When dealing with weighted networks, a useful generalization of this quantity is represented by the weighted-density k_w. This measure evaluates the intensity of each link rather than the simple absence/presence of connections i.e., 0/1.

3.2   NODE DEGREE

In the same way, the simplest attribute of a node is its connectivity degree, which is the total number of connections with other vertices. This quantity has to be split into in-degree d_{in} and out-degree d_{out}, when directed relationships are being considered. The formulation of the in-degree index d_{in} can be introduced as follows:

$$d_{in}(i) = \sum_{j \in V} a_{i,j} \ . \tag{3.2}$$

It represents the total amount of links incoming to the vertex i. V is the set of the available nodes and a_{ij} indicates the presence of the arc from the point j to the point i. In a similar way, for the out-degree:

$$d_{out}(i) = \sum_{j \in V} a_{j,i} \ . \tag{3.3}$$

It represents the total amount of links outgoing from the vertex i.

Indegrees and outdegrees have obvious functional interpretations. A high indegree indicates that a neural unit is influenced by a large number of other units, while a high outdegree indicates a large number of potential functional targets. For most neural structures, indegrees and outdegrees of neural units are subject to constraints due to growth, tissue volume or metabolic limitations. Connections cannot be attached or emitted beyond the limits imposed by these constraints (Sporns, O., 2002).

In a weighted graph, the natural generalization of the degree of a node i is the node *strength* or node weight or weighted-degree. The strength index integrates the information of the links' number (degrees) with the connections' weight, thus representing the total amount of outgoing intensity from a node or incident intensity into it.

3.3   DEGREE DISTRIBUTIONS

The arithmetical average of all nodes' degree $< d >$ is called mean degree of the graph. Indeed, this mean value gives little information about the behavior of degree within the system. Hence, it is

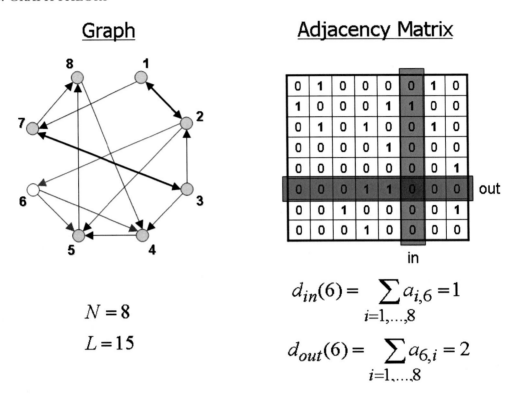

$$N = 8$$
$$L = 15$$

$$d_{in}(6) = \sum_{i=1,\dots,8} a_{i,6} = 1$$

$$d_{out}(6) = \sum_{i=1,\dots,8} a_{6,i} = 2$$

Figure 3.2: Example of in-degree and out-degree computation. At the left, the edges outgoing from the node 6 are colored in blue, while those incoming in the same node are colored in red. The rows and columns involved in the computation are colored in the adjacency matrix, too.

useful to introduce $P(d)$, the fraction of vertices in the graph that have degree d. Equivalently, $P(d)$ is the probability that a vertex chosen uniformly at random has degree d. A plot of $P(d)$ for any given network can be constructed by making a histogram of the degrees of vertices. This histogram is the degree distribution for the graph, and it allows better understanding of the degree allocation in the system.

If the graph is directed, the in-degree d_{in} and the out-degree d_{out} for each node must be considered separately (see Figure 3.3).

For large networks both distributions $P(d_{in})$ and $P(d_{out})$ over the entire digraph may be inspected for scale ¬ free attributes such as power laws (Albert and Barabási, 2002). Besides, contrasting biological degree distributions with those obtained from same-sized random digraphs could reveal interesting differences.

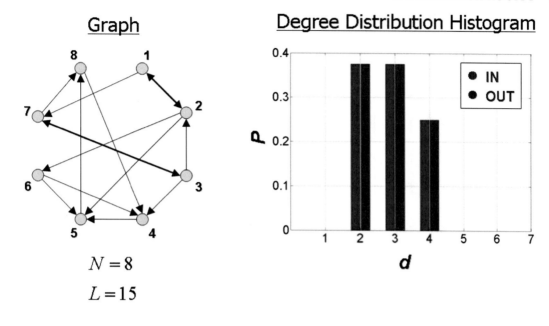

Figure 3.3: Example of in-degree distribution and out-degree distribution computation. The histogram at the right of the figure indicates the percentage of nodes in the graph that holds "d" number of incoming (red bar) or outgoing (blue bar) connections.

For a weighted graph, it is useful to introduce $R(s)$ as the fraction of vertices in the graph that have strength equal to s. In the same way, $R(s)$ is the probability that a vertex chosen uniformly at random has weight $= s$.

3.4   LINK RECIPROCITY

In a directed network, the analysis of *link reciprocity* reflects the tendency of vertex pairs to form mutual connections between each other [44]. The correlation coefficient index ρ proposed by Garlaschelli and Loffredo (2004) measures whether double links (with opposite directions) occur between vertex pairs more or less often than expected by chance. The correlation coefficient can be written as follows:

$$\rho(A) = \frac{r(A) - k(A)}{1 - k(A)} \ .$$

(3.4)

In this formula, r is the ratio between the number of links pointing in both directions and the total number of links, while k is the connection density that equals the average probability of finding a reciprocal link between two connected vertices in a random network. As a measure of reciprocity, ρ is an absolute quantity that directly allows one to distinguish between reciprocal ($\rho > 0$) and anti-reciprocal ($\rho < 0$) networks, with mutual links occurring more and less often than random, respectively. The neutral or areciprocal case corresponds to $\rho = 0$. Note that if all links occur in reciprocal pairs one has $\rho = 1$, as expected.

Graph   ## Adjacency Matrix

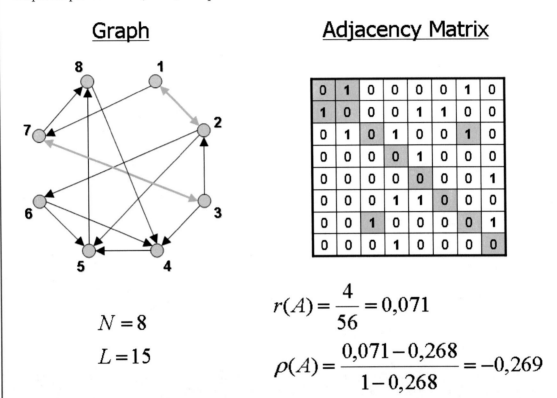

$$r(A) = \frac{4}{56} = 0{,}071$$

$$N = 8$$

$$L = 15$$

$$\rho(A) = \frac{0{,}071 - 0{,}268}{1 - 0{,}268} = -0{,}269$$

Figure 3.4: Example of reciprocity computation. At the left of the figure, the reciprocal edges in the graph are colored in green. At the right of the figure, the reciprocal elements that are located in symmetrical positions in the adjacency matrix are colored in green, while the elements in main diagonal are colored in gray.

In the brain, reciprocal connections are found in many systems and at many levels of scale, from local circuits to pathways between brain areas (where they are very abundant). Evaluating $\rho(A)$ provides a first-order estimate of the extent of reciprocal dynamical coupling present within the entire network. Typical large-scale cortical connection matrices have $\rho(A)$ of around 0.7-0.8 (perhaps higher, given that reciprocity of connection pathways may be underestimated due to missing anatomical information). This is much higher than the $\rho(A)$ of random networks of corresponding size (N, L) (Sporns, O., 2002).

In general, the existence of reciprocal anatomical connections does not imply a symmetrical functional relationship between the linked neural units. For example, connection pathways in the cortex show characteristic laminar termination patterns that differ for feedforward and feedback connections (Felleman and Van, 1991) and that may have different functional impact on their neuronal targets.

3.5   MOTIFS

By motif it is usually meant a small connected graph of M vertices and a set of edges forming a subgraph of a larger network with $N > M$ nodes (Milo et al., 2002). For each N, there are a limited number of distinct motifs. For $N = 3$, 4, and 5, the corresponding numbers of directed motifs is 13, 199, and 9364. In the present book, we focus on directed motifs with $N = 3$. The 13 different 3-node directed motifs are shown in Figure 3.5.

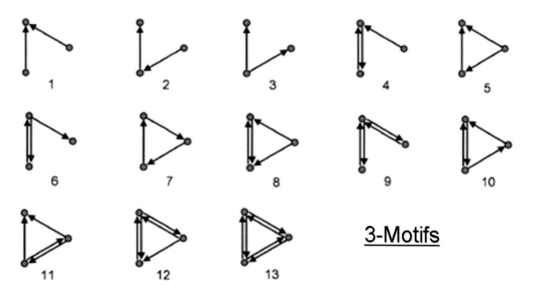

Figure 3.5: The 13 possible schemes of connectivity that can be achieved with a sub-graph of 3 nodes.

Counting how many times a motif appears in a given network yields a frequency spectrum that contains important information on the network basic building blocks. Eventually, one can looks at those motifs within the considered network that occur at a frequency significantly higher than in random graphs.

The functional motif frequency spectrum provides a sophisticated way of characterizing subtypes of such networks geared at more specific functional modes of information processing.

An innovative study showed that the functional motif number of a variety of real brain networks is very high compared to equivalent random networks. Motif number can yield networks that resemble real brain networks in several structural characteristics, including their motif frequency spectra,

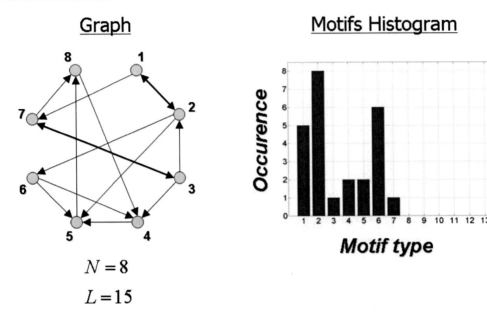

Figure 3.6: Example of 3-motif spectrum computation. At the right of the figure, the thirteen possible motif types are displayed on the x-axes, while the occurrence of each motif is displayed on y-axes.

motifs that occur in significantly increased numbers, and small-world measures (Sporns and Kotter, 2004).

3.6   NETWORK STRUCTURE

Two measures are frequently used to characterize the local and global structure of unweighted graphs: the average shortest path PL and the clustering index C (Watts and Strogatz, 1998). The former measures the efficiency of the passage of information among the nodes, the latter indicates the tendency of the network to form highly connected clusters of vertices.

Recently, a more general setup has been examined in order to investigate weighted networks. In particular, Latora and Marchiori (2001) considered weighted networks and defined the efficiency coefficient e of the path between two vertices as the inverse of the shortest distance between the vertices (note that in weighted graphs the shortest path is not necessarily the path with the smallest number of edges). In the case where a path does not exist, the distance is infinite and $e = 0$. The average of all the pair-wise efficiencies e_{ij} is the global-efficiency E_g of the graph. Thus, global-efficiency can be defined as:

$$E_g(A) = \frac{1}{N(N-1)} \sum_{i \neq j \in V} \frac{1}{d_{i,j}} \tag{3.5}$$

where N is the number of vertices composing the graph. Since the efficiency e also applies to disconnected graphs, the local properties of the graph can be characterized by evaluating for every vertex i the efficiency coefficients of Ai, which is the sub-graph composed by the neighbors of the node i. The local-efficiency E_l is the average of all the sub-graphs global-efficiencies:

$$E_l(A) = \frac{1}{N} \sum_{i \in V} E_{glob}(A_i) . \tag{3.6}$$

Since the node i does not belong to the sub-graph Ai, this measure reveals the level of fault-tolerance of the system, showing how the communication is efficient between the first neighbors of i when i is removed.

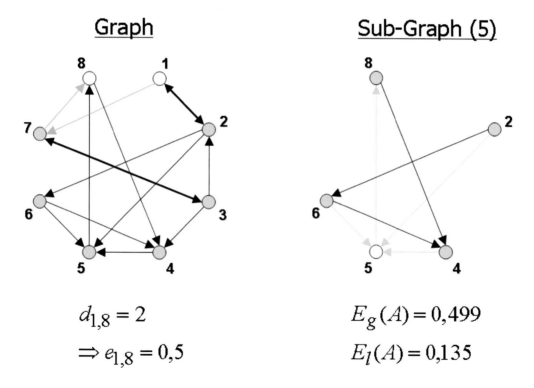

Figure 3.7: Example of efficiency computation. At the left of the figure, the shortest path between the nodes 1 and 8 is colored in green. At the right of the figure, the sub-graph obtained by removing the node 5 from the original graph is illustrated.

In general, the length of the shortest path in anatomical brain networks is indicative of the potential strength of functional interactions. However, distance *per se* in graphs make no reference to the physical distances between neuronal units located in metric space. It is possible for one pair of neuronal units to be separated by $d_{ij} = 3$ and be 200 μm apart, while another pair is directly connected ($d_{ij} = 1$), but separated by a metric distance of *20 mm*. It seems reasonable to assume that, in many cases, graph distance is a better predictor of the degree and strength of functional interactions than metric distance.

The cluster index for a network expresses the extent to which the units within the network share common neighbours that "talk" among each other, an attribute that has been called the "cliquishness" of the network. A high cluster index C points to a global organizational pattern consisting of groups of units that mutually share structural connections and can thus be surmised to be functionally closely related. However, the cluster index does not provide information about the number or size of these groups and only captures local connectivity patterns involving the direct neighbours of the central vertex (Sporns, O., 2002).

Global- (E_g) and local-efficiency (E_l) were demonstrated to reflect the same properties of the inverse of the average shortest path $1/PL$ and the clustering index C. In addition, this new definition is attractive since it takes into account the full information contained in the weighted links of the graph and provides an elegant solution to handle disconnected vertices.

CHAPTER 4

High-Resolution EEG

Nowadays, it is well understood that brain activity generates a variable electromagnetic field that can be detected quite accurately by using scalp electrodes as well as by superconductive magnetic sensors. Electroencephalography (EEG) and magnetoencephalography (MEG) are therefore useful techniques for the study of brain dynamics and functional cortical connectivity because of their high temporal resolution i.e., milliseconds (Nunez, P., 1981).

Electroencephalography reflects the activity of cortical generators oriented both in tangential and radial way with respect to the scalp surface, whereas MEG reflects mainly the activity of the cortical generators oriented tangentially with respect to the magnetic sensors. However, the different electrical conductivity of the brain, skull and scalp markedly blurs the EEG potential distributions and makes the localization of the underlying cortical generators through this technique a problematic issue.

To overcome this problem, the high-resolution EEG (HR-EEG) technology was introduced during the last decade, and it was shown to greatly improve the spat)ial resolution of the conventional EEG (Nunez, P., 1995; Gevins et al., 1991, 1999; Babiloni et al., 2000). Such technology includes (i) the sampling of the spatial distribution of scalp potential with a larger number of surface electrodes (typically 64–256); (ii) the use of sequential magnetic resonance images (MRI) to describe mathematically the different conductivity effects of the inner head structures between scalp and cortex; (iii) the solution to a linear inverse problem to estimate the original EEG signals generated in the cortex from the measured scalp signals.

4.1 HEAD MODELS AND REGIONS OF INTEREST

In order to quantify the different conductivity effects that blur the original EEG signal, realistic head models reconstructed from T1-weighted MRIs can be used. The main layers scalp, skull, dura mater and cortex are segmented as compartments from MRIs and tessellated with a large number of triangles. The use of an accurate model for the cortical sources typically includes a large number of triangles. Actually, each triangle represents the electrical dipole of a particular neuronal population and the estimation of its current density is computed by solving the linear inverse problem (see next paragraph). The electrical dipole model is largely used in literature since it can describe accurately the activities of small patches of the cortical tissue. Each dipole placed inside the volume conductor model of the head has a unitary strength and different direction, according to the local cortical geometry or to the adopted reference coordinate system. There is no limitation for the number of sources placed inside the head model, which depends on the modeling capabilities of the used computational system.

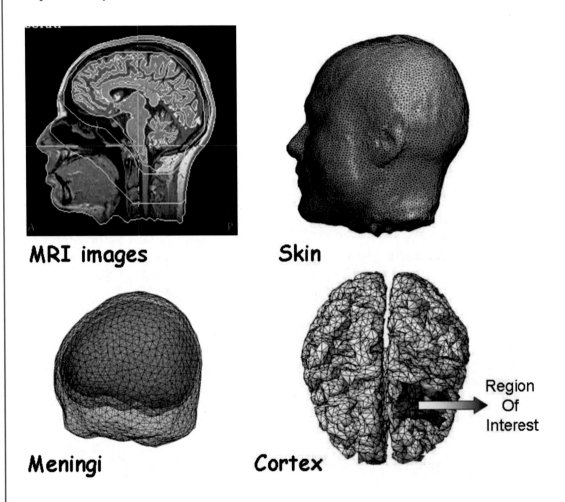

MRI images   Skin

Meningi   Cortex   Region Of Interest

Figure 4.1: Tesselation of the realistic head models from MRI images.

In the following, N indicates the number of dipoles whose activity is to be estimated from the M-dimensional measurement vector b. The typical values for N are between 1000 and 7000, while the values for M are in the range 64-256. The variable x is the N-dimensional vector of the unknown current strengths for the dipoles.

Eventually, particular cortical regions of interest (ROIs) can be drawn by a neuroradiologist on the computer-based cortical reconstruction of the individual head model. By following a Brodmann's criterion the whole cortex can be parcellated in several macro-areas consisting of hundreds of triangles. In this way, the electrical activity of particular Regions Of Interest (ROIs) can be obtained by averaging the current strengths of the various dipoles within the considered cortical area.

4.2   ESTIMATION OF CORTICAL SOURCE CURRENT DENSITY

In the estimation of neuronal activity from non-invasive measurements, a mathematical model should be used for the description of the propagation of the potential distribution from each modeled sources to the sensors positions. In other words, it is necessary to compute the potential distribution occurring on the set of the M sensors over the head model due to the ith unitary dipole placed at the ith cortical location.

In the following, A_i indicates the potential distribution over the M sensors due to the unitary ith cortical dipole. The collection of all the M-dimensional vectors A_i, $(i = 1, \ldots, N)$ describes how each dipole generates the potential distribution over the head model. This collection is called the lead field matrix A.

With the definitions provided above, the estimate of the strength of the modeled dipolar source strength x from the non invasive set of measurement b, is obtained by solving the following linear system:

$$Ax = b \tag{4.1}$$

where A is the $M \times N$ lead field matrix, x is the N-dimensional array of the unknowns cortical strengths and b is the M-dimensional vector of the instantaneous (electrical or magnetic) measurements.

This is a strongly underdetermined linear system, in which the number of the unknown variables (i.e., the dimension of the vector x), is greater than the number of measurement b of about one order of magnitude. In this case, from the linear algebra there are infinite solutions for the vector of dipole strengths x. All these solutions explain in the same way the data vector b. Furthermore, the linear system is ill conditioned as results of the substantial equivalence of the several columns of the electromagnetic lead field matrix A. In fact, each column of the lead field matrix arise from the potential distribution generated by the dipolar sources that are located in similar position and orientation along the used cortical model.

Regularizing the inverse problem consists in attenuating the oscillatory modes generated by vectors associated with the smallest singular values of the lead field matrix A, introducing supplementary and a priori information on the sources to be estimated.

In the following, the term "solution space" represents the "best" current strength solution x will be found. The "measurement space" is the vectorial space in which the vector b of the gathered data is considered. The solution of the linear problem in Eq. (4.1) with the variation approach is based on the idea of selecting metrics from the solution space and the measurement space, respectively. These two metrics are characterized by symmetric matrices and express our idea of closeness in the same spaces.

With this approach, the minimization function is composed of two terms, one that evaluates how well the solution explains the data, and the other that measures the closeness of the solution to an a priori selected.

The formulation of the problem expressed in the Eq. (4.1) now becomes the following:

$$\xi = \arg\min_{x} \left(\|Ax - b\|_{W_M}^2 + \lambda^2 \|x\|_{W_N}^2 \right) \tag{4.2}$$

where the matrices W_M and W_N are associated with the metrics of the measurement and source space, respectively, and λ is the Lagrangian parameter. The source estimate ξ was hence the source distribution that between the infinite possible solutions to the undetermined problem described in the Eq. (4.2) explains the EEG data with a minimum amount of energy (weighted minimum norm solution). By setting the matrices W_M and W_N to the identity, the minimum norm estimation was obtained.

The solution of the problem described in Eq. (4.2) has the following the form:

$$\xi = Gb \tag{4.3}$$

where under the hypothesis that the metric W_N and W_N are invertible the pseudo-inverse matrix G is given by:

$$G = W_N^{-1} A^T \left(A W_N^{-1} A^T + \lambda W_M^{-1} \right)^{-1}. \tag{4.4}$$

Hamalainen and Ilmoniem proposed the estimation of the cortical activity with the minimum norm solution in 1984.

However, it was recognized that in this particular application, the solutions obtained with the minimum norm constraints were biased toward those dipoles that are located nearest to the sensors. In fact, there is a dependence of the distance in the law of potential (and magnetic field) generation and this dependence tends to increase the activity of the more superficial dipoles while depress the activity of the dipoles far from the sensors.

The solution to this bias was obtained by taking into account a compensation factor, for each used dipole, that equalized the visibility of the dipole from the sensors. This technique, called column norm normalization, was used in the linear inverse problem by Pascual-Marqui, R. (1995) and then adopted largely by the scientist in this field.

With the column norm normalization, a diagonal $N \times N$ matrix W was formed, whose generic ith term on the diagonal is equal to:

$$W_{ii} = \|A_i\|^{-2} \tag{4.5}$$

representing the L2-norm of the ith column of the lead field matrix A. In his way, dipoles near to the sensors (with a large $\|A_i\|$) will be depressed in the solution of Eq. (4.2), since their activations are not convenient from the point of view of the functional cost. In fact, dipoles with low visibility from the sensors (with a low $\|A_i\|$) have an associate cost function convenient to use. The use of this W_N matrix in the source estimation is known as weighted minimum norm solution (Grave de Peralta et al., 1997).

Another question of interest in the solution of linear inverse problem is the setting of the Lagrangian parameter λ that regulates the presence of the a priori information inside the solution of the problem. How it is possible to set such parameter in an "optimal?

An optimal regularization of the linear system described in Eq. (4.2) was obtained through the Tickhonov and L-curve approach (Hansen P., 1992). This curve plots the residual norm versus the solution norm at different values of the regularization parameter lambda. It is worth of notice that the optimal regularization value can be selected automatically. In fact, this value is located at the "corner" of the L-curve plot.

Figure 4.2 illustrates the effect of the linear inverse problem's solution. From a scalp potential distribution, one can estimate accurately the original cortical potential.

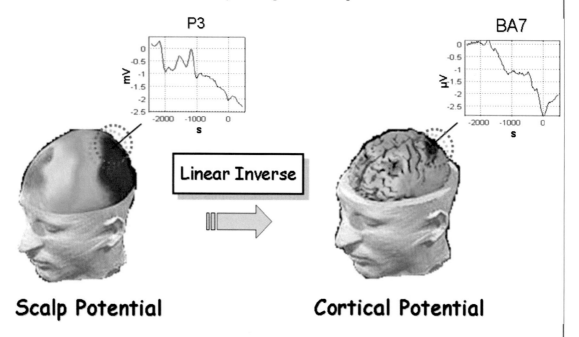

Figure 4.2: Electrical activity estimation in the Brodmann area 7 from the scalp measurement in the parietal sensor P3.

CHAPTER 5

Cortical Networks in Spinal Cord Injured Patients

In this first application of the methodology developed in the previous chapters, we would like to analyze the structure of cortical connectivity during the attempt to move a paralyzed limb by a group of spinal cord injured (SCI) patients. The question is whether the "architecture" of the functional connectivity in SCI patients, evaluated by graph analysis, may differ from healthy behavior. In particular, we wonder if SCI patients could show a more efficient cortical network in order to compensate the altered behavior of their primary motor areas because the spinal injury. By using tools derived from graph theory (see Chapter 3), the indexes related to the topology of the cortical networks estimated were derived.

The main experimental questions investigated in this work are the following:

1. Is the efficiency index significantly different in the cortical connectivity networks estimated from normal and SCI subjects during the performance of the same task?

2. If it does exist, is such a difference dependent on the frequency contents of the cortical activity?

3. Are the efficiency values estimated in the two populations' networks different from those obtained in "random" graphs having the same dimensions?

5.1 EXPERIMENTAL DESIGN

All experimental subjects participating in the study were recruited by advertisement. Informed consent was obtained from each subject after the explanation of the study, which was approved by the local institutional ethics committee.

The healthy group consisted of five volunteers (age, 26– 32 years; five males). They had no personal history of neurological or psychiatric disorder, were not taking medication, and were not abusing alcohol or illicit drugs. The SCI group consisted of five patients (age, 22–25 years; two females and three males). Spinal cord injuries were of traumatic aetiology and located at the cervical level (C6 in three cases, C5 and C7 in two cases, respectively); patients had not suffered a head or brain lesion associated with the trauma leading to the injury.

For EEG data acquisition, subjects were comfortably seated on a reclining chair, in an electrically shielded, dimly lit room. They were asked to perform a brisk protrusion of their lips (lip pursing) while they were performing (healthy subjects) or attempting (SCI patients) the right foot movement.

The choice of this joint movement was suggested by the possibility to trigger the SCI patients' attempt at foot movement. In fact, patients are not able to move their limbs; however, they could move their lips. By attempting a foot movement associated with a lips protrusion, they provided an evident trigger after the volitional movement activity. Such a trigger has been recorded to synchronize the period of analysis for both the populations considered.

The task was repeated every 6–7 s, in a self-paced manner, and the 100 single trials recorded were used for the estimate of functional connectivity by means of the Directed Transfer Function (DTF, see Chapter 2). A 96-channel system (BrainAmp, Brainproducts, Germany) was used to record EEG and EMG electrical potentials by means of an electrode cap and surface electrodes, respectively. The electrode cap was built according to an extension of the 10-20 international system to 64 channels. Structural MRIs of the subject's head were taken with a Siemens 1.5T Vision Magnetom MR system (Germany).

5.2   CORTICAL ACTIVITY AND FUNCTIONAL CONNECTIVITY

Cortical activity from high resolution EEG recordings was estimated using realistic head models and cortical surface models with an average of 5,000 dipoles, uniformly disposed. Estimation of the current density strength, for each one of the 5000 dipoles, was obtained by solving the linear inverse problem, according to techniques described in the Chapter 4.

By using the passage through the Tailairach coordinates system, twelve Regions Of Interest (ROIs) were then obtained by segmentation of the Brodmann areas (B.A.) on the accurate cortical model utilized for each subject. Bilateral ROIs considered in this analysis are the primary motor areas for foot (MIF) and lip movement (MIL), the proper supplementary motor area (SMAp), the standard pre-motor area (BA6), the cingulated motor area (CMA) and the associative area (BA7). As an example for the different steps involved in the generation of the high resolution EEG model employed in this study, Figure 5.1 presents a superimposition of the electrode montage with actual head structures as well as the cortical areas employed as regions of interest.

For each EEG time point, the magnitude of the 5,000 dipoles composing the cortical model was estimated by solving the associated linear inverse problem. Then average activity of dipoles within each ROI was computed. In order to study the preparation for an intended foot movement, a time segment of 1.5 s before the lips pursing was analyzed; lips movement was detected by means of an EMG.

Although motor responses could be analyzed at the same way, we would concentrate on the "intention-to-move" time interval in order to have results that could be used successively in the Brain Computer Interface context. BCI is a recent field of research in which brain signals related to movement intention can be suitably treated to control external devices. This fact would improve the condition of SCI patients in a next future.

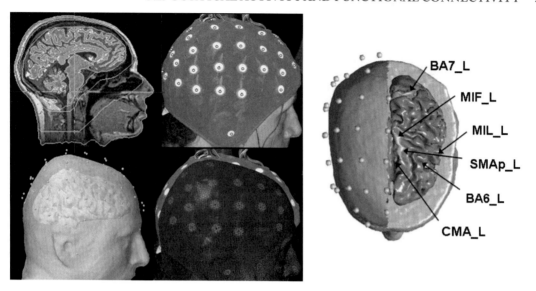

Figure 5.1: Left: Four steps involved in the generation of the Lead Field matrix for the estimation of cortical current density from EEG recordings. From left to right, from top to the bottom: MRI images from a healthy subject, generation of the head models and superimposition with electrodes cap. On the right of the figure the Regions of Interest (ROIs), taken into account for successive connectivity estimations, are illustrated. Cortical activity was estimated on the cortical areas of interest from the high resolution EEG recordings performed in both the populations.

Resulting cortical waveforms, one for each predefined ROI, were then simultaneously processed for the estimation of functional connectivity by using the Directed Transfer Function (see Chapter 2). Application of this method to the ROIs waveforms returns a cortical network for each frequency band of interest: (Theta 4–7 Hz, Alpha 8–12 Hz, Beta 13–29 Hz, 30–40 Hz). Only estimated DTF connections that are statistically significant (at $P < 0.001$) after a contrast with the surrogate distribution of DTF values on the same ROIs obtained with a Montecarlo procedure were considered for the network to be analyzed with graph theory's tools. This procedure enables us to consider only those functional links that are not due to chance.

As an example of the networks estimated for the two populations analyzed, Figure 5.2 shows the average cortical network estimated in the Beta frequency band for the SCI group and for the healthy group during the motor attempt/execution of the task.

The figure shows the average intensity of 30% of the greater connections belonging to two experimental subjects at least. The cortex of one particular subject was used for display purposes, being the computations performed on the realistic ROIs of each individual cortex.

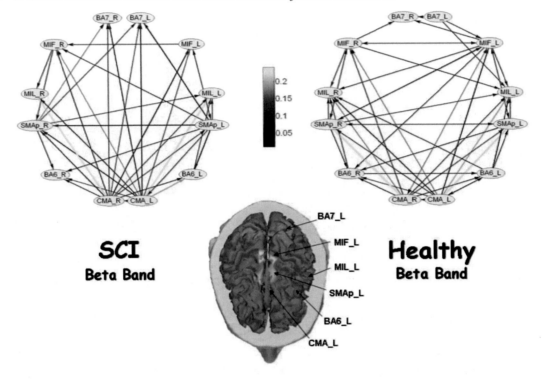

Figure 5.2: Average connectivity networks among ROIs for SCI group (upper left) and healthy group (upper right), obtained from DTF in the Beta frequency band during the foot-lip task. They show the 30% of the most powerful edges. Only edges shared by at least two experimental subjects are shown. Nodes follow the disposition of ROIs on the cortex, represented at the bottom of the figure. Head is seen from above, with nose towards the bottom of the page; left hemisphere is on the right part of the figure. Flows direction is represented by an arrow, while intensity is coded by its color and size. Each node is labeled with the ROI acronym.

One arrow from the cortical region X to the cortical region Y describes the existence of a stable causal relation between them. Estimated cortical waveform in the X region "Granger"-causes the estimated cortical waveform in the Y region, in the particular frequency range.

5.3   GRAPH THEORETICAL MODELING

In order to achieve topological features, a cortical network has to be converted into a directed unweighted graph (digraph). In this study, connection matrix contains DTF values for each directed pair of ROIs and can be converted to an adjacency matrix A by considering a threshold T that represents the number of the most powerful connections to be considered. If the number of links in the DTF matrix exceeds T, less powerful connections will be removed until that threshold is reached.

Also, T can be expressed as connection density, that is, the ratio between the number of all the effective connections and the number of all possible connections within the graph. In order to study the topology of the networks at different connection densities or costs (Latora and Marchiori, 2003), a range of values (0.1, 0.15, 0.2, 0.25, 0.3, 0.35, 0.4, 0.45, 0.5) was explored.

Once the cortical network has been converted, it is possible to characterize the digraph in terms of its degrees, degree distributions, global, and local efficiency.

5.4   RESULTS

All results displayed are relative to networks with a threshold applied equals to 0.3. This means that all digraphs analyzed have the same number of connections representing the 30% of the most powerful links within the network. This particular value represents the median of an interval of thresholds (from 0.1 to 0.5) for which results remain significantly stable. In all the networks investigated values of degrees for each node increase or decrease proportionally to the threshold selected. Thus, all differences among graphs' degrees are maintained. In particular, the global- and local-efficiency of the two populations for each of the following thresholds were statistically compared. It has been seen that the only significant ($p < 0.05$) differences were due to local efficiency in a sub-interval ranging from 0.2 to 0.4. The stability of results is then clear in a range of $\pm 10\%$ around the chosen value. Trends of average local efficiencies at different values of threshold in the two experimental populations for a representative frequency band are shown in Figure 5.3.

Figure 5.4 shows results relative to the average incoming connections (in-degree) and the average outgoing connections (out-degree) for the ROIs of the SCI population in four different frequency bands analyzed. Level of involvement has to be considered separately for incoming connections and outgoing connections.

Contrast between degrees "in" and "out" of the normal population, during the preparation to the movement execution, is presented in Figure 5.5.

Direct comparisons of the data presented in Figures 5.4 and 5.5 shows that for all the frequency bands, there is a strong involvement of SMAp areas in SCI population during the attempt to move the paralyzed limb, which is not so evident in the healthy subjects. The absolute level of incoming connections in each single ROI seems rather similar in the two populations. Anyway some differences appear in BA7 areas, where in-degree is higher in the SCI group irrespectively of the frequency band.

The number of outgoing connections for cingulate motor areas (CMA) is very high for both of the two experimental groups, while no connections seem to come from the primary motor areas of the lips (MIL). In particular, in the Beta frequency band, healthy subjects present a remarkable flow coming from their primary motor areas (MIF), while a large number of links in the SCI patients come out from the SMAp areas.

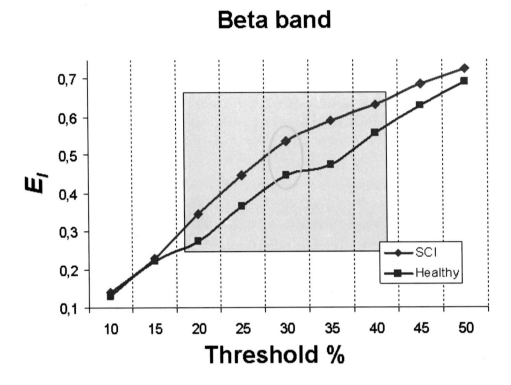

Figure 5.3: Average values of local efficiency at different values of threshold in the two experimental populations for the representative Beta frequency band. Red line represents values for SCI networks while blue line refers to Healthy networks. The square inside the figure displays the sub-interval of thresholds for which E_l remains significantly ($P < 0.05$) different between the two groups. Yellow ellipse localizes the threshold used (30%) in order to illustrate all following results.

As outlined in Chapter 3, the mean degrees are just indicative of the global behavior of a digraph's connectivity. For a more detailed analysis, it is necessary to compute two degree distributions Pin and Pout for each of the networks obtained.

At the top of Figure 5.6 average trends of the degree distributions are shown for SCI and Healthy groups, in a representative frequency band.

Histogram values were normalized to the size of the digraph, which is the number of the elements within the network (12 ROIs). An interesting result is that in-degree and outdegree distributions show different trends within each group. Right-skew tails of out-degree distribution indicates the presence of few nodes with a very high level of outgoing connections, while for the in-degree distribution there are no ROIs in the network with more than six (seven in some other bands) incoming connections. This unsettling, found in each of the two experimental groups and

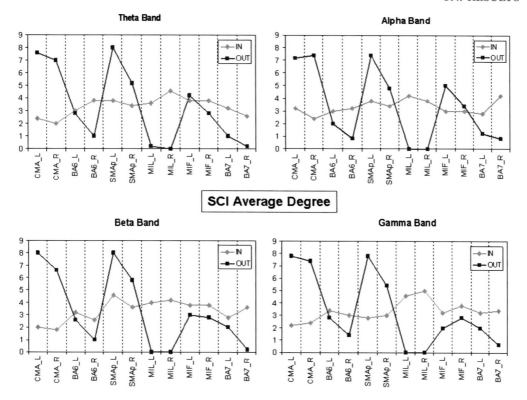

Figure 5.4: Average degrees of SCI group in the frequency bands analyzed. Light blue line represents degree-in, while dark blue refers to degree-out. On the abscissas ROIs label are displayed; on the ordinates, there are degree values.

frequency band, is an attribute of the real networks that cannot be observed in random graphs. At the bottom of Figure 5.6, average degree distributions obtained from a set of five random digraphs are shown; it is evident the absence of "hubs" either for efferent and afferent flows.

In order to catch inter-individual variance of the results, values of indexes employed (E_g, E_l) were computed for cortical networks estimated on each experimental subject (spinal cord injured and healthy), in the four frequency bands. Successively, a contrast between values of the two populations was addressed by using the Analysis of Variance (ANOVA) that is known to be robust with respect to the departure of normality and homoscedasticity of data being treated (Zar, 1984).

Separate ANOVAs were conduced for each of the two variables E_g and E_l, computed in each frequency band relevant for this study. Statistical significance was put at 0.05 and main factors of the ANOVAs were the "between" factor GROUP (with two levels: SCI and Healthy) and the "within" factor BAND (with four levels: Theta, Alpha, Beta, and Gamma). Greenhouse and Geisser

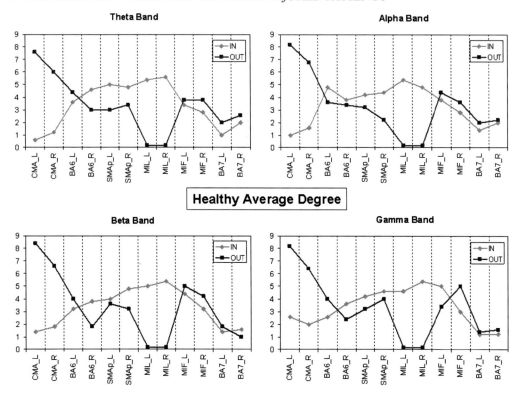

Figure 5.5: Average degrees of the healthy group in all the frequency bands analyzed. Same conventions than in Figure 5.4.

correction was used for the protection against the violation of the sphericity assumption in the repeated measure ANOVA. Besides, post-hoc analysis with the Duncan's test and significance level at 0.05 was performed. All the statistical analysis was performed with the software Statistica', StatSoft, Inc.

Average values of the SCI and healthy population are then reported in the Figures 5.7 and 5.8, respectively, in the four frequency bands.

ANOVA performed on the Eg variable showed no significant differences for the main factors GROUP and BAND. In particular, "between" factor GROUP was found having an F value of 0.83, $p = 0.392$ while the "within" factor BAND was found having an F value of 0.002 and $p = 0.99$.

ANOVA performed on the El variable revealed a strong influence of the between factor group (F = 32.67, $p = 0.00045$); while the BAND factor and the interaction between GROUP X BAND were found not significant (F = 0.21 and F = 0.91, respectively, p values equal to 0.891 and 0.457). Post-hoc tests revealed a significant difference between the two examined experimental groups (SCI,

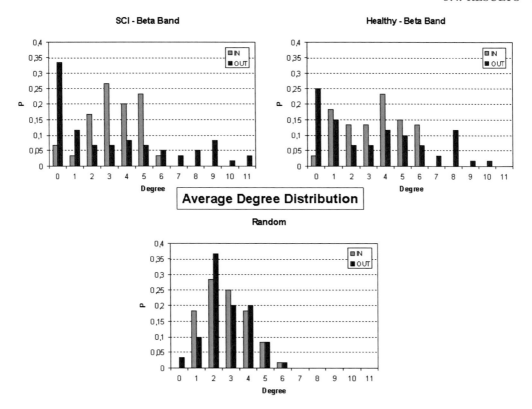

Figure 5.6: Average trends of degree distributions are shown for SCI (upper-left) and healthy (upper-right) group in the Beta frequency band. In-degree distribution Pin is represented by light blue bars, while dark blue bars refer to out-degree distribution Pout. In order to have comparable results, the histogram values were normalized to the number of the elements within the network (12 ROIs). Average degree distributions of five random digraphs are shown in the bottom part of the figure with the same previous conventions.

Healthy) in Theta, Alpha, and Beta band (p = 0.006, 0.01, 0.03, respectively). It can be observed (Figure 5.8) that the average values of the local efficiency in the SCI subjects are significantly higher than those obtained in the Healthy group, for the three frequency bands.

A hypothesis testing procedure was employed in order to detect significant differences between average values from random digraphs and average values from experimental digraphs. The contrast between functional networks was obtained from the two experimental groups and random digraphs having the same number of nodes and connections of the cortical networks was investigated. These graphs were generated distributing a fixed number of connections between randomly chosen couple of nodes. A set of 1000 random digraphs was collected and the respective distributions of global and

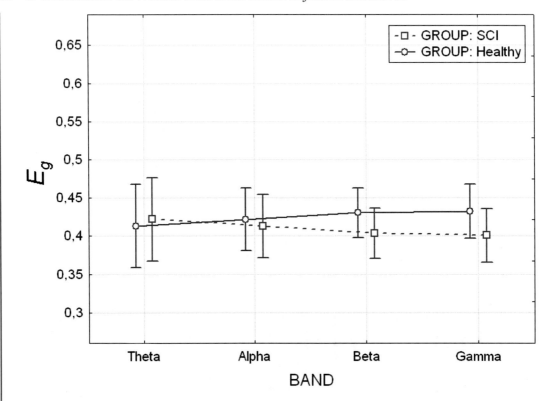

Figure 5.7: Average values of global efficiency (Eglob) for the levels of "within" factor BAND, grouped by SCI and healthy subjects. No statistically significant differences were found between normal subjects and SCI patients. Vertical bars denote 0.95 confidence intervals.

local efficiency values were calculated. The large amount of random values assures the normality of their distributions, then it was separately performed a z-test for each average value of global and local efficiency gathered from the two populations in each frequency band.

Significance level was posed at 0.05 (Bonferroni corrected for multiple comparisons), and it was determined whether an average value obtained from an experimental group in a particular band could belong to a normal distribution of 1000 values with a known standard deviation.

Figure 5.9 shows the contrast between the values obtained for global and local efficiency in the two populations studied with those obtained in a set of 1000 random digraphs, having the same number of nodes and arcs.

To state the statistical significance of the differences observed in the mean values of those indexes, separate Z-tests, at the significance level of 0.05 (Bonferroni-corrected for multiple comparisons), were performed and are summarized in the Table 5.1.

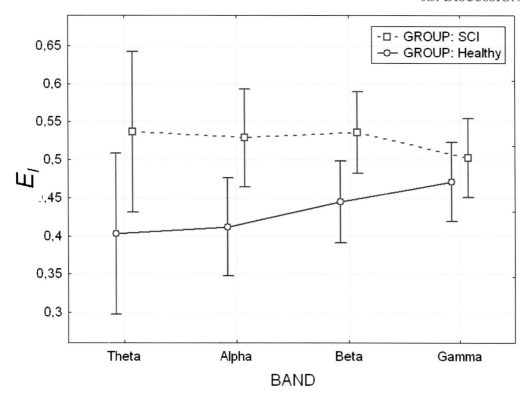

Figure 5.8: Average values of local efficiency (Eloc) for the levels of "within"factor BAND, grouped by SCI and Healthy subjects. A statistically significant difference ($P < 0.05$) was noted between normal subjects and SCI patients. Vertical bars denote 0.95 confidence intervals.

Results show that for both of the experimental groups (SCI and healthy) and in all the frequency bands employed, global efficiency is significantly lower than the random mean value. Instead, local efficiency for the SCI group in every band is significantly higher than the random one. The same behavior appears for the healthy population in each frequency band except for Theta and Alpha that contrarily show lower values of local efficiency when compared with the random distribution.

5.5   DISCUSSION

Analysis performed on the cortical networks estimated from the group of normal and SCI patients revealed that both groups present few nodes with a high out-degree value. This property is valid in the networks estimated for all the frequency bands investigated. In particular, cingulate motor areas (CMAs) ROIs act as "hubs" for the outflow of information in both groups, SCI and healthy.

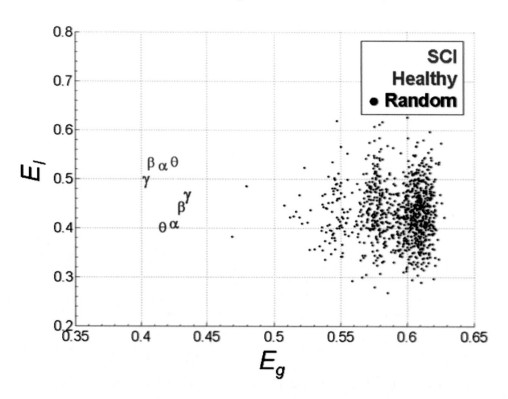

Figure 5.9: Scatter plot of global and local efficiency. All the average values are grouped by SCI patients (red symbols) and healthy subjects (blue symbols) while black dots represent the distribution of 1000 random digraphs. Greek symbols characterize the frequency bands involved in the study.

This means that removal of CMAs from the estimated patterns will cause a collapsing of the whole cortical network, thus corrupting the characteristic behavior of the preparation to the effecting of this experimental task. In addition, while SCI patients show a remarkable flow outgoing from their SMAp areas in the Beta frequency band, healthy subjects show a relevant outflow from the MIF areas in the same frequency band.

Although the presence of "hubs" in the out-degree distributions of all the cortical digraphs could suggest a power-law trend, we cannot formally assert their scale-free properties, according to actual procedures (Boccaletti et al., 2006) because the small size of the networks involved prevents us from achieving a reliable degree distributions.

Table 5.1: Statistical Comparison Between Cortical Networks and Random Networks.

Z	SCI-θ	SCI-α	SCI-β	SCI-γ	Heal.-θ	Heal.-α	Heal.-β	Heal.-γ
E_g	237.1	250.5	262.1	267.3	249.8	238.2	225.9	223.4
E_l	57.7	53.3	57.0	38.9	15.9	11.1.	7.2	21.7

All contrasts were significant (P < 0.001); so respective percentiles are not reported. Positive/negative z-values indicate that the mean value of the random distribution is higher/lower than the average value of an experimental group in a particular frequency band.

Results suggest that spinal cord injuries affect the functional architecture of the cortical network sub-serving the volition of motor acts mainly in its local feature property. In fact, SCI patients have shown significant differences from healthy subjects in this index; this could be due to a functional reorganization phenomenon, generally known as brain plasticity (Raineteau and Schwab, 2001). The higher value of local efficiency E_l suggests a larger level of the internal organization and fault tolerance (Sivan et al., 1999). In particular, this difference can be observed in three frequency bands, Theta, Alpha and Beta, which are already known for their involvement in electrophysiologic phenomena related to the execution of foot movements (Pfurtscheller and Lopes da Silva, 1999). A high local efficiency implies that the network tends to form clusters of ROIs which hold an efficient communication. These efficient clusters, noticed in the SCI group, could represent a compensative mechanism as a consequence of the partial alteration in the primary motor areas (MIF) due to the effects of the spinal cord injury. Instead, it seems that the global level of integration between the ROIs within the network do not differ significantly from the healthy behavior. This could mean that spinal cord injuries do not affect the global efficiency of the brain which attempts to preserve the same external properties observed during the foot-lip task in the cortical networks of healthy subjects.

By perusing data presented in both Table 5.1 and Figure 5.9, it is clear that cortical networks estimated in this study are not structured like random networks. Instead, well ordered properties arise from most of the digraphs obtained from each experimental group and frequency band. In fact, they show similar values of global and local efficiency and more precisely fault tolerance is privileged with respect to global communication. These results indicate that cortical networks behave globally in the same way as they behave locally. Moreover, these real digraphs show a lower global efficiency and a higher local efficiency than respective values obtained from random digraphs. This fact suggests during the experimental task brain networks tend to follow an ordered spatial topology rather than a small-world or a random architecture. Here, random digraphs are generated with the same number of nodes and edges of the connectivity patterns obtained. Anyway, in order to have more robust comparisons algorithms that also preserve the degree distributions are available and recently applied to cortical networks (Sporns and Zwi, 2004).

On the basis of these experimental results obtained from the application of graph theory tools to the functional con-nectivity networks estimated by using advanced high resolution EEG and DTF algorithms, we have possible answers to the experimental questions posed in the introduction section.

In particular:

1. It seems that there are significant differences in the cortical functional connectivity networks between SCI patients and healthy subjects. Such differences are related to the internal organization of the network and its fault tolerance, which in SCI patients appears to be higher than in normal subjects, as suggested by the significant increase of the local efficiency.

2. Differences in this functional connectivity networks are higher in the Theta, Alpha and Beta frequency bands, which are already known to be involved in the phenomena related to the execution of motor acts.

3. All functional connectivity networks extracted from the two experimental groups showed ordered properties and significant differences from "random" networks having the same characteristic sizes.

CHAPTER 6

Cortical Networks During a Lifelike Memory Task

The second study proposed as example for the application of the methodology developed in the chapters I-III investigates the brain behavior by estimating the functional connectivity changes during the visualization of television (TV) commercials (see Figure 6.1).

Figure 6.1: Picture representing the methods and technologies employed in the present study.

In particular, the aim was to elucidate if TV commercials that will be remembered by the subjects several days after their showing generate different cortical networks when compared to those related to the TV spots that will be forgotten.

To achieve this, EEG recordings were first performed in a healthy group of subjects and the respective cortical activity was achieved by using advanced high-resolution EEG methods including realistic MRI-constructed head models (see Chapter 4). Then, we estimated the functional connectivity patterns among the Regions of Interest (ROIs) for this cognitive task by means of the Partial Directed Coherence (see Chapter 2). Finally, we evaluated the topological-weighted properties of the cortical networks by calculating their global- and local-efficiency indexes (see Chapter 3), which return an appropriate measure for the inverse of the average shortest path length $1/PL$ and clustering index C, respectively.

In particular, the main questions we wanted to address are:

1. Does the structure of cortical network present significant differences during the visualization of the TV spots that will be remembered and those that will be forgotten?

2. If existing, is such difference depending on the spectral content of the cortical activity?

3. How the obtained results can be interpreted with respect to a random benchmark?

6.1   EXPERIMENTAL DESIGN

Ten voluntary and healthy subjects participated in the study (age, 26-32 years; five males). They had no personal history of neurological or psychiatric disorder and they were free from medications, or alcohol or drugs abuse.

For the EEG data acquisition, subjects were comfortably seated on a reclining chair in an electrically shielded room. They were exposed to the vision of a film of about 30 minutes for five consecutive days.

Each film consisted in a different neutral documentary. Three interruptions have been generated at the beginning, at the middle and at the end of the documentary. Each interruption was composed by six commercial video-clips of about 30 seconds. Eighteen commercials were showed during the whole documentary. The TV spots were relative to standard international brands of commercial products - like cars, food, etc... - and no profit associations that have been never broadcasted in the country in which the experiment have been performed. Hence, the advertising material was new to the subject as well as the documentary observed. Each day the documentary changed as well as the order of presentation of the video-clips within it. The eighteen TV spots presented along all these five days remained unchanged.

After a period of ten days, each experimental subject was contacted and an interview was performed. In the interview, the subjects were asked to recall spontaneously which clips they remembered. Then, the EEG signals were segmented and classified according to the periods relative to the video-clips remembered (RMB) and forgotten (FRG).

The Figure 6.2 illustrates the structure of the film and the timing of the experimental design.

A 96-channel system with a sampling frequency of 200 Hz (BrainAmp, Brainproducts GmbH, Germany) was used to record the EEG electrical potentials by means of an electrode cap. The electrode cap was built according to an extension of the 10-20 international system to 64 channels. The EEG signals were referred to linked A1 and A2 electrodes. The structural MRIs of each

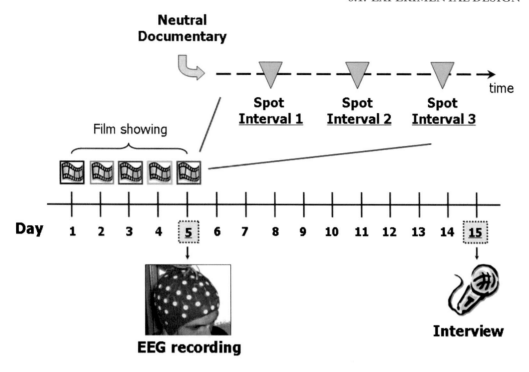

Figure 6.2: Representation of the experimental design followed in this study. Each film consisted of a neutral documentary of about 30 min. Three TV-spot intervals were inserted at the beginning, at the middle and at the end of the documentary. Each interval was composed by a slot of six television commercials. The experimental subjects watched five different films along five consecutive days. Every day, the commercials within the spot intervals were randomized. The EEG recordings were performed during the fifth day. After 10 days, an interview to the subject was generated in order to correlate the recorded EEG signals to the TV spots that were remembered or forgotten.

subject's head were taken with a Siemens 1.5T Vision Magnetom MR system (Germany). The three-dimensional electrode positions were obtained by using a photogrammetric localization technique (Photomodeler, Eos Systems Inc., Canada) with respect to the anatomic landmarks: nasion and the two pre-auricular points. Trained neurologists visually inspected EEG data, and they rejected trials containing artifacts. Subsequently, the EEG signals were baseline-adjusted and low-pass filtered at 45 Hz.

6.2   CORTICAL ACTIVITY AND FUNCTIONAL CONNECTIVITY

High-resolution EEG technologies were used to estimate the cortical activity from the scalp measurements (see Chapter 4). In this study, fourteen regions of interest were manually segmented from the cortical model of each subject. The ROIs considered for the left (_L) and right (_R) hemisphere are the Brodmann areas 10 (10_L and 10_R), 9 (9_L and 9_R), 8 (8_L and 8_R) and the anterior cingulate areas (AC_L and AC_R). The bilateral Brodmann areas 40 (40_L and 40_R), 5 (5_L and 5_R) and 7 (7_L and 7_R) were considered, too.

The estimate of functional connectivity was addressed through the partial directed coherence index (see Chapter 2) from the obtained cortical waveforms. This measure returned a weighted and directed cortical network for each frequency band of interest: Theta (3-6 Hz), Alpha (7-12 Hz), Beta (13-29 Hz), and Gamma (30-40 Hz).

In order to consider only the significant functional connections in each cortical network, a surrogate data-generation procedure was performed. This method consists in transforming each signal to the frequency domain by means of a Fourier transform and by multiplying each amplitude by $e^{i\varphi}$, where φ is the phase of the signal independently chosen for each frequency from the interval [0, 2π] (Theiler et al., 1992). In the present work, φ is obtained by randomly shuffling the phases of the original signal. The inverse transform represents the surrogate data that has the same power spectrum, but random phases. In order to extract a significant threshold, we first performed this randomization procedure 1000 times for each EEG dataset. Then, we computed the respective cortical network and collected the distribution of 1000 values of PDC for each pair of ROIs and for each frequency band. In this way, we obtained a threshold for each functional link by considering the 99th percentile of the respective distribution.

According to the previous methods, we obtained a functional connectivity pattern among fourteen ROIs from high-resolution EEG data, for each experimental subject, condition (FRG and RMB) and frequency band (Theta, Alpha, Beta, and Gamma). In general, the intensity of the overall connectivity was higher in the RMB situation irrespectively of the spectral content.

This difference can be observed at the left side of Figure 6.3, where the functional links of a representative subject in the Alpha band are illustrated on the realistic head model for the FRG and RMB task. The lighter is color of the arrow; the higher is intensity of the functional flow. At the right side of the figure, the same functional flows among the cortical areas are represented as weighted and directed graphs.

6.3   GRAPH THEORETICAL MODELING

The structural properties of the cortical networks were analyzed through the global-efficiency E_g based on shortest paths length and through the local-efficiency E_l based on clustering properties were employed. In this study, a common number of connections (or density) was considered in each weighted graph in order to analyze the structure of the cortical networks across all the subjects,

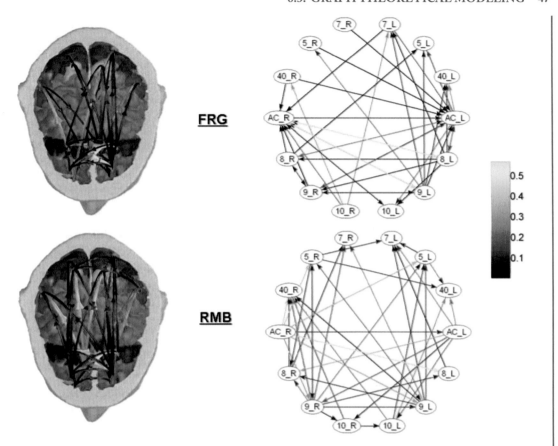

Figure 6.3: Cortical networks in the Alpha frequency band for a representative subject during the different conditions of spot-remembered (RMB) and spot-forgotten (FRG). In the left part of the figure, the estimated functional connectivity among the considered ROIs is shown on the realistic model of the subject's head. The bilateral ROIs are depicted with different colors on the grey cortex. In the right part of the figure, the connectivity patterns are represented as graphs. Each arrow represents a directed information flow whose intensity is coded by its color. Nodes stand for cortical areas, disposed according a pseudo-anatomical order (nose toward the top of the figure).

frequency bands and time samples. In fact, when seeking a common behavior among different networks, the mathematical indexes that evaluate their properties could suffer in robustness when they apply to graphs with different densities. Consequently, this study employed 42 connections – i.e., a connection density equal to 0.23 – for each network, by removing the weakest links from each weighted graph. The chosen density of the connections assured the best operative conditions for the indexes of the network structure E_g and E_l. In fact, with such connection density it has been found that these indexes keep their usual independency, i.e., their ability to detect global and local properties in a network, even in a small 14 nodes-graph like those employed in the present study (data not shown here).

Eventually, all the efficiency indexes calculated on the cortical networks were evaluated with respect to the distributions obtained from random graphs. E_g and E_l were compared with the theoretical values from networks generated by randomly shuffling the edges of the original cortical pattern. One hundred random networks were generated for each subject and frequency band. The mean $E_g - r$ and $E_l - r$ were calculated from all these random patterns as reference values for the global- and local-efficiency. In this work, the scaled values $E_g/E_g - r$ and $E_l/E_l - r$ were considered.

In order to evaluate whether the functional structure of the cortical network changes during the observation of the remembered TV spots the analysis of variance – ANOVA – with a significance level of 0.05 was employed. In particular, the main factors of the two ANOVAs performed for the global- and local-efficiency indexes were the factor TASK with two levels (spot-forgotten, FRG; spotremembered, RMB) and the factor BAND with four levels related to the frequency bands adopted (Theta, Alpha, Beta, and Gamma). Post hoc comparisons between different levels of the investigated factors were performed with the Scheffe's test, at the 0.05 of statistical significance.

6.4 RESULTS

Figure 6.4 shows the average profiles of the scaled global-efficiency E_g/E_{g-r} grouped by the TASK factor and across the frequency bands investigated (factor BAND). Irrespectively of the spectral content, the average value during the RMB condition ($E_g/E_{g-r} \sim 0.7$) is significantly lower than the FRG situation ($E_g/E_{g-r} \sim 0.83$).

In particular, the performed ANOVA indicates that the main factor TASK significantly affected the scaled global-efficiency of the cortical network (F(1, 18) = 168.97, p = 0.00001), as well as the BAND factor that was also statistically significant (F(3, 54) = 4.42, p = 0.007) for the same index. However, there is no evidence of a synergistic or interaction effect of the two independent variables, since their joint factor TASK x BAND was not significant (F(3, 54) = 1.81, p = 0.156).

Scheffe's post hoc tests performed at 5% significance level returned a significant decrease of the scaled global-efficiency index $E_g/E_g - r$ in the RMB condition when compared to the FRG one for the Beta and Gamma frequency bands, with a significance level of p = 0.0004 and p = 0.0235, respectively. The comparisons performed in the Theta and Alpha band are not statistically significant (p = 0.27 and p = 0.38, respectively).

Figure 6.5 shows the average profiles of the scaled local-efficiency $E_l/E_l - r$ according to the conventions as in Figure 6.4.

In a similar way, the average value during the RMB condition ($E_l/E_l - r \sim 1.38$) was lower than the FRG situation ($E_l/E_l - r \sim 1.62$), irrespectively of the frequency band. The analysis of variance (ANOVA) for the scaled local-efficiency values returned a significant effect of the main

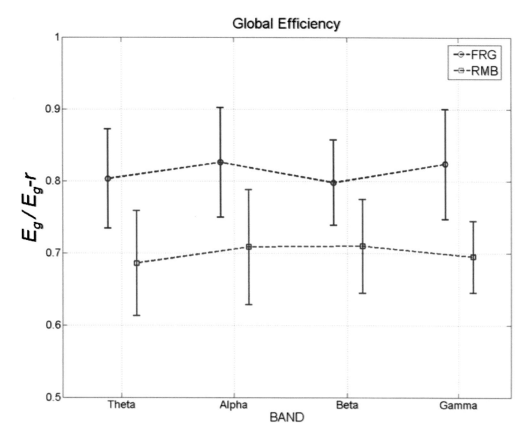

Figure 6.4: Average values of the global-efficiency in the four frequency BANDs grouped by the TASK factor. The efficiency indexes $E_g / E_g - r$ are scaled to the mean value from a distribution of 100 random graphs. The red line represents a "Spot-Remembered" condition (RMB), while the blue one stands for a "Spot-Forgotten" situation (FRG). Vertical bars denote 0.95 confidence intervals.

factors TASK ($F(1, 18) = 14.369$, $p = 0.0013$), and BAND ($F(3, 54) = 22.8$, $p = 0.00001$) as well as for their interaction TASK x BAND ($F(3, 54) = 1.54$, $p = 0.0001$).

Scheffe's post hoc tests performed at 5% significance level returned a significant decrease of the scaled local-efficiency index in the RMB condition when compared to the FRG only for the Alpha frequency band, with a significance level of $p = 0.00001$. The comparisons performed in the Theta ($p = 0.96$), Beta ($p = 0.17$) and Gamma ($p = 0.94$) bands were not statistically significant.

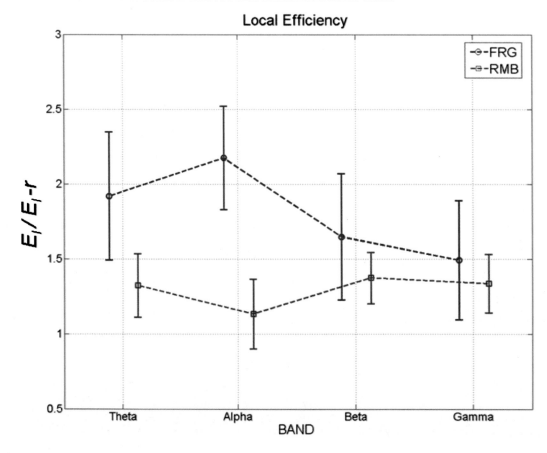

Figure 6.5: Average values of the scaled local-efficiency $E_l/E_l - r$ in the four frequency BANDs grouped by the TASK factor. The red line represents a "Spot-Remembered" condition (RMB), while the blue one stands for a "Spot-Forgotten" situation (FRG). Vertical bars denote 0.95 confidence intervals.

6.5   DISCUSSION

This application of the graph theory methods aimed to evaluate the changes in the global and local properties of the estimated cortical networks during the successful or unsuccessful memorization of several TV commercials. Due to the experimental design adopted, the changes in these indexes could be related both to memory coding activity as well as to increase/decrease of attentive state of the subjects. A graph theoretical approach was employed in order to evaluate the structural properties of the estimated cortical networks. In particular, we dealt with two characteristic measures to capture the global and local efficiency of the cerebral networks (Latora and Marchiori, 2001). The scaling procedures was adopted for the evaluation of the global-efficiency E_g and local-efficiency E_l indexes, because the randomization method does not preserve the degree of distribution of the original graph (Sporns and Zwi, 2004), because the small number of nodes could prevent any resulting significant difference. It is worth to mention that among the possible network configurations, random graphs have the highest values of global-efficiency and simultaneously the lowest values of local-efficiency. High values of E_g are then obtained when compared to random networks if the respective contrast $E_g/E_g - r$ is greater than 0.75 – i.e., threshold for a 50% decrease – while high values of E_l are considered if the contrast $E_l/E_l - r$ is greater than 1.5 – i.e., threshold for a 50% increase.

Results show that the successful encoding of TV spots significantly ($p < 0.00001$) affects the organization of the functional flows of the cortical network, together with a statistical dependence of such flow also by the frequency bands analyzed. In fact, both the average values of global- and local-efficiency present an evident decrease during the spot-remembered (RMB) task, as illustrated in Figures 6.5 and 7.1. Irrespectively of the spectral content, the average values of $E_g/E_g - r \sim 0.84$ and $E_l/E_l - r \sim 1.62$ during the spot-forgotten FRG condition suggest a small-world configuration of the respective cortical network. Instead, during the RMB situation, the same indexes appear significantly lower ($E_g/E_g - r \sim 0.73$ and $E_l/E_l - r \sim 1.38$) and the estimated cortical networks rather present a weak pattern of communication characterized by a non-homogeneous distribution of the functional links and few clustering connections between the considered ROIs.

Many scientific studies have shown that the cerebral networks exhibited a small-world architecture during no-tasks and/or rest periods (Salvador et al., 2005; Achard and Bullmore, 2007; Stam, C., 2004). In the present experimental condition, the small-world characteristic was found during a period of time in which an unsuccessful memorization of TV commercials is performed by the analyzed population (FRG dataset). Besides the local properties, also the global-efficiency index $E_g/E_g - r$ showed a significant decrease between the FRG and RMB conditions in the Beta and Gamma bands. It is interesting to note that the Gamma spectral content has been already found to be an important channel for the memory formation as revealed by an increasing Event Related Potential activity during the encoding phase (Sederberg et al., 2003; Gruber et al., 2004). In this study, the estimated cortical networks returned further information about the complex interactions among the considered cortical areas. In fact, during the RMB situation, the functional network in the Beta and Gamma band presented a significant non-homogeneous allocation of the involved information flows and a consequent reduction of the efficiency in the overall communication between the network nodes. In the Beta and Gamma frequency bands, the respective reduction of global-efficiency as well as the reduction of local-efficiency for the Alpha band of the cortical network communication could represent a predictive measure for the accurate recall of the commercials that will be remembered.

On the base of the obtained results from the application of theoretical graph indexes to the functional networks estimated by using advanced high-resolution EEG, we have possible answers to the experimental questions posed at the beginning of the present chapter.

In particular:

1. The structure of cortical network presents significant differences during the visualization of the TV spots that were remembered with respect to those that were forgotten by the analyzed

population. In particular, the successful encoding of TV spots produces an evident decrease of the average values of global- and local-efficiency in the estimated cortical networks.

2. The main differences between the two analyzed conditions (FRG and RMB) appear in the Beta and Gamma bands for the global-efficiency index and in the Alpha band for the local-efficiency index.

3. In the light of the comparison performed with random graphs, the optimal level of global and local communication – i.e., small-world network – during the visualization of the spots that will not be remembered (FRG) reflects a low level of neural engagement. Instead, the presence of attentive and semantic processes during the visualization of the video-clips that will be remembered (RMB) leads to a weak pattern of communication among the considered ROIs.

CHAPTER 7

Application to Time-varying Cortical Networks

The aim of this third application of the graph methodology to the experimental neuroscience is to evaluate the functional time-varying dynamics of the cortical network during the preparation and the execution of the foot movement.

This perspective sounds relevant since the large part of the current methods for the estimate of the functional connectivity return static relationships between fixed temporal windows of brain activity. This stationarity assumption could bias the physiologic interpretation of the obtained results. In fact, the transitory transfer of causal information could remain hidden because a unique model is estimated on the whole time interval.

On the contrary, the change of direction and intensity of the functional links could point out some new insights about the transient processes that are supposed to occur among different specialized cortical areas and that cannot be observed by means of static or low-time resolution methods.

In this study, the cortical connectivity is expected to change rapidly during the preparation and the execution of the foot movement. According to this hypothesis, the theoretical graph approach aims at extracting the time-varying properties of the evolving cortical network.

7.1 EXPERIMENTAL DESIGN

Five voluntary and healthy subjects participated to the study (age, 26–32 years; five males). They had no history of neurological or psychiatric disorder, and they were free from medications, or alcohol or drug abuse. The informed consent statement was signed by each subject after the explanation of the study, which was approved by the local institutional ethics committee. For the EEG data acquisitions, the participants were comfortably seated on a reclining chair in an electrically shielded and dimly lit room. They were asked to perform a dorsal flexion of their right foot, whose preference was previously attested by simple questionnaires (Chapman et al., 1987).

The movement task was repeated every 8 s, in a self-paced manner and 200 single trials were recorded by using 200 Hz of sampling frequency. A 96-channel system (BrainAmp, Brainproducts GmbH, Germany) was used to record EEG signals by means of an electrode cap and EMG electrical potentials through surface electrodes. In accordance to an extension of the 10–20 international system, the electrode cap was composed by 64 channels. The structural MRIs of the subject's head were taken with a Siemens 1.5T Vision Magnetom MR system (Germany).

Three-dimensional electrode positions were obtained by using a photogrammetric localization (Photomodeler, Eos Systems Inc., Canada) with respect to the anatomic landmarks: nasion and the two pre-auricular marks. Some trained electroencephalographists visually inspected EEG data and all the trials containing artefacts were rejected. Subsequently, they were baseline adjusted and low-pass filtered at 45 Hz.

7.2   CORTICAL ACTIVITY AND FUNCTIONAL CONNECTIVITY

High-resolution EEG technologies were used to estimate the cortical activity from the scalp measurements (see Chapter 4). The present experiment was based on sixteen regions of interest that were segmented from the cortical model of each subject.

The ROIs considered for the left (_L) and right (_R) hemisphere are the primary motor areas of the foot (MF_L and MF_R), the proper supplementary motor areas (SM_L and SM_R) and the cingulate motor areas (CM_L and CM_R). The bilateral Brodmann areas 6 (6_L and 6_R), 7 (7_L and 7_R), 8 (8_L and 8_R), 9 (9_L and 9_R) and 40 (40_L and 40_R) were also considered. In order to inspect the brain dynamics during the preparation and the execution of the studied movement, a time segment of 2 s was analyzed, after having centred it on the onset detected by a tibial EMG. The most interesting cerebral processes concerning the detected movement are actually thought to occur within this interval (Pfurtscheller and Lopes da Silva, 1999).

In the present study, the time-varying formulation of the PDC based on adaptive MVAR models (see Chapter 2) was thus employed. This measure is very interesting because it estimates a connectivity pattern for each time-point and it maintains the same time resolution of the employed brain-imaging technique. The time-dependent parameter matrices were estimated by means of the recursive least squares algorithm with forgetting factor (RLS), as described in Chapter 2. The multivariate Akaike's criterion (Akaike, H., 1974) was applied on each time sample and the highest order obtained was then utilised in all the recursive estimations. The order of the used aMVAR models ranged from 14 to 16 for all the five experimental subjects.

The rough connectivity estimation produces a full connected weighted and asymmetric matrix, representing the Granger-causal influences (Granger, C., 1969) among all the cortical regions of interest. In order to consider only the task-related connections, a filtering procedure based on statistical validation was adopted. In each trial, a rest period of 2 s preceding the movement was selected as an element of contrast (from −4 to −2 s before the onset). The connection intensities regarding the pairs of ROIs for each time sample were collected in order to obtain, a distribution of values belonging to the rest period. A threshold range was then extracted from the values of the rest-distribution by considering a percentile of 0.01 and 0.99, respectively, for the lowest and highest edge, with the aim of testing the significance of the estimated connections within the period of interest. After statistical filtering, the remaining weighted connections represent the significant relationships among the sixteen ROIs considered in the experimental task.

Figure 7.1 illustrates the locations of the regions of interest (ROIs) on the left hemisphere of the cortex model together with their estimated temporal activity. At the bottom, the time-varying cortical network in the Beta frequency band is shown for a representative subject. In particular, three instants are highlighted: one second before the onset, the onset itself and one second after the onset.

In the present work, we focused our analysis on two particular spectral ranges related to the movement, specifically the Alpha and Beta band. In fact, those frequency bands have been suggested to be the most responsive channels to the preparation and execution of a simple limb movement (Pfurtscheller and Lopes da Silva, 1999).

7.3   GRAPH THEORETICAL MODELING

A common number of connections (or density) has been considered in each weighted graph in order to analyze the structure of the cortical networks across all the subjects, frequency bands and time samples. In fact, when seeking a common behaviour among different networks, the mathematical indexes that evaluate their properties could suffer in robustness when they apply to graphs with different densities.

Consequently, this study will take into consideration 48 connections (i.e., density = 0.2) for each network obtained by removing the weakest links from each weighted graph. The choice of this connection-density was surely the most favourable condition for the significance of the indexes of the network structure (E_g and E_l). At a more specific analysis, it has been found that these indexes keep their usual independency—characterized by their ability to detect global and local properties—even in a small 16 nodes-graph (data not shown here).

All the indexes calculated on the cortical networks were standardized by considering their Z-score with respect to the distribution obtained from 50 random graphs. For each frequency band and time sample, random patterns were generated from the cerebral network of each subject, by shuffling the connections and maintaining the same in-and out-degree, but not the same strength (Sporns and Zwi, 2004).

7.4   RESULTS

Figure 7.2 (a) shows the in-strength values for the average network during three moments of interest that presented significant differences from random networks. Among all the cortical regions, the supplementary motor areas of both hemispheres (SM_L and SM_R) show the highest values of in-strength index. In the time points that precedes the onset movement (−560 ms) also the right and left primary motor areas of the foot (MF_L and MF_R) present a considerable number of incoming functional links. Figure 7.2 (b) shows the average values of out-strength obtained during the three time points of interest. In this particular case, it is evident that the large part of the cortical areas does not produce outgoing edges, while the bilateral cingulate motor region (CM_L and CM_R) presents very high out-strength values. All the indexes calculated on the cortical networks were

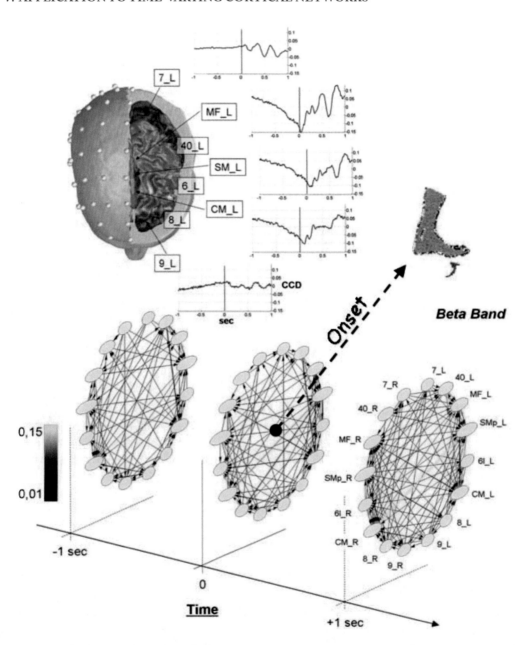

Figure 7.1: (Up). Realistic head model for a representative subject and cortical activity for the ROIs in the left hemisphere. (Bottom). Three-dimensional representation of the estimated time-varying network in the Beta band for the same subject.

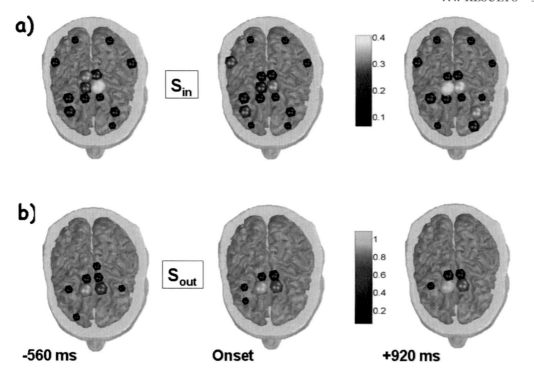

Figure 7.2: Average in- and out-strength in the Beta band during three significant moments. The little spheres are located in correspondence of each ROI. The size and color of each sphere encodes the degree value.

standardized by considering their Z-score with respect to the distribution obtained from 50 random graphs.

Figure 7.3 shows the average Z values in the analyzed population for the time-varying in-strength R_{in} and out-strength R_{out} distributions in the representative Beta frequency band. An interesting result is that in-strength (R_{in}) and out-strength (R_{out}) distributions show different characteristics. The high Z-scores in correspondence with the high values of S_{out} (i.e., the "right tail" of the distribution) suggest the presence of few ROIs with a very high level of outgoing flows, which makes them act as cortical "hubs." In particular, the intensity of their outgoing links seems to increase as time elapses from the movement preparation to the movement execution, as revealed by the respective shift of the significant Z values towards high levels of out-strength.

The level of organization in the time-varying cortical networks during the foot movement was analyzed by computing the efficiency indexes E_g and E_l. The E_g and E_l indexes estimated in every subject from the respective cortical networks were contrasted with the ones obtained from the respective random structures.

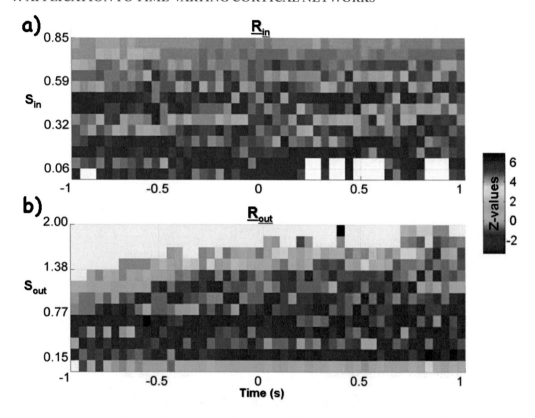

Figure 7.3: (a) Average time-varying in-strength distributions for the Beta band. The latency from the movement onset is shown on the x-axes; the in-strength (S_{in}) values on the y-axes. The colour encodes the group-averaged intensity of the R_{in} Z-score. (b) Average time-varying out-strength distributions for the Beta band. Same conventions as above.

Figure 7.4 shows the average Z-scores of the time-varying E_g (solid line) and E_l - (dotted line) of the connectivity patterns in the Beta frequency band. In particular, one second before the onset (from about -1 to -0.5 s), the cortical networks mostly show low values of E_g and E_l, reflecting a weak pattern of communication characterized by long average distances and few clustering connections between the ROIs. Throughout the period closer to the execution of the movement (from about -0.5 s to the onset), both the global and local properties increase and in correspondence with it, we observe high values of E_g and E_l.

Consequently, the structure of the cortical networks tends to maximize the interplay between the global integration and its local interactions. This particular structure represents one of the best way in which the cortical areas communicate since the relevant network presents simultaneously short links between each pair of ROIs and highly connected clusters (i.e., small-world architecture).

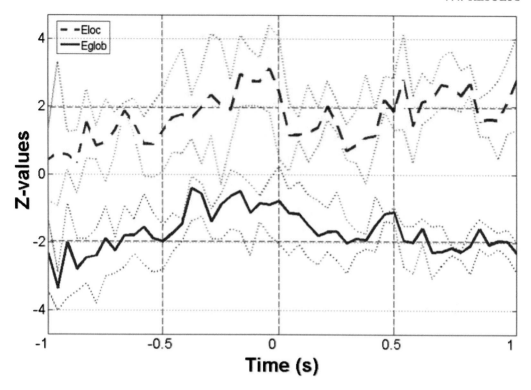

Figure 7.4: Average time-varying efficiency indexes. The lighter lines around the mean value indicate the time courses of the 25th and 75th percentile. The latency from the movement onset is shown on the x-axes.

After the onset (from the onset to +0.5 s), the estimated cortical networks show a typical random organization of the functional links, with a high E_g and a low E_l, reflecting the dense presence of wide-scope interactions among the ROIs, but a low tendency of the same cortical regions to form functional clusters. In the last period of the movement execution (from about +0.5 to +1 s), the estimated cortical networks mainly show high E_l values and low E_g values. The resulting structure is known to reflect the properties of regular and ordered graphs in which the local property of clustering is privileged with respect to the overall communication. Figure 4.1 (a) shows the average time-varying course of the weighted-density k_w in the Beta band during the analyzed period of interest.

Table 7.1 summarizes in a schematic way the average behaviour of these indexes obtained during the task performance of the analyzed population.

Figure 7.5 (a) shows the average time-varying course of the weighted-density k_w in the Beta band during the analyzed period of interest. The average intensity of the network links during the

Table 7.1: Global- and Local- Efficiency Variation.			
$-1, 5\,\mathrm{s} \rightarrow -0, 5\,\mathrm{s}$	$-0, 5\,\mathrm{s} \rightarrow$ **onset**	**onset** $\rightarrow +0, 5\,\mathrm{s}$	$+0, 5 \rightarrow +1\,\mathrm{s}$
E_g ↓ *	↑	↑	↓ *
E_l ↓	↑ *	↓	↑ *

Average profiles of E_g and E_l during the different stages of the foot movement. The pointing upward arrow indicates an increase of the respective value of the parameter and vice versa. The asterisk means that the increase/decrease is significantly different from random networks ($p < 0.05$).

preparation (from -0.5 s to the onset) is relatively low if compared with its maximum value reached in the following movement execution. In correspondence with this period, the network structure presents the most efficient pattern of communication, as revealed by the estimated small-world characteristic. Therefore, it is interesting to note that the optimal organization of the functional links among the cortical areas during the preparation of the foot movement is not correlated to the need of a high level of overall connectivity.

The analysis of the average time-varying reciprocity index revealed an interesting behavior during the preparation (from about -1 to 0 s) of the movement in the Beta frequency band. In such a period, the functional network moved from a reciprocal ($\rho = 0.1$) to an anti-reciprocal ($\rho = -0.1$) state. This aspect emphasizes the role of the early preparation in which a high level of mutual exchange of information is required to speed up the cortical process in expectation of the execution. Moreover, by tracking the evolving involvement of each single reciprocal connection (see Fig. 7.6 (a)), it is possible to observe their "persistence" during the entire period of interest. In particular, the persistent bilateral links between the cingulate motor areas and the supplementary motor areas (they correspond to the rows 58 and 69) in the Beta band reveals a novel aspect of such a connection that anyway was expected in a self-paced modality of movement generation, as in our experimental condition.

In Figure 7.6 (b), we compared the 3-motif properties of real brain networks with random networks and we identified some motif classes that occurred more frequently during particular stages of the movement. Of particular interest is the involvement of the feed-forward-loop motif (the fifth in the Figure 3.1) that tends to significantly ($p < 0.01$) increase during the proper movement execution (from about 0 to $+1$ s). This type of building block is known to play an important functional role in information processing. In fact, one possible function of this circuit is to activate output only if the input signal is persistent and to allow a rapid deactivation when the input goes off.

In the cortical context, a possible interpretation of such a motif would make a particular ROI act as a "switch" for the communication between the others two ROIs composing the triad. Another interesting aspect was revealed by the significant ($p << 0.01$) "persistence" of the single-input motif (the third in the Figure 3.1) that represented the highest recurrent pattern of interconnections during the entire evolution of the foot movement. The main function of this motif is known to involve the

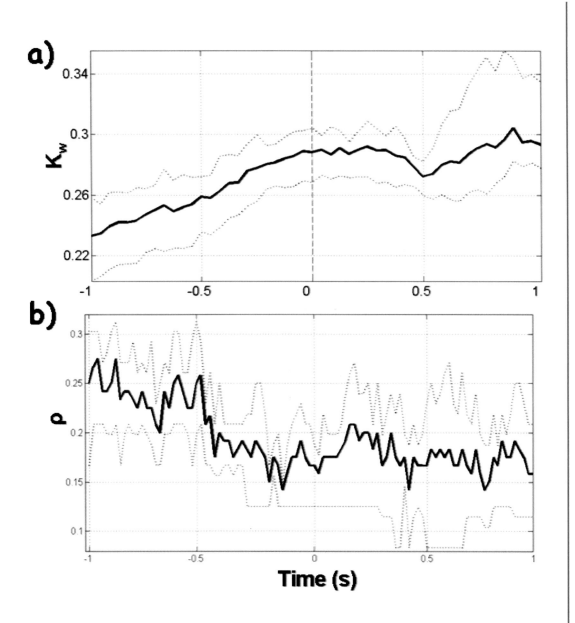

Figure 7.5: (a) Average time-varying "weighted-density" in the Beta band. (b) Average time-varying "reciprocity" during the period of interest in the Beta band. On y-axes the correlation coefficient ρ while time in seconds on x-axes.

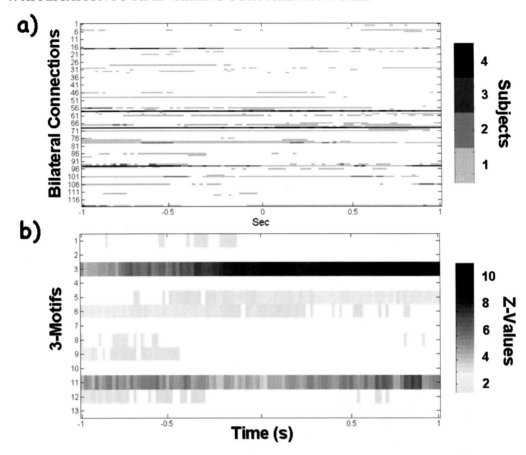

Figure 7.6: (a) Time-varying persistence of the bilateral connections in the cortical network. On *y*-axes all the 120 possible reciprocal connections while time in seconds on *x*-axes. The colour of the line corresponding to a particular link codes the number of subjects that actually hold such a connection. (b) Average time-varying 3-motif spectra. On *y*-axes all the 13 possible directed 3-motifs are listed while time in seconds is displayed on *x*-axes.

"activation" of several parallel pathways by a single activator. Thus, since the single-input only differs from the feed-forward-loop motif for the functional link between the two areas activated, we can claim the privileged scheme of communication within the functional networks estimated consists in a parallel activation from a particular ROI of two other distinct areas, whose communication seems to increase significantly only during the proper movement execution.

7.5   DISCUSSION

The possibility to deal explicitly with weighted and asymmetric relationships as well as the observation of transient couplings would increase the analytical power to observe the specific cortical network dynamics during the task. In order to limit the discussion of the results, the present study has analyzed the cortical networks in the Alpha (7–12 Hz) and Beta (13–29 Hz) frequency band representing the spectral contents principally involved in the preparation and the execution of simple motor acts (Pfurtscheller and Lopes da Silva, 1999). However, this methodology is not limited to a particular frequency band or a particular set of ROIs since it can be adapted to investigate experimental tasks in any spectral content.

The successful execution of a simple movement relies on the integrated neural activity in spatially distributed networks, which encompass several frontal pre-motor, primary sensorimotor and subcortical structures (Ohara et al., 2001). As regards the Alpha and Beta band, the strong functional links between the cingulate motor (CM_L and CM_R) areas and the supplementary motor (SM_L and SM_R) areas are expected to be involved in a self-paced modality of movement generation, exactly as in our experimental condition (Gerloff et al., 1998). Indeed, the bilateral SM in the Alpha and Beta band and the CM areas in the Beta band showed a persistent involvement throughout the entire considered period. The significant ($p < 0.01$) differences with respect to a random re-distribution of the links reveal the characteristic role of such regions. In particular, the SM areas represent the main target for a large number of functional connections taking origin from the other investigated cortical areas. In addition, the CM_L and CM_R represent the main sources of outgoing flows towards all the other ROIs, as revealed by the high statistical significance of their out-strength indexes.

Beyond the simple evaluation of the estimated functional flow intensities in different cortical areas, the strength distribution gave additional information about the allocation of these flows within the modelled cortical system. The average time-varying profiles reveal a different behaviour between the distributions of the incoming and outgoing strength indexes in the Alpha and Beta band. In fact, the out-strength distribution only indicates a significant ($p < 0.01$) presence of few cortical areas - i.e., the nodes of the modelled graph - acting as "hubs," characterized by a very high level of outgoing functional flows. Looking more closely at the strength values of the ROIs previously analyzed, the cingulate motor areas (CM_L and CM_L) are found to act as the centre of outgoing flows for the estimated cortical functional network. This fact suggests a central role of these regions in both the considered spectral contents, since their removal would immediately corrupt the organization of the estimated functional network by reducing the overall level of connectivity. It is interesting to remark that only in the Beta band this specific involvement increases during the execution of the foot movement, which indicates a predominant role of the CM areas in this period and spectral content.

The study of the structural properties of the graph, evaluated by the clustering index C and the inverse of the average shortest path $1/PL$ was performed by calculating the local-efficiency E_l and the global-efficiency E_g.

The results in the Beta band revealed the succession of different functional architectures during the preparation and the execution of the foot movement. In particular, 1 s before the onset movement (from about -1 to -0.5 s), the cortical networks mostly show low values of $1/PL$ and C, which reflects a weak pattern of communication characterized by long average distances and few clustering connections between the ROIs.

The closer the execution of the movement is (from about -0.5 s to the onset), the more both $1/PL$ and C increase. Consequently, the structure of the cortical networks tends to maximize the interplay between the global integration and its local interactions. This particular structure represents one of the best way in which the cortical areas communicate, since the relevant network presents simultaneously short links between each pair of ROIs and highly connected clusters — i.e., small-world architecture (Watts and Strogatz, 1998).

After the onset (from the onset to $+0.5$ s), the estimated cortical networks show a typical random organization of the functional links (Watts and Strogatz, 1998), with a high $1/PL$ and a low C, reflecting the dense presence of wide-scope interactions among the ROIs, but a low tendency of the same cortical regions to form functional clusters.

In the last period of the movement execution (from about $+0.5$ to $+1$ s) the estimated cortical networks mainly show high C values and low $1/PL$ values. The resulting structure is known to reflect the properties of regular and ordered graphs (Watts and Strogatz, 1998) in which the local property of clustering is privileged with respect to the overall communication.

In the Alpha band, the network structure seems to follow a stationary profile, since the inverse of average shortest path and the clustering index remain quite constant during the entire analyzed period. In particular, the functional networks hold a regular configuration independently from the cortical dynamics that characterized the task, which indicates that main network changes are encoded in the higher spectral frequencies (Beta band).

In addition to the structural analysis, the evaluation of the average weighted-density k_w associated to the evolution of the cortical network returned further information regarding the varying level of overall connectivity. In the Beta band, the average intensity of the network links during the preparation (from -0.5 s to the onset) is relatively low if compared with its maximum value reached in the following movement execution. In correspondence with this period the network structure presents the most efficient pattern of communication, as revealed by the estimated small-world characteristic. Therefore, it is interesting to note that the optimal organization of the functional links among the cortical areas during the preparation of the foot movement is not correlated to the need of a high level of overall connectivity.

The analysis of the average time-varying reciprocity index revealed the significant presence of mutual links within the cortical networks during the entire period analysed.

In particular, during the preparation (from about -1 to 0 s) of the movement in the Beta frequency band the functional network moved from a high ($\rho > 0.25$) to a lower ($\rho < 0.17$) reciprocal state. This aspect emphasizes the role of the early preparation in which a higher level of mutual exchange of information is required to speed up the cortical process in expectation of the execution. Although the cortical networks seem not much correlated during their maximum state of reciprocity, this result does not diverge too much from other empirical results obtained for networks of neuron classes in which values ranged around 0.17–0.18 (Garlaschelli and Loffredo, 2004).

Moreover, by tracking the evolving involvement of each single reciprocal connection, it was possible to observe their 'persistence' during the entire period of interest. Interestingly, the permanence of stable mutual links seems a peculiar characteristic of the estimated functional networks, which instead cannot be observed in any sequence of random graphs. In particular, the persistent bilateral links between the cingulate motor areas (CM left and right) and the supplementary motor areas (SM left and right) in the Beta band reveal a novel aspect of such a connection that anyway was expected in a self-paced modality of movement generation, as under our experimental condition (Gerloff et al., 1998).

In brain networks, motifs are those basic building blocks consisting of a set of cortical areas and pathways that can potentially engage in different patterns of interactions depending on their degree of activation, the surrounding neural context (Sporns and Kotter, 2004). The analysis of the average time-varying spectra of the 3-motifs revealed the basic rules governing the structure of the complex networks estimated during the task performance in the Beta frequency band. We compared the 3-motif properties of real brain networks with those of random networks, and we identified some motif classes that occurred more frequently during particular stages of the movement.

Of particular interest is the involvement of the feed-forward-loop motif that tends to significantly ($p < 0.01$) increase during the proper movement execution (from about 0 to +1 s). This type of building block is known to play an important functional role in information processing. In fact, one possible function of this circuit is to activate output only if the input signal is persistent and to allow a rapid deactivation when the input goes off (Shen-Orr et al., 2002).

In the cortical context, a possible interpretation of such a motif would make a particular ROI act as a 'switch' for the communication between the others two ROIs composing the triad. Another interesting aspect was revealed by the significant ← ($p < 0.01$) → "persistence" of the single-input motif that represented the highest recurrent pattern of interconnections within the cortical network during the entire evolution of the foot movement. The main function of this motif is known to involve the *activation* of several parallel pathways by a single activator (Shen-Orr et al., 2002). Thus, since the single input only differs from the feed-forward-loop motif for the lack of a functional link between the two areas activated, we can claim that the privileged scheme of communication within the functional networks estimated consists in a parallel activation from a particular ROI of two other distinct areas, whose communication seems to increase significantly only during the proper movement execution.

CHAPTER 8

Conclusions

The entire work exposed in the previous pages suggests that it is possible to treat the brain activity from neuroelectric measurements with mathematical instruments and methodologies derived from other fields of science. In particular, the present book explored the use of graph theory indexes on the assessment of particular brain functions, such for instance memory or movement tasks.

This book presents and describes different brain networks that have no particular "physical" or "anatomical" support. In other words, the brain networks estimated and described by the indexes of the graph theory in the previous pages are really "functional" brain networks. Such "functional" brain networks could change in topology and properties according to the specific subject's behavior.

In some sense, with the proposed methodology, it is possible to describe externally the flow of the brain networks that sub-serve particular macroscopic behavioral function, such as memory or the movement planning. Since these behavioral functions are likely change in time, also this "functional" brain networks change their topology and appearance during the time. The possibility to estimate such flow of brain networks sub-serving the different functions in the human here explored it is really promising for a generation of a better understanding of the brain functions.

The fact that such investigations are possible with the low-cost technique of the high resolution EEG is also a strong conclusion of this book; that suggests how it is possible to estimate the activity and the functional connectivity at the level of the Brodmann areas of the cortex by using only electrodes placed on the scalp.

This methodology then allows the representation of the graph nodes as particular regions of interest on the cortex. And it gives to the researcher a "window" to access the brain functions in a different perspective than the usual techniques encountered in the neuroscience literature.

In fact, the development of brain imaging devices (such as the functional Magnetic Resonance Imaging (fMRI), but also the high-resolution EEG technology) often give to the scientist a series of colored hot-spots in the brain that sub-serve the functions performed by the subject during a particular task.

Actually, if we look at thousands of fMRI studies, a possible impression is that a specific cortical area gets "activated" during the performance of whatever cognitive or motor operation. In this scenario of modern "color phrenology," the study of functional cortical connectivity suggests an image of the brain as a system of objects that rapidly changes the way in which they are intermingled, according to the complexity and to the dynamics of the task proposed to the subject.

It is the opinion of the authors of the present book that the perspective offered by the use of graph theory for functional cortical connectivity networks estimated from standard EEG recordings could be a promising way to approach the brain functioning from a modern point of view.

Bibliography

Achard, S. and Bullmore, E. (2007). Efficiency and cost of economical brain functional networks, *PloS Comp. Biol.*, 3(2):e17. DOI: 10.1371/journal.pcbi.0030017 1, 51

Akaike, H. (1974). A new look at statistical model identification. IEEE Trans Automat Control, AC-19:716–723. DOI: 10.1109/TAC.1974.1100705 54

Albert, R. and Barabási, L. (2002). Statistical mechanics of complex networks. *Rev. Mod. Phys.*, 74, 47–97. DOI: 10.1103/RevModPhys.74.47 16

Amaral, L., Scala, A., Barthelemy, M. and Stanley, H. (2000). Classes of small-world networks, *Proc. Natl. Acad. Sci.*, 97111, 49–52. DOI: 10.1073/pnas.200327197 2

Astolfi, L., Cincotti, F., Mattia, D., De Vico Fallani, F., Tocci, A., Colosimo, A., Salinari, S., Marciani, M.G., Hesse, W., Witte, H., Ursino, M., Zavaglia, M. and Babiloni, F. (2008). Tracking the time-varying cortical connectivity patterns by adaptive multivariate estimators. *IEEE Trans. Biomed. Eng.*, 55(3):902–13. DOI: 10.1109/TBME.2007.905419 9

Astolfi, L., Cincotti, F., Mattia, D., Marciani, M.G., Baccalà, L., De Vico Fallani, F., Salinari, S., Ursino, M., Zavaglia, M., Ding, L., Edgar, J.C., Miller, G.A., He, B. and Babiloni, F. (2006). A comparison of different cortical connectivity estimators for high resolution EEG recordings. *Human Brain Mapping*, 28(2):143–57. DOI: 10.1002/hbm.20263 1, 8

Babiloni, F., Babiloni, C., Locche, L., Cincotti, F., Rossini, P.M. and Carducci, F. (2000). High resolution EEG: source estimates of Laplacian-transformed somatosensory-evoked potentials using a realistic subject head model constructed from magnetic resonance images. *Med. Biol. Eng. Comput.* 38:512–9. DOI: 10.1007/BF02345746 23

Babiloni, F., Cincotti, F., Babiloni, C., Carducci, F., Basilisco, A., Rossini, P.M., Mattia, D., Astolfi, L., Ding, L., Ni, Y., Cheng, K., Christine, K., Sweeney, J. and He, B. (2005). Estimation of the cortical functional connectivity with the multimodal integration of high resolution EEG and fMRI data by Directed Transfer Function. *NeuroImage*, 24(1):118–3. DOI: 10.1016/j.neuroimage.2004.09.036 1

Baccalà, L.A. and Sameshima, K. (2001). Partial Directed Coherence: a new concept in neural structure determination. *Biol. Cybern.*, 84:463–474. DOI: 10.1007/PL00007990 5, 6, 8

Baccalà, L.A. (2001). On the efficient computation of partial coherence from multivariate autoregressive model, in Callaos, N., Rosario, D. and Sanches, B. (Eds.), *Proceedings of the 5th World Conference Cybernectis Systemics and Informatics SCI 2001*, Orlando. 5

Barabási, A.L. and Albert, R. (1999). Emergence of scaling in random net works. *Science*, 286:509–512. DOI: 10.1126/science.286.5439.509 2

Bartolomei, F., Bosma, I., Klein, M., Baayen, J.C., Reijneveld, J.C., Postma, T.J., Heimans, J.J., van Dijk, B.W., de Munck, J.C., de Jongh, A., Cover, K.S. and Stam, C.J. (2006). Disturbed functional connectivity in brain tumour patients: evaluation by graph analysis of synchronization matrices. *Clin. Neurophysiol.*, 117:2039–2049. DOI: 10.1016/j.clinph.2006.05.018 2

Bassett, D.S., Meyer-Linderberg, A., Achard, S., Duke, Th. and Bullmore, E. (2006). Adaptive reconfiguration of fractal small-world human brain functional networks. *PNAS*, 103:19518–19523. DOI: 10.1073/pnas.0606005103 2

Bendat, J.S. and Piersol, A.G., *Engineering Applications of Correlation and Spectral Analysis*, Wiley, 1993.

Boccaletti, S., Latora, V., Moreno, Y., Chavez, M. and Hwang, D.U. (2006). Complex networks: Structure and dynamics. *Phys. Rep.*, 424:175–308. DOI: 10.1016/j.physrep.2005.10.009 40

Bressler, S.L. (1995). Large-scale cortical networks and cognition. *Brain Res. Rev.*, 20(3):288–304. DOI: 10.1016/0165-0173(94)00016-I 5

Brovelli, A., Ding, M., Ledberg, A., Chen, Y., Nakamura, R. and Bressler, S.L. (2004). Beta oscillations in a large-scale sensorimotor cortical network: directional influences revealed by Granger causality. *Proc. Natl. Acad. Sci., USA, 2004, June 29*, 101(26):9849–54. DOI: 10.1073/pnas.0308538101 5

Brovelli, A., Lachaux, J.P., Kahane, P. and Boussaoud, D. (2005). High gamma frequency oscillatory activity dissociates attention from intention in the human premotor cortex. *NeuroImage*, July 12. DOI: 10.1016/j.neuroimage.2005.05.045

Buchel, C. and Friston, K.J. (1997). Modulation of connectivity in visual pathways by attention: cortical interactions evaluated with structural equation modeling and fMRI. *Cereb. Cortex*, 7(8):768–78. DOI: 10.1093/cercor/7.8.768 5

Chapman, J.P., Chapman, L.J. and Allen, J.J. (1987). The measurement of foot preference. *Neuropsychologia*, 25(3), 579–584. DOI: 10.1016/0028-3932(87)90082-0 53

Clifford, C.G. (1987). Coherence and time delay estimation. *Proc. IEEE*, 75:236 –255. 1

David, O., Cosmelli, D. and Friston, K.J. (2004). Evaluation of different measures of functional connectivity using a neural mass model. *NeuroImage*, vol. 21, pp. 659–673. DOI: 10.1016/j.neuroimage.2003.10.006 1

De Vico Fallani, F., Astolfi, L., Cincotti, F., Mattia, D., Marciani, M.G., Salinari, S., Kurths, J., Gao, S., Cichocki, A., Colosimo, A. and Babiloni, F. (2007). Cortical Functional Connectivity Networks In Normal And Spinal Cord Injured Patients: Evaluation by Graph Analysis. *Human Brain Mapping*, 28:1334–46. DOI: 10.1002/hbm.20353

De Vico Fallani, F., Astolfi, L., Cincotti, F., Mattia, D., Marciani, M.G., Tocci, A., Salinari, S., Soranzo, R., Colosimo, A. and Babiloni, F. (2008). Structure of the cortical networks during successful memory encoding in TV commercials. *Clinical Neurophysiology*, 119: 2231–2237. DOI: 10.1016/j.clinph.2008.06.018

De Vico Fallani, F., Astolfi, L., Cincotti, F., Mattia, D., Marciani, M.G., Tocci, A., Salinari, S., Soranzo, R., Colosimo, A. and Babiloni, F. (2008). Structure of the cortical networks during successful memory encoding in TV commercials. *Clinical Neurophysiology*, 119: 2231–2237. DOI: 10.1016/j.clinph.2008.06.018

De Vico Fallani, F., Astolfi, L., Cincotti, F., Mattia, D., Marciani, M.G., Tocci, A., Salinari, S., Witte, H., Hesse, W., Gao, S., Colosimo, A. and Babiloni, F. (2008). Cortical network dynamics during foot movements. *Neuroinformatics*, DOI 10.1007/s12021–007-9006-6. DOI: 10.1007/s12021-007-9006-6

Ding, L., Bressler, S.L., Yang, W., Liang Ding, H. (2000). Short-window spectral analysis of cortical event-related potentials by adaptive multivariate autoregressive modeling: data preprocessing, model validation, and variability assessment. Bio. Cybern. 83: 35–45. DOI: 10.1007/s004229900137 8

Eguiluz, V.M., Chialvo, D.R., Cecchi, G.A., Baliki, M. and Apkarian, A.V. (2005). Scale-free brain functional networks. *Phys. Rev. Lett.*, 94:018102. DOI: 10.1103/PhysRevLett.94.018102 1

Felleman, D.J. and Van Essen, D.C. (1991). Distributed hierarchical processing in the primate cerebral cortex. *Cerebral Cortex*, 1, 1–47. DOI: 10.1093/cercor/1.1.1-a 19

Garlaschelli, D. and Loffredo, M.I. (2004). Patterns of link reciprocity in directed networks, *Phys. Rev. Lett.*, 93268701. DOI: 10.1103/PhysRevLett.93.268701 2, 17, 65

Gerloff, C., Richard, J., Hadley, J., Schulman, A.E., Honda, M. and Hallett, M. (1998). Functional coupling and regional activation of human cortical motor areas during simple, internally paced and externally paced finger movements. *Brain*, 1211513–31 63, 65

Gevins, A.S., Cutillo, B.A., Bressler, S.L., Morgan, N.H., White, R.M., Illes, J. and Greer, D.S. (1989). Event-related covariances during a bimanual visuomotor task. II. Preparation and feedback. *Electroencephalogr. Clin. Neurophysiol.*, 74:147–160. DOI: 10.1016/0168-5597(89)90020-8 1, 5

Gevins, A., Brickett, P., Reutter, B., Desmond, J. (1991). Seeing through the skull: advanced EEGs use MRIs to accurately measure cortical activity from the scalp. *Brain Topogr.*, 4:125–31. DOI: 10.1007/BF01132769 23

Gevins, A., Le, J., Leong, H., McEvoy, L.K. and Smith, M.E. (1999). Deblurring. *J. Clin. Neurophysiol.*, 16(3):204–13. DOI: 10.1097/00004691-199905000-00002 23

Granger, C.W.J. (1969). Investigating causal relations by econometric models and cross-spectral methods. *Econometrica*, 37, 424–438. DOI: 10.2307/1912791 5, 54

Grave de Peralta, R., Hauk, O., Gonzalez Andino, S., Vogt, H. and Michel, C.M. (1997). Linear inverse solution with optimal resolution kernels applied to the electromagnetic tomography, *Human Brain Mapping*, 5, 454–67. DOI: 10.1002/(SICI)1097-0193(1997)5:6%3C454::AID-HBM6%3E3.0.CO;2-2 26

Gross, J., Kujala, J., Hämäläinen, M., Timmermann, L., Schnitzler, A. and Salmelin, R. (2001). Dynamic imaging of coherent sources: studying neural interactions in the human brain. *Proc. Natl. Acad. Sci. USA*, **98** 2,694–699. 5

Gross, J., Timmermann, L., Kujala, J., Salmelin, R. and Schnitzler, A. (2003). Properties of MEG tomographic maps obtained with spatial filtering. *NeuroImage*, 19:1329–1336. DOI: 10.1016/S1053-8119(03)00101-0 5

Gruber, T., Tsivilis, D., Montaldi, D., Muller, M.M. (2004). Induced gamma band responses: an early marker of memory and retrieval. Neuroreport 2004; 15: 1837–41. 51

Hämäläinen, M. and Ilmoniemi, R. (1984). Interpreting measured magnetic field of the brain: estimates of the current distributions. *Tech. Rep. TKKF-A559*, Espoo (Finland): Helsinki University of Technology.

Hansen, P.C. (1992). Analysis of discrete ill-posed problems by means of the L-curve. *SIAM Rev.*, 34:561–80. DOI: 10.1137/1034115 27

Hesse, W., Möller, E., Arnold, M. and Schack, B. (2003). The use of timevariant EEG Granger causality for inspecting directed interdependencies of neural assemblies. *J. Neurosci. Methods*, vol. 124, pp. 27–44. DOI: 10.1016/S0165-0270(02)00366-7 9

Horwitz, B. (2003). The elusive concept of brain connectivity. *NeuroImage*, 19, 466–470. DOI: 10.1016/S1053-8119(03)00112-5 1, 5

Inouye, T., Iyama, A., Shinosaki, K., Toi, S. and Matsumoto, Y. (1995). Intersite EEG relationships before widespread epileptiform discharges. *Int. J. Neurosci.*, 82:143–153. DOI: 10.3109/00207459508994298 1

Jancke, L., Loose, R., Lutz, K., Specht, K. and Shah, N.J. (2000). Cortical activations during paced finger-tapping applying visual and auditory pacing stimuli. *Cogn. Brain Res.*, 10(1–2):p. 51-66 DOI: 10.1016/S0926-6410(00)00022-7 5

Jeong, H., Tombor, B., Albert, R., Oltvai, Z.N. and Barab'asi, A.L. (2000). The large-scale organization of metabolic networks. *Nature*, 407651–4. DOI: 10.1038/35036627 2

Kaminski, M. and Blinowska, K. (1991). A new method of the description of the information flow in the brain structures. *Biol. Cybern.*, 65:203–210. DOI: 10.1007/BF00198091 5, 8

Kaminski, M., Ding, M., Truccolo, W.A. and Bressler, S. (2001). Evaluating causal relations in neural systems: Granger causality, directed transfer function and statistical assessment of significance. *Biol. Cybern.*, 85:145–157. DOI: 10.1007/s004220000235 5, 8

Kus, R., Kaminski, M. and Blinowska, K.J. (2004). Determination of EEG activity propagation: Pair-wise versus multichannel estimate, *IEEE Trans. Biomed. Eng.*, vol. 51, no. 9, pp. 1501–1510, Sep. 2004. DOI: 10.1109/TBME.2004.827929 6, 8

Lago-Fernandez, L.F., Huerta, R., Corbacho, F. and Siguenza, J.A. (2000). Fast response and temporal coherent oscillations in small-world networks, *Phys. Rev. Lett.*, 84:2758–61. DOI: 10.1103/PhysRevLett.84.2758 1

Latora, V. and Marchiori, M. (2001). Efficient behaviour of smallworld networks. *Physical Review Letters*, 87, 198701. DOI: 10.1103/PhysRevLett.87.198701 20, 50

Latora, V. and Marchiori, M. (2003). Economic small-world behaviour in weighted networks. *Eur. Phys. JB*, 32:249–263. DOI: 10.1140/epjb/e2003-00095-5 33

Lee, L., Harrison, L.M, Mechelli, A. (2003). The functional brain connectivity workshop: report and commentary. Neuroimage 2003; 19: 457–465. DOI: 10.1088/0954-898X/14/2/201 5

Micheloyannis, S., Pachou, E., Stam, C.J., Vourkas, M., Erimaki, S. and Tsirka, V. (2006), Using graph theoretical analysis of multi channel EEG to evaluate the neural efficiency hypothesis. *Neuroscience Letters*, 402:273–277. DOI: 10.1016/j.neulet.2006.04.006 2

Milgram, S. (1967). The Small World Problem. *Psychology Today*, 1, 60-67. 1

Milo, R., Shen-Orr, S., Itzkovitz, S., Kashtan, N., Chklovskii, D. and Alon, U. (2002). Network motifs: simple building blocks of complex networks. *Science*, 298824–7. DOI: 10.1126/science.298.5594.824 2, 19

Moeller, E., Schack, B., Arnold, M. and Witte, H. (2001). Instantaneous multivariate EEG coherence analysis by means of adaptive high-dimensional autoregressive models. *J. Neurosci. Methods*, vol. 105, pp. 143–158. DOI: 10.1016/S0165-0270(00)00350-2 8, 9

Nunez, P., *Electric Fields of the Brain*. New York, Oxford University Press, 1981. 23

Nunez, P.L., *Neocortical Dynamics and Human EEG Rhythms*, New York, Oxford University Press, 1995. 1, 23

Ohara, S., Mima, T., Baba, K., Ikeda, A., Kunieda, T. and Matsumoto, R. (2001). Increased synchronization of cortical oscillatory activities between human supplementary motor and primary sensorimotor areas during voluntary movements. *Journal of Neuroscience*, 21(23),9377–9386. 63

Pascual-Marqui, R.D. (1995). Reply to comments by Hämäläinen, Ilmoniemi and Nunez. In *ISBET Newsletter*, N .6, December 1995. Ed. W. Skrandies., 16–28. 26

Pfurtscheller, G. and Lopes da Silva, F.H. (1999). Event-related EEG/MEG synchronization and desynchronization: basic principles. *Clin. Neurophysiol.*, 110(11):1842–57. DOI: 10.1016/S1388-2457(99)00141-8 5, 41, 54, 55, 63

Raineteau, O. and Schwab, M. (2001). Plasticity of motor systems after incomplete spinal cord injury. *Nat. Rev. Neurosci.*, 2:263–273. DOI: 10.1038/35067570 41

Salvador, R., Suckling, J., Coleman, M.R., Pickard, J.D., Menon, D and Bullmore, E. (2005). Neurophysiological Architecture of Functional Magnetic Resonance Images of Human Brain. *Cereb. Cortex*, 15(9):1332–42. DOI: 10.1093/cercor/bhi016 1, 51

Sederberg, P.B., Kahana, M.J., Howard, M.W., Donner, E.J., Madsen, J.R. (2003). Theta and gamma oscillations during encoding predict subsequent recall. J. Neurosci 2003; 26: 10809–14. 51

Shen-Orr, S., Milo, R., Mangan, S. and Alon, U. (2002). Network motifs in the transcriptional regulation network of Escherichia coli. *Nat. Genet.*, 31:64–68. DOI: 10.1038/ng881 65

Singer, W. (1999). Neuronal synchrony: a versatile code for the definition of relations. *Neuron.*, 24(1):49–65. DOI: 10.1016/S0896-6273(00)80821-1 2

Sivan, E., Parnas, H. and Dolev, D. (1999). Fault tolerance in the cardiac ganglion of the lobster. *Biol. Cybern.*, 81:11–23. DOI: 10.1007/s004220050541 41

Sporns, O., Chialvo, D.R., Kaiser, M. and Hilgetag, C.C. (2004). Organization, development and function of complex brain networks. *Trends Cogn. Sci.*, 8:418–25. DOI: 10.1016/j.tics.2004.07.008 2

Sporns, O. and Zwi, J.D. (2004). The small world of the cerebral cortex. *Neuroinformatics*, 2:145–162. DOI: 10.1385/NI:2:2:145 2, 41, 51, 55

Sporns, O. and Honey, C.J. (2006). Small worlds inside big brains. *PNAS 2006*, 103(51):19219–19222. DOI: 10.1073/pnas.0609523103 2

Sporns, O., Tononi, G. and Edelman, G.E. (2000). Connectivity and complexity: the relationship between neuroanatomy and brain dynamics. *Neural Netw.* 13:909–922. DOI: 10.1016/S0893-6080(00)00053-8 1

Sporns, O. (2002). Graph theory methods for the analysis of neural connectivity patterns. Kötter, R. (Ed.), *Neuroscience Databases. A Practical Guide*, pp. 171–186, Klüwer, Boston, MA. 15, 18, 22

Sporns, O. and Kotter, R. (2004). Motifs in brain networks. *PLoS Biol*, 2e369. DOI: 10.1371/journal.pbio.0020369 20, 65

Stam, C.J. and van Dijk, B.W. (2002). Synchronization likelihood: an unbiased measure of generalized synchronization in multivariate data sets. *Physica D*, 163:236–251. DOI: 10.1016/S0167-2789(01)00386-4 1

Stam, C.J., Jones, B.F., Nolte, G., Breakspear, M. and Scheltens, Ph. (2007). Small-world networks and functional connectivity in Alzheimer's disease. *Cereb. Cortex*, 17:92–99. DOI: 10.1093/cercor/bhj127 1, 2

Stam, C.J. (2004). Functional connectivity patterns of human magnetoencephalographic recordings: a 'small-world' network? *Neurosci. Lett.*, 355:25–28. DOI: 10.1016/j.neulet.2003.10.063 2, 51

Stam, C.J. and Reijneveld, J.C. (2007). Graph theoretical analysis of complexnetworks in the brain. *Nonlinear Biomed. Phys.*, 1:3 epub. DOI: 10.1186/1753-4631-1-3 1

Stephan, K.E., Hilgetag, C.-C., Burns, G.A.P.C., O'Neill, M.A., Young, M.P. and Kotter, R. (2000). Computational analysis of functional connectivity between areas of primate cerebral cortex. *Phil. Trans. R. Soc. Lond. B*, 355:111–126. DOI: 10.1098/rstb.2000.0552 1

Strogatz, S.H. (2001). Exploring complex networks. *Nature*, 410:268-276. DOI: 10.1038/35065725 1, 2

Taniguchi, M., Kato, A., Fujita, N., Hirata, M., Tanaka, H., Kihara, T., Ninomiya, H., Hirabuki, N., Nakamura, H., Robinson, S.E., Cheyne, D. and Yoshimine, T. (2000). Movement-related desynchronization of the cerebral cortex studied with spatially filtered magnetoencephalography. *NeuroImage* 12(3):298–306. DOI: 10.1006/nimg.2000.0611 5

Theiler, J., Eubank, S., Longtin, A., Galdrikian, B. and Farmer, D. (1992). Testing for nonlinearity in time series: the method of surrogate data. *Physica D*, 58:77–94. DOI: 10.1016/0167-2789(92)90102-S 46

Tononi, G., Sporns, O. and Edelman, G.M. (1994). A measure for brain complexity: relating functional segregation and integration in the nervous system. *Proc. Natl. Acad. Sci. USA*, 91:5033–5037. DOI: 10.1073/pnas.91.11.5033 1

Urbano, A., Babiloni, C., Onorati, P. and Babiloni, F. (1998). Dynamic functional coupling of high resolution EEG potentials related to unilateral internally triggered one-digit movements. *Electroencephalogr. Clin. Neurophysiol.*, 106(6):477–87. DOI: 10.1016/S0013-4694(97)00150-8 1, 5

Varela, F., Lachaux, J.P., Rodriguez, E. and Martinerie, J. (2001). Thebrainweb: Phase synchronization and large-scale integration. *Nature Reviews Neuroscience*, 2:229–239. DOI: 10.1038/35067550 2

Watts, D.J. and Strogatz, S.H. (1998). Collective dynamics of 'small-world' networks. *Nature,* 393:440–442. DOI: 10.1038/30918 2, 20, 64

Wasserman, S. and Faust, K., *Social Network Analysis*, Cambridge: Cambridge University Press, 1994.

Zar, J.H. (1984). *Biostatistical Analysis*. Englewood Cliffs, NJ: Prentice Hall. 35

Authors' Biographies

FABRIZIO DE VICO FALLANI

Fabrizio DeVico Fallani is a Research fellow in the Department of Human Physiology, University of Rome "Sapienza" (Subject - Development and application of brain network analyses in healthy and diseased human subjects). He is also a term-contract worker at the Neuroeletrical Imaging and Brain Computer Interface Laboratory. IRCCS "Fondazione S.Lucia" (Subject - Development and application of brain network analyses in healthy and diseased human subjects during cognitive and motor tasks). In 2005 he got his master's degree in Computer Science Engineering at the University of Rome "Sapienza" (Thesis title - Advanced Methods for the Estimation of the Cortical Connectivity from high resolution EEG recordings in a group of Spinal Cord Injured Patients). In 2009, he got his PhD degree in Biophysics at the University of Rome "Sapienza". (Thesis title - Theoretical Graph Approach to Brain Functional Networks from High-Resolution EEG). He also received the best PhD thesis prize on Biomedical signal processing and imaging awarded by the Italian National Group of Bioengineering. He has participated with both oral and poster presentations at more than 45 international conferences and has received several travel grants and visiting research grants. He is the author of more than 30 scientific articles in peer-reviewed journals and his H-index is 6. His current interests are in the field of graph theoretical approaches, cortical connectivity estimation, and Brain Computer Interface.

FABIO BABILONI

Fabio Babiloni was born in Rome in 1961. He received his the master's degree in Electronic Engineering at the University of Rome "La Sapienza"and the PhD in Computational Engineering at the Helsinki University of Technology, Helsinki in the 2000 with a dissertation on the multimodal integration of EEG and fMRI. He is currently Associate Professor of Human Physiology at the Faculty of Medicine of the University of Rome "La Sapienza", Rome, Italy. He is the author of more that **120** papers on bioengineering and neurophysiological topics in international peer-reviewed scientific journals, and more than **250** contributions to conferences and book chapters. His total impact factor is more than **290** and his H-index is **26**. His current interests are in the field of multimodal integration of EEG, MEG and fMRI data, cortical connectivity estimation and Brain Computer Interface. Prof. Babiloni is currently grant reviewer for the National Science Foundation (NSF) USA, the Academy of Finland, Finland, the Austrian Fund of Research, Austria and the European Union through the FP6 and FP7 research programs. Prof. Babiloni is president of the International Society of Functional Source Imaging, member of the Italian Society of Physiology and the Italian Society of Clinical Neurophysiology. He is an Associate Editor of four scientific Journals *"Frontiers in Neuroscience,"* *"International Journal of Bioelectromagnetism,"* *"IEEE Trans. On Neural System and Rehabilitation Engineering,"* and *"Computational Intelligence and Neuroscience."*

Printed in the United States
by Baker & Taylor Publisher Services

Super Resolution of
Images and Video

Super Resolution of Images and Video
Aggelos K. Katsaggelos, Rafael Molina and Javier Mateos

ISBN: 978-3-031-01115-3   paperback
ISBN: 978-3-031-02243-2   ebook

DOI: 10.1007/978-3-031-02243-2

A Publication in the Springer series

SYNTHESIS LECTURES ON IMAGE, VIDEO, AND MULTIMEDIA PROCESSING #7

Lecture #7
Series Editor: Alan C. Bovik, University of Texas at Austin

Library of Congress Cataloging-in-Publication Data

Series ISSN: 1559-8136   print
Series ISSN: 1559-8144   electronic

First Edition
10 9 8 7 6 5 4 3 2 1

Super Resolution of Images and Video

Aggelos K. Katsaggelos
Department of Electrical Engineering and Computer Science,
Northwestern University, Evanston, Illinois, USA

Rafael Molina
Departamento de Ciencias de la Computación e Inteligencia Artificial,
Universidad de Granada, Granada, Spain

Javier Mateos
Departamento de Ciencias de la Computación e Inteligencia Artificial,
Universidad de Granada, Granada, Spain

SYNTHESIS LECTURES ON IMAGE, VIDEO, AND MULTIMEDIA PROCESSING #7

ABSTRACT

This book focuses on the super resolution of images and video. The authors' use of the term *super resolution* (**SR**) is used to describe the process of obtaining a *high resolution* (**HR**) image, or a sequence of HR images, from a set of *low resolution* (**LR**) observations. This process has also been referred to in the literature as *resolution enhancement* (**RE**). SR has been applied primarily to spatial and temporal RE, but also to hyperspectral image enhancement. This book concentrates on motion based spatial RE, although the authors also describe motion free and hyperspectral image SR problems. Also examined is the very recent research area of SR for compression, which consists of the intentional downsampling, during pre-processing, of a video sequence to be compressed and the application of SR techniques, during post-processing, on the compressed sequence.

It is clear that there is a strong interplay between the tools and techniques developed for SR and a number of other inverse problems encountered in signal processing (e.g., image restoration, motion estimation). SR techniques are being applied to a variety of fields, such as obtaining improved still images from video sequences (video printing), high definition television, high performance color Liquid Crystal Display (LCD) screens, improvement of the quality of color images taken by one CCD, video surveillance, remote sensing, and medical imaging. The authors believe that the SR/RE area has matured enough to develop a body of knowledge that can now start to provide useful and practical solutions to challenging real problems and that SR techniques can be an integral part of an image and video codec and can drive the development of new coder-decoders (codecs) and standards.

KEYWORDS

Image processing, Video processing, Image compression, Image enhancement, Resolution enhancement, Super resolution, High resolution, SR, LR, RE.

Contents

List of Figures

List of Abbreviations

1D	one-dimensional
2D	two-dimensional
3D	three-dimensional
AM	alternating minimization
AR	autoregression
BD	blind deconvolution
CCD	couple charge device
CFA	color filter array
CG	conjugate gradient
CIF	common intermediate format
codec	coder–decoder
CR	Cramer-Rao
EM	expectation–maximization
DCT	discrete cosine transform
DFD	displaced frame difference
DFT	discrete Fourier transform
DVD	digital versatile disc
DWT	discrete wavelet transform
ETM	enhanced thematic mapper
FT	Fourier transform
GGMRF	generalized Gaussian Markov random field
GOI	group of images
GOP	group of pictures
GOV	group of videoobjectplane
GPS	global positioning system
HDTV	high definition television
HR	high resolution

ICM	iterated conditional modes
KL	Kullback-Leibler
LCD	liquid crystal display
LR	low resolution
MAP	maximum *a posteriori*
MC	Markov chain
MCMC	Markov chain Monte Carlo
ML	maximum likelihood
MMSE	minimum mean square error
MSE	mean squared error
MPEG	Motion Pictures Expert Group
MRI	magnetic resonance imaging
MRF	Markov random field
NTSC	National Television System Committee
PCA	principal component analysis
pdf	probability density function
PET	positron emission tomography
POCS	projection onto convex sets
PSF	point spread function
PSNR	peak signal-to-noise ratio
QCIF	quarter common intermediate format
RANSAC	random sample consensus
RE	resolution enhancement
RGB	red, green, and blue
SAR	simultaneous autoregressive
SDTV	standard definition television
SQCIF	sub-quarter common intermediate format
SNR	signal to noise ratio
SPECT	single photon emission computed tomography
SR	super resolution
SVC	scalable video coding

Preface

The topic of super resolution (SR) or resolution enhancement (RE) as described in this book first appeared in the early 1980s, with one of the first papers in the signal processing community, the paper by Tsai and Huang [190]. Since then the topic has been active, with some of the early good results appearing in the 1990s. The last five or so years however have witnessed an enormous resurgence in SR activity. This assessment is supported by the number of the recent papers on the topic but also with the ongoing R&D projects on super resolution in various segments of the industry and various research laboratories. In parallel with the activities in the signal processing community increased activity was observed recently in the area of applying Bayesian modeling and inference to the SR task as well as in pattern recognition and computer vision communities in applying learning techniques to the SR problem. The comprehensive list of publications in the book therefore spans quite a range of journals and consequently makes the SR arena a perfect environment for cross-fertilization between scientific bodies of knowledge. We believe that the SR/RE area has matured enough to develop a body of knowledge that can now start to provide useful and practical solutions to challenging real problems. This indeed provides an explanation for the renewed interest in it.

We have been working on the SR topic for over 15 years and on the closely related topic of image and video restoration and on solving inverse problems for considerably longer than that. We felt that due to the mathematical challenges presented by the topic but also due to the recent interest and activities in it, it was appropriate to undertake this project and write this monograph. It is also clear that there is a strong interplay between the tools and techniques developed for SR and a number of other inverse problems encountered in signal processing (e.g., image restoration, motion estimation). Another motivation for writing this book is our strong believe, as described in the last chapter of the book, that SR techniques can be an integral part of an image and video codec and they can drive the development of new coder–decoders (codecs) and standards.

We were, therefore, glad to undertake this task when we were contacted by the Series Editor Prof. Al Bovik to write a short textbook creating expert treatises or Notes on a topic of special and timely interest. We provide our personal perspective on the topic of super resolution through the filter of Bayesian inference. We present in a systematic way the building blocks of the Bayesian framework, which is also used as a reference in reviewing and comparing SR approaches which have appeared in the literature.

It is our hope and expectation that this monograph will serve as a good reference to the graduate student who would like to work in this area, but also to the practicing engineer, and to the person completely outside of this research area who would like to "just find out what is this excitement all about!" As we already alluded above the material in this monograph can also serve well the scientist in applying some of the tools and results to other related problems.

This has been a very stimulating and enjoyable (most of the time) project for the authors. There were plenty of bits that crossed the Atlantic in the form of emails and CDs and faxes, but also enjoyable meetings in Granada, Chicago, and a recent intense week in Athens, Greece in trying to bring this project to completion (not even enough time to visit the Acropolis!).

There are a number of people we would like to thank for helping us bring this project to completion. First of all, Al for asking us to undertake the project and for feedback along the way and Joel Claypool for all the input but also his patience. We would like to thank Prof. P. Milanfar for providing us with the software used to run the Bilateral Total Variation example in Sections 4.5 and 5.1 and Javier Martín, undergraduate student at the University of Granada, for implementing as a final year project the Baker-Kanade algorithm used in Section 5.1. We would also like to thank the Ph.D students Derin Babacan and Bruno Amizic (Northwestern University) and Dácil Barreto (University de las Palmas de Gran Canaria), and Tom Bishop (University of Edinburgh), for their comments on the book (the last two while visiting the University of Granada). Thank you is also due for the comments and images to the Ph.D students at the University of Granada, L.D. Alvarez and Jaime Cancino. Finally, we would like to thank our families for all their love and support (and also patience) over years: Ereni, Zoe, Sofia, and Sally (AKK), Conchi and Paloma (RM), Marian, David, and Cristina (JM).

An example in this book uses the FERET database of facial images collected under the FERET program, sponsored by the DOD Counterdrug Technology Development Program Office [142, 143].

This work was supported by the "Comisión Nacional de Ciencia y Tecnología" under contract TIC2003-00880, by the Greece-Spain Integrated Action HG2004-0014, and by the "Instituto de Salud Carlos III" project FIS G03/185.

Chicago   Aggelos Katsaggelos
La Herradura   Rafael Molina
Granada   Javier Mateos
August, 2006

CHAPTER 1

Introduction

1.1   WHAT IS SUPER RESOLUTION OF IMAGES AND VIDEO?

We use the term super resolution (SR) to describe the process of obtaining an high resolution (HR) image or a sequence of HR images from a set of low resolution (LR) observations. This process has also been referred to in the literature as resolution enhancement (RE) (henceforth, we will be using the two terms interchangeably; we will also use the terms image(s), image frame(s), and image sequence frame(s) interchangeably). SR has been applied primarily to spatial and temporal RE, but also to hyperspectral image enhancement (a comprehensive classification of spatiotemporal SR problems is provided in [22, 24]). In this book we mainly concentrate on motion-based spatial RE (see Fig. 1.1), although we also describe motion-free [42] and hyperspectral [115, 118] image SR problems. Furthermore, we examine the very recent research area of SR for compression [13, 112], which consists of the intentional downsampling (during pre-processing) of a video sequence to be compressed and the application of SR techniques (during post-processing) on the compressed sequence.

In motion-based spatial RE, the LR-observed images are under-sampled and they are acquired by either multiple sensors imaging a single scene (Fig. 1.1(a)) or by a single sensor imaging a scene over a period of time (Fig. 1.1(b)). For static scenes the observations are related by global subpixel displacements (due, for example, to the relative positions of the cameras, and to camera motion, such as panning or zooming), while for dynamic scenes they are related by local subpixel displacements due to object motion, in addition to possibly global displacements. In both cases the objective again of SR is to utilize either the set of LR images (Fig. 1.1(a)) or a set of video frames (Fig. 1.1(b)) to generate an image of increased spatial resolution.

Increasing the resolution of the imaging sensor is clearly one way to increase the resolution of the acquired images. This solution, however, may not be feasible due to the increased associated cost and the fact that the shot noise increases during acquisition as the pixel size becomes smaller [135]. Furthermore, increasing the chip size to accommodate the larger number of pixels increases the capacitance, which in turn reduces the data transfer rate. Therefore, signal processing techniques, like the ones described in this book, provide a clear alternative for increasing the resolution of the acquired images.

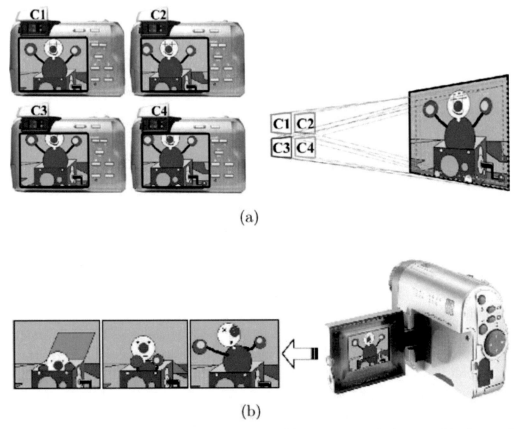

FIGURE 1.1: (a) Several cameras acquire still images of the same scene which are combined to produce an HR image; (b) a video camera records a dynamic scene

There are various possible models in describing the systems depicted in Fig. 1.1. The block diagram of a general such model is shown in Fig. 1.2. According to it, there is a source discrete-space HR image which can represent the source continuous two-dimensional (2D) scene, that is, it resulted from the Nyquist sampling of the continuous 2D scene. This HR image is then warped and blurred before being downsampled. The warping represents the translation or rotation or any sophisticated mapping that is required to generate either the four still images shown in the example of Fig. 1.1(a), or a set of individual frames in the example of Fig. 1.1(b), from the original HR image. Registration is another term used to describe the estimation of the parameters of the warping model which will allow the re-alignment of the individual images or frames. The blur is modeling the response of the camera optics or the sensors or it can be due to the motion of the object or the out-of-focus acquisition of the scene. An interesting modeling question is the order in which these two operations—blurring and

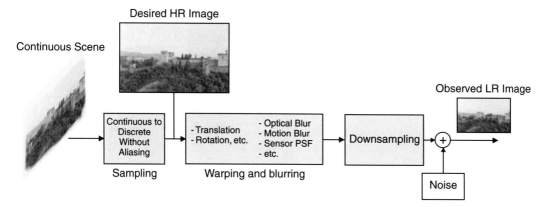

FIGURE 1.2: Block diagram of an SR system

warping—are applied. Both resulting systems, the so-called warp–blur and blur–warp models, have been used and are analyzed in detail in Chapter 3. Similarly, there are various ways to describe the downsampling block in Fig. 1.2, which will generate the aliased LR images of the original HR image. Finally, noise is added to the output of the system introducing the downsampling, resulting in the observed LR image. The noise component can model various elements in the imaging chain, such as, the misregistration error and the thermal or electronic noise, or the errors during storage or transmission. In the book we also consider the case when the available LR images have been compressed. In this case, one more block representing the compression system is added at the end of the block diagram of Fig. 1.2. Another noise component due to quantization needs to be considered in this case, which is typically considered to be the dominant noise component, especially for low bit-rates. As is clear from the block diagram of Fig. 1.2, in a typical SR problem there are a number of degradations that need to be ameliorated, in addition to increasing the resolution of the observations. The term RE might be therefore more appropriate in describing the type of problems we are addressing in this book.

Image restoration, a field very closely related to that of SR of images, started with the scientific work in the Space programs of the United States and the former Soviet Union [11, 116]. It could be said that the field of SR also started in the sky with the launching of Landsat satellites. These satellites imaged the same region on the Earth every 18 days. However, the observed scenes were not exactly the same since there were small displacements among them [190]. Since then, a wealth of research has considered modeling the acquisition of the LR images and provided solutions to the HR problem. Recent literature reviews are provided in [22, 26, 41, 85, 123]. More recent work, e.g. [108, 133, 134, 166], also addresses the HR problem when the available LR images are compressed using any of the numerous

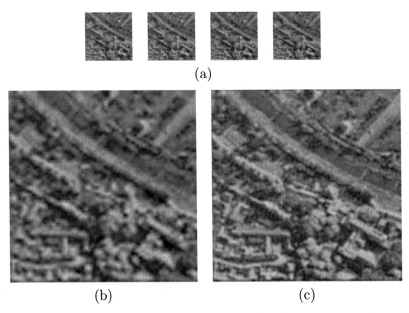

FIGURE 1.3: (a) Observed LR images with global subpixel displacement among them; (b) upsampled image of the top left LR image using the *Corel® Paint Shop Pro® X* smart size filter; (c) HR image obtained by combining the information in the LR observations using the algorithm in [193]

image and video compression standards or a proprietary technique. In the book we address both cases of uncompressed and compressed LR data.

To illustrate the effectiveness of SR methods consider, for example, Fig. 1.3. In Fig. 1.3(a), four LR images of the same scene are shown. These observed images are under-sampled and they are related by global subpixel shifts which are assumed to be known. In Fig. 1.3(b), the HR image obtained by the smart size filter in *Corel® Paint Shop Pro® X* on one of the observed images is shown. In Fig. 1.3(c), the super-resolved image obtained with the use of the algorithm in [193] (to be described later) that combines the four LR images is shown. As can be seen, a considerable improvement can be obtained by using the information contained in all four images.

1.2 WHY AND WHEN IS SUPER RESOLUTION POSSIBLE?

A fundamental question is what makes SR possible. We address this question graphically with the aid of Fig. 1.4. Figure 1.4(a) shows a representation of a couple charge device (CCD) of a certain physical size consisting of 8×8 cells or pixels, and Fig. 1.4(b) shows a CCD of the same physical size but with 16×16 pixels. The second CCD will obviously produce higher spatial resolution images. We now consider the case of having two 8×8 pixel CCDs

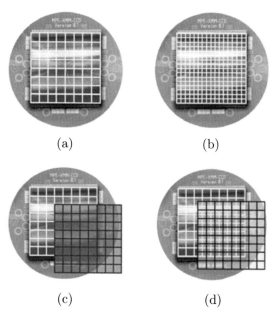

FIGURE 1.4: (a) A CCD with 8×8 pixels; (b) a CCD of the same size with 16×16 pixels; (c) pixel shift of the second CCD does not provide additional information about the scene; (d) subpixel shift of the second CCD provides additional information about the scene

imaging the same scene. Had the two CCDs been arranged as shown in Fig. 1.4(c), i.e., shifted by an integer number of pixels, we could not have improved the spatial resolution of the acquired 8×8 images. However, an LR subpixel shift, as the one shown in Fig 1.4(d), provides additional information about the scene, thus allowing the potential resolution enhancement.

Intuitively, each LR-observed image represents a subsampled (i.e., aliased) version of the original scene. The aliasing could not be removed if we were to process one image only. Due to the subpixel shifts, however, each observed image contains complementary information. With exact knowledge of the shifts, the observed images can be combined to remove the aliasing and generate a higher resolution image. If we assume that the resolution of this HR image is such that the Nyquist sampling criterion is satisfied, this HR image then represents an accurate representation of the original (continuous) scene. Aliased information can therefore be combined to obtain an alias-free image. Figure 1.5 shows the relationship among the subsampled LR images and the HR image in this degradation-free example. Assuming that the image in Fig. 1.5(e) is available, that is, assuming that the subpixel shifts are known, the SR problem becomes a resampling problem, that of converting an arbitrarily sampled image (Fig. 1.5(e)) to a uniformly sampled one (Fig. 1.5(f)) .

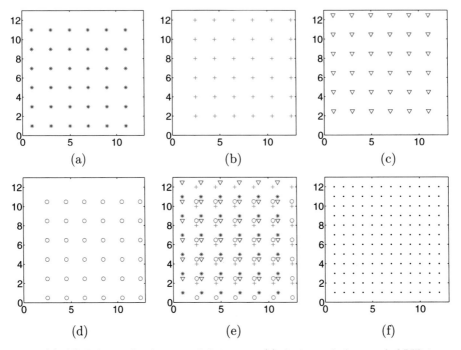

FIGURE 1.5: (a)–(d) Relationship between LR images; (e) the irregularly sampled HR image; and (f) the regularly sampled HR image

Let us now provide a more specific explanation of why SR is possible by following [190] (see also [183]). For simplicity, we are using a one-dimensional (1D) example, with neither blur nor noise included in the observation. Let us consider a 1D continuous signal $f(x)$ and p versions of it, each shifted by an amount δ_k, $k = 0, \ldots, p - 1$, that is,

$$f_k(x) = f(x + \delta_k), \qquad (1.1)$$

with $\delta_0 = 0$.

Consider now the sampled version of f_k, that is,

$$f_{kn} = f(nT + \delta_k), \qquad (1.2)$$

where $n = 0, \ldots, N - 1$, and T represents the sampling period. That is, in SR terminology we have p LR images each of size N.

Taking the continuous Fourier transform (FT) of both sides of Eq. (1.1) and using its shifting property, we have that

$$F_k(w) = e^{j2\pi \delta_k w} F(w), \qquad (1.3)$$

where $F_k(w)$ and $F(w)$ are the FTs of $f_k(x)$ and $f(x)$, respectively.

Let us denote by F_{kl} the N-point discrete fourier transform (DFT) of the kth-sampled image f_{kn}, that is,

$$F_{kl} = \sum_{n=0}^{N-1} f_{kn} \quad \exp\left[-j2\pi\frac{ln}{N}\right], \quad l = 0, \ldots, N-1. \tag{1.4}$$

Then, the FT of $f_k(x)$ and the DFT of its sampled version are related by the sampling theorem [153], i.e.,

$$F_{kl} = \frac{1}{T} \sum_{m=-\infty}^{m=\infty} F_k\left(\frac{l}{NT} + m\frac{1}{T}\right)$$

$$= \frac{1}{T} \sum_{m=-\infty}^{m=\infty} e^{j2\pi\delta_k(\frac{l}{NT}+m\frac{1}{T})} F\left(\frac{1}{T}\frac{l+mN}{N}\right) \text{ (using Eq. (1.3))} \tag{1.5}$$

for $l = 0, \ldots, N-1$.

Let us now examine how many different sampled signals with the sampling period of T we need to recover the original band limited signal f. Since f is bandlimited, given the sampling period T we can always find L such that

$$F(w) = 0 \text{ for } |w| \geq L/T. \tag{1.6}$$

Note that in order to avoid aliasing we should have sampled with $T' = T/2L$ and so for $L = 1$ we should have sampled at $T/2$ or for $L = 2$ the sampling period should have been $T/4$ instead of T.

Using Eq. (1.5), each F_{kn}, $k = 0, \ldots, p - 1$, $n = 0, \ldots, N - 1$, can be written as a combination of $2L$ samples of $F(w)$. Note that the same samples of $F(w)$ are used to define the combination producing F_{kn} with n fixed and $k = 0, \ldots, p - 1$. From the above discussion it is clear that we have a set of $p \times N$ linear equations with $2L \times N$ unknown. Thus, the linear system can be solved if $p \geq 2L$. The $2L \times N$ calculated samples of $F(w)$ can now be used to estimate $f(x)$ from $x = 0, \ldots, (N - 1)T$, with spacing $T/2L$ using $k = 0$ and replacing N by $2LN$ and T by $T/2L$ in Eq. (1.5). The resolution is thus increased by a factor of $2L$.

The above example and its theoretical justification apply to the simplest scenario and an ideal situation. There are several issues that need to be addressed in a practical situation. For example, the displacements among images may not be exactly known and therefore need to be estimated from the available data. In addition, they may be more complicated than a subpixel translation, and the motion may have both a global and a local component. Furthermore, the blurring and downsampling processes have to be taken into account together with the noise involved in the observation process. In the case of compressed data, information provided by the bitstream, such as motion vectors and quantization step sizes, should also be utilized in performing SR.

We finally mention here that an important building block of SR methods is the conversion of arbitrarily sampled data to evenly spaced data, that is, a system which performs interpolation. Critical theoretical aspects therefore of SR methods are related to results on interpolation, such as, the work by Papoulis [131].

1.3   APPLICATIONS

SR techniques are being applied to a variety of fields, such as obtaining improved still images from video sequences (video printing), high definition television, high performance color liquid crystal display (LCD) screens, improvement of the quality of color images taken by one CCD, video surveillance, remote sensing, and medical imaging.

Negroponte [121] describes early work at Media Lab on the problem of obtaining a "salient still" from video sequences. This SR problem arises when we need to create an enhanced resolution still image from a video sequence, as when printing stills from video sources. The human visual system requires a higher resolution still image than that of a frame in a sequence of frames, in order to experience the same perceptual quality. National Television System Committee (NTSC) video yields at most 480 vertical lines, while more than twice the number of lines are required to print with reasonable resolution at standard size on modern printers. An extension of this problem is encountered when a standard definition television (SDTV) signal is to be displayed on an HR display (i.e., a high definition television (HDTV) display). The problem is exacerbated when the video sequence has been compressed [5].

The advances in cell phones, digital still and video cameras, media players, portable digital versatile disc (DVD) players and portable global positioning system (GPS) devices are fueling the demand for higher performance color LCDs . Some of these LCDs are now being based on treatment of the red, green, and blue (RGB) information at the subpixel level [145]. Most of the available nonprofessional still and video cameras use one CCD to capture all three RGB bands and henceforth they spatially subsample the three observed bands. This subsampling leads to unpleasant effects, which SR techniques can be used to remove (this represents the so-called demosaicking problem)[194]. As an example, Fig. 1.6(a) shows an original color image, while Fig. 1.6(b) the captured raw image by a camera equipped with a single CCD and a color filter array (CFA). The available data shown in Fig. 1.6(b) are processed using bilinear interpolation and the SR-based algorithm in [70] to generate, respectively, the images in Figs. 1.6(c) and 1.6(d).

As described by Chaudhuri and Taur [43], several video surveillance products are available on the market for office or home security and remote surveillance. A home, an office, or a location of interest, is monitored by capturing motion events using webcams or camcorders and

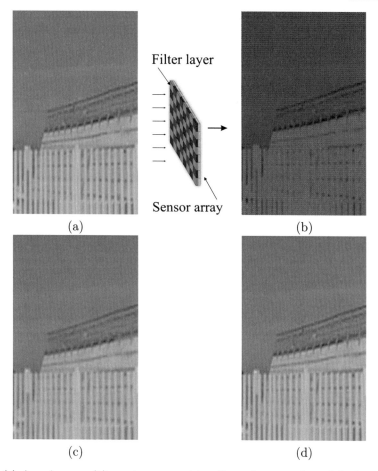

FIGURE 1.6: (a) A real scene; (b) raw image resulting from the capturing of the image in (a) using a single CCD color camera and the filter layer shown. Results of processing the image in (b) by (c) bilinear interpolation; and (d) the algorithm in [70]

detecting abnormalities. The visual data is saved into compressed or uncompressed video clips, and the system triggers various alerts such as initiating an ftp session, sending an e-mail (in case of a webcam), or setting off a buzzer. Some of these products also have video streaming capabilities, which enable one to watch the webcam over the Web from any location. Due to bandwidth constraints, the frames are often captured at a lower rate. Hence, there is need to both temporally and spatially super resolve the data.

With the current emphasis on security and surveillance, intelligent personal identification and authentication (biometrics) is also an important area of application of SR techniques. The iris has been widely studied for personal identification and authentication applications because of its extraordinary structure and SR techniques have been applied to this problem [47].

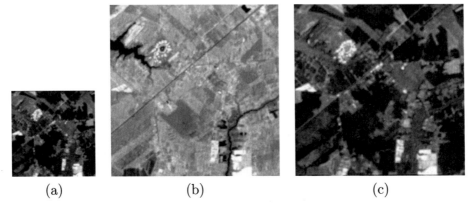

(a)   (b)   (c)

FIGURE 1.7: (a) Observed LR Landsat ETM+ image (only bands 1 to 3 are shown); (b) panchromatic HR image; (c) enhanced version of (a) using (b) and the SR technique in [115]

Improving the quality of images of faces is also an area of application of SR techniques (see [9], and also [73] following the line of work by Capel and Zisserman on super-resolving text [32]).

SR techniques are also being applied to multispectral images when several multispectral cubes of the same scene have been observed and such cubes are shifted at the subpixel level [1]. The first work on SR was aimed at improving the resolution of Landsat images [190]. When dealing with multispectral images, the term SR (or pansharpening) is also used to define the problem of combining HR images which are observed with panchromatic sensors with LR images observed at a given wavelength (see [115] for a short review with applications to Landsat enhanced thematic mapper (ETM) + imagery). Figure 1.7 shows an example of a super-resolved Landsat ETM+ image. In Fig. 1.7(a), the image of bands 1–3 of an LR Landsat ETM + image is shown. In Fig. 1.7(b), the HR panchromatic image is shown and in Fig. 1.7(c), the HR image generated by the algorithm in [115] utilizing the images in Figs. 1.7(a) and 1.7(b). Related to remote sensing and use of multispectral images, SR techniques have also been applied to astronomical images [44, 64, 107].

Medical image processing represents another area where SR has been recently applied, in particular, in magnetic resonance imaging (MRI). In this case, the thickness of the 2D slices is attempted to be reduced [141] and 2D slides have been used to obtain three-dimensional (3D) images [69]. Unfortunately not much work has been reported on the application of SR techniques to positron emission tomography (PET) or single photon emission computed tomography (SPECT), although we believe that this is a potential area of application of SR techniques.

In the discussion so far SR is utilized as a means to ameliorate the undesirable reduction in resolution introduced by the imaging system. However, the intentional downsampling during pre-processing of a video sequence to be compressed and the application of SR techniques

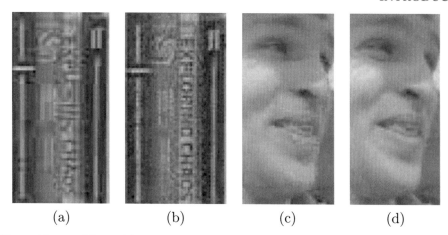

(a)	(b)	(c)	(d)

FIGURE 1.8: Figures (a) and (c) show details of regions of the "Deadline" and "Foreman" sequences, respectively, compressed at 512 kbps using MPEG-4; (b) and (d) the same regions obtained by spatially subsampling the original video sequences before compression and upsampling the compressed video sequence, also at 512 kbps, at the decoder (see [13])

during post-processing of the compressed sequence can be utilized as a mechanism to optimize the overall quality of the HR reconstructed video in the rate-distortion sense. Details of two images in two video sequences compressed at 512 kbps using Motion Pictures Expert Group (MPEG)-4 are shown in Figs. 1.8(a) and 1.8(c). The same regions obtained by spatially subsampling the original video sequences before compression and upsampling the compressed video sequence, also at 512 kbps, at the decoder are shown in Figs. 1.8(b) and 1.8(d) (see [13]). Considerable detail is gained in Figs. 1.8(b) and 1.8(d) while some of the compression artifacts are being removed.

There are a number of additional areas of application of SR techniques; of particular interest is the motion-free SR problem, where cues are used instead of motion to super resolve a scene [42] (for instance, observing the same scene with different blurs [150]). We will only comment on motion-free SR techniques when referring to their relation to motion-based SR techniques.

1.4   BOOK OUTLINE

Once we have described the SR problem, examined why it is possible to improve the resolution of an image or sequence of images and also described some of the areas of application of SR techniques, the goals in this book can be summarized as follows:

1. to review the SR techniques used in the literature from the point of view of Bayesian modeling and inference;

2.   to present applications of SR techniques;

3.   to introduce the use of SR techniques as tools to compress video sequences for storage and transmission; and

4.   to comment on open problems in the field of SR.

The rest of the book is organized as follows. First, in Chapter 2, we present in its basic form the Bayesian formulation of the problem of reconstructing HR images and motion vectors from LR observations. In Chapter 3, we model the process for obtaining the LR observations from the HR image we want to estimate. Based on this, in Chapter 4, we examine the prior distribution modeling of the motion vectors and study how the LR images are registered or motion compensated. Once the motion vectors are estimated or their priors described, we analyze, in Chapter 5, methods for modeling the HR image prior as well as for estimating the HR images. Chapter 6 describes SR Bayesian inference procedures that go beyond the sequential or alternate estimation of image and motion vectors, described in the previous chapters. In Chapter 7, we introduce the use of SR techniques as tools to compress video sequences for storage and transmission. Finally, we summarize the book and comment on SR problems that still remain open.

CHAPTER 2

Bayesian Formulation of Super-Resolution Image Reconstruction

Our aim in this monograph is to follow a Bayesian formulation in providing solutions to the SR problem. As will be described in the material that follows, a large number of techniques which have appeared in the literature can be derived from such a Bayesian formulation.

The formulation of a problem is usually coupled with the approach which will be followed in solving it. Consequently, in the following chapters we will provide the quantities required by the Bayesian paradigm for the SR problem, that is, the conditional and prior probabilities, as well as, the inference method. In this chapter we present such quantities, and we introduce the required notation.

2.1 NOTATION

Figure 2.1 provides a graphical description of the video spatial SR problem for the case of uncompressed observations. As shown in it, we denote by the vector $\mathbf{f}_k$ the (lexicographically) ordered intensities of the kth HR frame. A set of frames can be similarly ordered to form a new vector $\mathbf{f}$. We denote by the vector $\mathbf{g}_k$ the ordered intensities of the kth LR frame. Similarly, the set of LR frames corresponding to the HR frames in $\mathbf{f}$ can be ordered to form the vector $\mathbf{g}$. The block diagram in Fig. 1.2 describes the steps in obtaining $\mathbf{g}_k$ (or $\mathbf{g}$) from $\mathbf{f}$. We assume at this point that all the parameters required by the model in generating the LR observations from the HR images are known, except for the parameters of the warping model. We denote by the vector $\mathbf{d}_{l,k}$ the warping model parameters in mapping frame $\mathbf{f}_k$ to $\mathbf{f}_l$. In the context of dynamic video sequences, the difference between two frames is typically due to the motion of the camera and the objects in the scene. The vector $\mathbf{d}_{l,k}$ therefore contains the motion parameters for compensating frame $\mathbf{f}_l$ from frame $\mathbf{f}_k$ (this means that each pixel value in frame $\mathbf{f}_l$ can be predicted through the motion $\mathbf{d}_{l,k}$ from a pixel in frame $\mathbf{f}_k$). The motion vectors have subpixel accuracy and therefore the pixel values in $\mathbf{f}_k$ need to be interpolated in carrying out

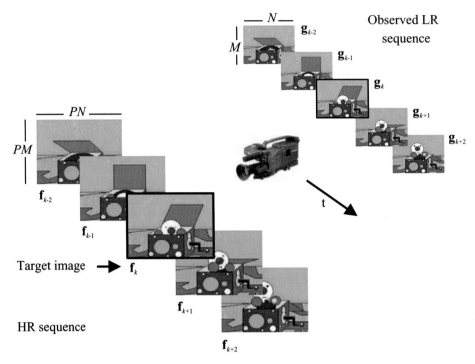

FIGURE 2.1: LR video acquisition model for the uncompressed case. The process to obtain the LR images from the HR ones has to account for sensor and optical distortions, motion, nonzero aperture time, downsampling, and noise. The HR images are denoted by $\mathbf{f}$ and the LR observations by $\mathbf{g}$. The target HR image is denoted by $\mathbf{f}_k$

the motion compensation. A set of vectors $\mathbf{d}_{l,k}$ can be ordered to form the vector $\mathbf{d}$. This same model and notation can be applied to the SR problem of still images (Fig. 1.1(a)). In this case, there is only one HR image $\mathbf{f}$ and a number of LR observations $\mathbf{g}_k$, which are related by LR warp parameters or shifts $\mathbf{d}$. The problem at hand in both cases is to estimate $\mathbf{f}$ and $\mathbf{d}$ (so that it is used in estimating $\mathbf{f}$) based on the observations $\mathbf{g}$, which will be generically denoted by $\mathbf{o}$.

The situation is slightly different when dealing with compressed observations, as depicted in Fig. 2.2. In this case, the observations $\mathbf{g}_k$ of Fig. 2.1 are compressed generating the compressed observations $\mathbf{y}_k$. Motion vectors are generated during compression by the encoder relating pairs of LR frames. Such LR motion vectors are denoted by $\mathbf{v}_{l,k}$ (or $\mathbf{v}$ if a number of them are stacked), with the same interpretation of such vectors as with the $\mathbf{d}_{l,k}$ vectors mentioned earlier. The encoder, in addition to $\mathbf{v}$, provides to the bitstream other information (such as the quantizer step sizes), which can be of use during the SR process. In this case, the observation $\mathbf{o}$ consists of $\mathbf{y}$ and $\mathbf{v}$ and any additional information useful for the SR problem that can be found in the bitstream. The problem at hand, again, is to estimate $\mathbf{f}$ and $\mathbf{d}$ based on the observation $\mathbf{o}$. This

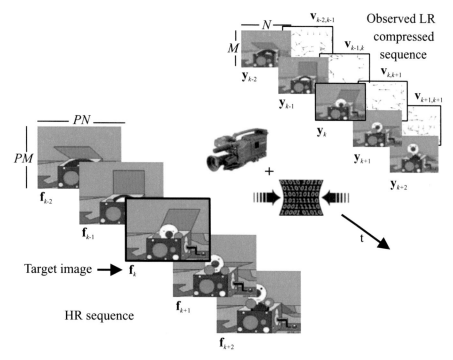

FIGURE 2.2: Graphical depiction of the acquisition model for the compressed case. The process for obtaining the LR images from high-resolution ones has to account for sensor and optical distortions, motion, nonzero aperture time, downsampling, compression, and noise

scenario depicted in Fig. 2.2 is also applicable to the case of still images with the appropriate interpretation of the various quantities.

2.2   BAYESIAN MODELING

We now proceed to cast the SR problem within the Bayesian framework. A fundamental principle of the Bayesian philosophy is to regard all parameters and observable variables as unknown stochastic quantities, assigning probability distributions based on subjective beliefs. Thus, in SR the original HR image f_k and motion vectors $\mathbf{d}$ are all treated as samples of random fields, with corresponding *prior* distributions that model our knowledge about the nature of the original HR image and motion vectors. The observation $\mathbf{o}$, which is a function of f_k and $\mathbf{d}$, is also treated as a sample of a random field, with corresponding *conditional* distribution that models the process to obtain $\mathbf{o}$ from f_k and $\mathbf{d}$. These distributions depend on parameters which will be denoted by Ω.

Assuming that Ω is known (its estimation will be studied in Chapter 6), the joint distribution modeling the relationship between the observed data and the unknown quantities of the SR problem becomes

$$\mathbf{P}(\mathbf{f}_k, \mathbf{d}, \mathbf{o}) = \mathbf{P}(\mathbf{f}_k, \mathbf{d})\mathbf{P}(\mathbf{o}|\mathbf{f}_k, \mathbf{d}), \qquad (2.1)$$

where $\mathbf{P}(\mathbf{o}|\mathbf{f}_k, \mathbf{d})$ is termed the *likelihood* of the observations.

2.3   BAYESIAN INFERENCE

Once the modeling has been completed, Bayesian inference is performed using the *posterior*

$$\mathbf{P}(\mathbf{f}_k, \mathbf{d}|\mathbf{o}) = \frac{\mathbf{P}(\mathbf{f}_k, \mathbf{d})\mathbf{P}(\mathbf{o}|\mathbf{f}_k, \mathbf{d})}{\mathbf{P}(\mathbf{o})}. \qquad (2.2)$$

There are a number of ways to estimate the HR image and motion vectors using Eq. (2.2). In its simplest form, Bayesian inference provides the *maximum* a posteriori (MAP) solution, represented by the values $\hat{\mathbf{f}}_k$ and $\hat{\mathbf{d}}$, that maximize the posterior probability distribution, that is,

$$\begin{aligned}
\hat{\mathbf{f}}_k, \hat{\mathbf{d}} &= \arg\max_{\mathbf{f}_k, \mathbf{d}} \mathbf{P}(\mathbf{f}_k, \mathbf{d}|\mathbf{o}) \\
&= \arg\max_{\mathbf{f}_k, \mathbf{d}} \left\{ \frac{\mathbf{P}(\mathbf{f}_k, \mathbf{d})\mathbf{P}(\mathbf{o}|\mathbf{f}_k, \mathbf{d})}{\mathbf{P}(\mathbf{o})} \right\} \\
&= \arg\max_{\mathbf{f}_k, \mathbf{d}} \left\{ \mathbf{P}(\mathbf{f}_k, \mathbf{d})\mathbf{P}(\mathbf{o}|\mathbf{f}_k, \mathbf{d}) \right\} \\
&= \arg\min_{\mathbf{f}_k, \mathbf{d}} \left\{ -\log \mathbf{P}(\mathbf{f}_k, \mathbf{d}) - \log \mathbf{P}(\mathbf{o}|\mathbf{f}_k, \mathbf{d}) \right\}. \qquad (2.3)
\end{aligned}$$

Depending on the prior and conditional probability density functions (pdfs), it may be difficult to find analytical solutions to Eq. (2.3), so numerical solutions are needed. Even with numerical solutions, a major problem in the optimization is the simultaneous estimation of the variables $\mathbf{f}_k$ and $\mathbf{d}$. The SR image and motion vector reconstruction literature has mainly concentrated so far on the following two alternating minimization (AM) methodologies which provide approximations to the solution of Eq. (2.3):

1. The first methodology (henceforth referred to as *alternate*) uses a cyclic coordinate descent procedure [102]. According to it, let $\hat{\mathbf{f}}_k^q$ be the current estimate of the HR image, where q is the iteration index. An estimate for the displacements is first found by solving

$$\hat{\mathbf{d}}^q = \arg\max_{\mathbf{d}} \mathbf{P}(\hat{\mathbf{f}}_k^q, \mathbf{d})\mathbf{P}(\mathbf{o}|\hat{\mathbf{f}}_k^q, \mathbf{d}). \qquad (2.4)$$

The HR image $\mathbf{f}_k$ is then estimated by assuming that the displacement estimates are exact, that is,

$$\hat{\mathbf{f}}_k^{q+1} = \arg \max_{\mathbf{f}_k} \mathbf{P}(\mathbf{f}_k, \hat{\mathbf{d}}^q)\mathbf{P}(\mathbf{o}|\mathbf{f}_k, \hat{\mathbf{d}}^q). \tag{2.5}$$

The iteration index q is then set to $q + 1$ and the solutions in Eqs. (2.4) and (2.5) are found again. The process iterates until convergence.

2. The second methodology (henceforth referred to as *sequential*) assumes that $\mathbf{d}$ is either known or estimated separately (not necessarily using Eq. (2.4)). With $\bar{\mathbf{d}}$ the known or previously estimated motion vector, Eq. (2.4) becomes $\hat{\mathbf{d}}^q = \bar{\mathbf{d}}, \forall q$, and the HR image is found with the use of

$$\hat{\mathbf{f}}_k = \arg \max_{\mathbf{f}_k} \mathbf{P}(\mathbf{f}_k, \bar{\mathbf{d}})\mathbf{P}(\mathbf{o}|\mathbf{f}_k\bar{\mathbf{d}}). \tag{2.6}$$

Using the fact that $\mathbf{P}(\mathbf{f}_k, \mathbf{d}) = \mathbf{P}(\mathbf{f}_k)\mathbf{P}(\mathbf{d}|\mathbf{f}_k)$, we proceed as follows in the subsequent chapters. We first study the distributions $\mathbf{P}(\mathbf{o}|\mathbf{f}_k, \mathbf{d})$ and $\mathbf{P}(\mathbf{d}|\mathbf{f}_k)$ in Chapters 3 and 4, respectively. With knowledge of these distributions, given an estimate of the HR original image we are able to estimate the original HR motion vectors for one iteration of the *alternate* SR methodology and also solve the motion estimation part of the *sequential* SR methodology. To complete our study of the vast majority of the SR methods presented in the literature, we analyze in Chapter 5 various forms of the distribution $\mathbf{P}(\mathbf{f}_k)$. With knowledge of this distribution we are able to estimate the original HR image at each iteration of the *alternate* procedure and also solve the image estimation part of the *sequential* SR methodology. We mention here that since the optimization steps in both the *alternate* and *sequential* methodologies are highly dependent on the pdfs used, they will not be discussed in the book. Note also that for some particular choices of the prior and conditional pdfs in Eq. (2.3) the associated optimization problem is typically seen as a non-Bayesian method and it is often described as a regularization-based SR solution.

2.4   HIERARCHICAL BAYESIAN MODELING AND INFERENCE

As we have already indicated, Bayesian inference provides us with a stronger and more flexible framework than the one needed to simply obtain estimates approximating the solution of Eq. (2.3). Together with the *simultaneous* estimation of the HR images and motion vectors, the Bayesian framework can also handle the estimation of the unknown parameter vector Ω involved in the prior and conditional pdfs. When Ω is unknown, we may adopt the *hierarchical* Bayesian framework in which case we also model our prior knowledge of their values. The probability distributions of the hyperparameters are termed *hyperprior* distributions. This hierarchical modeling allows us to write the joint global probability distribution as

$$\mathbf{P}(\Omega, \mathbf{f}_k, \mathbf{d}, \mathbf{o}) = \mathbf{P}(\Omega)\mathbf{P}(\mathbf{f}_k, \mathbf{d}|\Omega)\mathbf{P}(\mathbf{o}|\Omega, \mathbf{f}_k, \mathbf{d}). \tag{2.7}$$

Observe that assuming that the values of the hyperparameters are known is equivalent to using degenerate distributions (delta functions) for hyperpriors. A degenerate distribution on Ω is defined as:

$$\mathbf{P}(\Omega) = \delta\left(\Omega, \Omega_0\right) = \begin{cases} 1, & \text{if } \Omega = \Omega_0 \\ 0, & \text{otherwise} \end{cases}. \tag{2.8}$$

Under the hierarchical Bayesian framework, the SR task becomes one of performing inference using the posterior probability distribution

$$\mathbf{P}(\Omega, \mathbf{f}_k, \mathbf{d}|\mathbf{o}) = \frac{\mathbf{P}(\Omega)\mathbf{P}(\mathbf{f}_k, \mathbf{d}|\Omega)\mathbf{P}(\mathbf{o}|\Omega, \mathbf{f}_k, \mathbf{d})}{\mathbf{P}(\mathbf{o})}, \tag{2.9}$$

which in its simplest form can be carried out by calculating the maximum *a posteriori* (MAP) estimate of $\mathbf{P}(\Omega, \mathbf{f}_k, \mathbf{d}|\mathbf{o})$. However, the hierarchical Bayesian framework allows us to marginalize, for instance, $\mathbf{f}_k$, $\mathbf{d}$, or Ω to obtain, respectively, $\mathbf{P}(\Omega, \mathbf{d}|\mathbf{o})$, $\mathbf{P}(\Omega, \mathbf{f}_k|\mathbf{o})$ or $\mathbf{P}(\mathbf{f}_k, \mathbf{d}|\mathbf{o})$. Note that the variables we integrate upon are called *hidden variables* and that other marginalizations are also possible. This framework also allows us to calculate, approximate, or simulate the posterior distribution $\mathbf{P}(\Omega, \mathbf{f}_k, \mathbf{d}|\mathbf{o})$. The hierarchical Bayesian modeling and inference for SR will be studied in detail in Chapter 6, once the elements of the alternate and sequential methodologies have been explored in the next three chapters.

We finally note that in presenting the material in the following chapters we follow a kind of historical order that reflects the way the various results have appeared in the literature over time (one may argue that these results are presented from the less to the more complicated ones).

CHAPTER 3

Low-Resolution Image
Formation Models

As already explained in the previous chapter, our aim in this monograph is to follow a Bayesian formulation in providing solutions to the SR problem. In this chapter we describe the models used in the literature for obtaining the observed LR images from the HR-targeted source image. That is, we look into the details of the building blocks of the system in Fig. 1.2. The analysis will result in the determination of $\mathbf{P}(\mathbf{o}|\mathbf{f}_k, \mathbf{d})$ required by the Bayesian formulation of the SR problem (see Eq. (2.3)). We include both cases of recovering an HR static image and an HR image frame from a sequence of images capturing a dynamic scene. As we have already mentioned in the previous chapters, the LR sequence may be compressed. We therefore first describe the case of no compression and then extend the formation models to include compression.

Also in this chapter, we review the most relevant results on theoretically establish the degree of achievable resolution improvement, in other words, how much one can magnify an image as a function of the number of available LR images. This represents an important topic that has received considerable attention recently.

The chapter is organized as follows. In Section 3.1, we describe the image formation models for uncompressed LR observations, we consider two different models which depend on the order on which the warping and the blurring are applied in generating the LR observations (see Fig. 1.2). In Section 3.2, we extend those models to include compression. In Section 3.3, we describe some of the theoretical results on how much resolution enhancement can be theoretically achieved by the use of SR techniques.

3.1 IMAGE FORMATION MODELS FOR UNCOMPRESSED OBSERVATIONS

Let us denote by $f(x, y, t)$ the continuous in time and space dynamic scene which is being imaged. If the scene is sampled according to the Nyquist criterion in time and space, it is represented by the HR sequence $f_l(m, n)$, where $l = 1, \ldots, L$, $m = 0, \ldots, PM - 1$, and $n = 0, \ldots, PN - 1$, represent, respectively, the discrete temporal and spatial coordinates. For reasons that will become clear right away, the parameter P is referred to as the *magnification factor*. Note that although different magnification factors P_r and P_c can be used for rows and columns, respectively, for simplicity and without lack of generality, we use the same factor P for both directions. It is, however, important to observe that depending on the available images we may not be able to improve the spatial image resolution in both directions at the same degree.

Before we proceed, a matrix–vector representation of images and image sequences is introduced to be used in addition to the point-wise representation. Using matrix–vector notation, each $PM \times PN$ image can be transformed into a $(PM \times PN) \times 1$ column vector, obtained by ordering the image by rows (lexicographic ordering). The $(PM \times PN) \times 1$ vector that represents the lth image in the HR sequence is denoted by $\mathbf{f}_l$, with $l = 1, \dots, L$. If, in addition, all frames $\mathbf{f}_l$, $l = 1, \dots, L$, are lexicographically ordered, the vector $\mathbf{f}$ of dimensions $(L \times PM \times PN) \times 1$ is obtained.

The HR sequence $\mathbf{f}$ is input to the imaging system which generates the LR observations denoted by $\mathbf{g}$ (see Fig. 2.1). The objective of SR, as addressed in this book, is to obtain an estimate of *one* HR frame, $\mathbf{f}_k$, from the available LR observations. All of the described techniques, however, may be applied to the SR of video by using, for example, a sliding window approach, as illustrated in Fig. 3.1. Alternatively, temporally recursive techniques can be developed in estimating an SR sequence of images.

To obtain an estimate of $\mathbf{f}_k$, the imaging system and the temporal relationship of the HR and LR sequences need to be modeled and cast within the Bayesian framework, which is the subject of the subsequent sections.

A critical component in the system modeling the generation of the LR observations from HR source data is the warping system (see Fig. 1.2). We will discuss, in Chapter 4, the warp models used in the SR literature as well as techniques for the estimation of their parameters. For the majority of the published work, the sought after HR images $\mathbf{f}_1, \dots, \mathbf{f}_L$, are assumed to

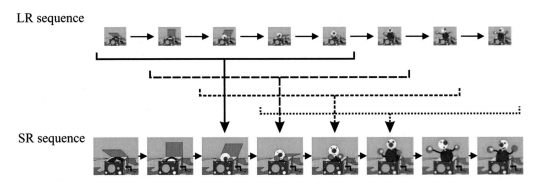

FIGURE 3.1: Obtaining a sequence of HR images from a set of LR observations (the sliding window approach)

satisfy

$$f_l(m, n) = f_k(m + d_{l,k}^x(m, n), n + d_{l,k}^y(m, n)) , \qquad (3.1)$$

where $d_{l,k}^x(m, n)$ and $d_{l,k}^y(m, n)$ denote respectively the horizontal and vertical components of the displacement, that is,

$$\mathbf{d}_{l,k}(m, n) = (d_{l,k}^x(m, n), d_{l,k}^y(m, n)). \qquad (3.2)$$

The model of Eq. (3.1) is a reasonable one under the assumption of constant illumination conditions in the scene. It leads to the estimation of the optical flow in the scene, not necessarily, to the estimation of the true motion. Notice that the above model applies to both local and global motion (the estimation of global motion is typically referred to as a registration problem). Notice also that there may exist pixels in one image for which no motion vector exists (occlusion problem), and pixels for which the displacement vectors are not unique. Finally, notice that we are not including noise in the above model, since we will incorporate it later when describing the process to obtain the LR observations.

Equation (3.1) can be rewritten using matrix–vector notation as

$$\mathbf{f}_l = \mathbf{C}(\mathbf{d}_{l,k})\mathbf{f}_k , \qquad (3.3)$$

where $\mathbf{C}(\mathbf{d}_{l,k})$ is the $(PM \times PN) \times (PM \times PN)$ matrix that maps frame $\mathbf{f}_l$ to frame $\mathbf{f}_k$, and $\mathbf{d}_{l,k}$ is the $(PM \times PN) \times 2$ matrix defined by lexicographically ordering the vertical and horizontal components of the displacements between the two frames. We will be using the scalar and matrix–vector notation interchangeably through this manuscript.

The motion estimation problem, as encountered in many video processing applications, consists of the estimation of $\mathbf{d}_{l,k}$ or $\mathbf{C}(\mathbf{d}_{l,k})$ given $\mathbf{f}_l$ and $\mathbf{f}_k$. What makes the problem even more challenging in SR is the fact that although the HR motion vector field is required, the HR images are not available, and therefore this field must be estimated utilizing the LR images (which might also be compressed). The accuracy of the $\mathbf{d}_{l,k}$ is of the outmost importance in determining the quality of the sought after HR images. To further illustrate the quantities and challenges involved in the motion estimation problem, let us consider a 1D example.

Figure 3.2(a) depicts the image $\mathbf{f}_l$ containing two partially overlapping objects, while Fig. 3.2(b) depicts the image $\mathbf{f}_k$ in which the one object has completely occluded the second one. The mathematical expressions describing the images are

$$f_l(x) = \begin{cases} 0 & 0 \leq x \leq 3 \quad \text{and} \quad 9 \leq x \leq 10 \\ 0.5 & x = 8 \\ 1 & 4 \leq x \leq 7 \end{cases} \qquad (3.4)$$

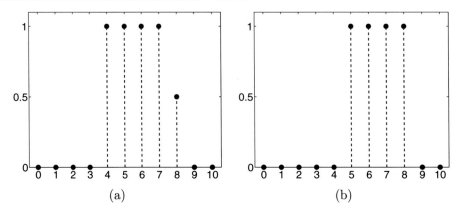

FIGURE 3.2: (a) Image $\mathbf{f}_l$ containing two partially overlapping objects; (b) image $\mathbf{f}_k$ containing two completely overlapping objects (the object with higher intensity has occluded the other object)

and

$$f_k(x) = \begin{cases} 0 & 0 \le x \le 4 \quad \text{and} \quad 9 \le x \le 10 \\ 1 & 5 \le x \le 8. \end{cases} \tag{3.5}$$

Note that in this case

$$d_{l,k}(x) = \begin{cases} 0 & 0 \le x \le 3 \quad \text{and} \quad 9 \le x \le 10 \\ \text{not defined} & x = 8 \\ 1 & 4 \le x \le 7 \end{cases} \tag{3.6}$$

and the motion compensation matrix has the form

$$\mathbf{C}(\mathbf{d}_{l,k}) = \begin{pmatrix} 1 & 0 & 0 & 0 & 0 & 0 & 0 & 0 & 0 & 0 & 0 \\ 0 & 1 & 0 & 0 & 0 & 0 & 0 & 0 & 0 & 0 & 0 \\ 0 & 0 & 1 & 0 & 0 & 0 & 0 & 0 & 0 & 0 & 0 \\ 0 & 0 & 0 & 1 & 0 & 0 & 0 & 0 & 0 & 0 & 0 \\ 0 & 0 & 0 & 0 & 0 & 1 & 0 & 0 & 0 & 0 & 0 \\ 0 & 0 & 0 & 0 & 0 & 0 & 1 & 0 & 0 & 0 & 0 \\ 0 & 0 & 0 & 0 & 0 & 0 & 0 & 1 & 0 & 0 & 0 \\ 0 & 0 & 0 & 0 & 0 & 0 & 0 & 0 & 1 & 0 & 0 \\ * & * & * & * & * & * & * & * & * & * & * \\ 0 & 0 & 0 & 0 & 0 & 0 & 0 & 0 & 0 & 1 & 0 \\ 0 & 0 & 0 & 0 & 0 & 0 & 0 & 0 & 0 & 0 & 1 \end{pmatrix} \tag{3.7}$$

where $*$ denotes that these matrix entries are not defined. Note that by considering only the signal values (grayscale values in an image) a number of other displacement matrices can describe the mapping between Figs. 3.2(a) and 3.2(b) like for example

$$
\mathbf{C}(\mathbf{d}'_{l,k}) =
\begin{pmatrix}
1 & 0 & 0 & 0 & 0 & 0 & 0 & 0 & 0 & 0 & 0 \\
0 & 1 & 0 & 0 & 0 & 0 & 0 & 0 & 0 & 0 & 0 \\
0 & 0 & 1 & 0 & 0 & 0 & 0 & 0 & 0 & 0 & 0 \\
0 & 0 & 0 & 1 & 0 & 0 & 0 & 0 & 0 & 0 & 0 \\
0 & 0 & 0 & 0 & 0 & 0 & 0 & 0 & 1 & 0 & 0 \\
0 & 0 & 0 & 0 & 0 & 0 & 1 & 0 & 0 & 0 & 0 \\
0 & 0 & 0 & 0 & 0 & 0 & 0 & 1 & 0 & 0 & 0 \\
0 & 0 & 0 & 0 & 0 & 1 & 0 & 0 & 0 & 0 & 0 \\
* & * & * & * & * & * & * & * & * & * & * \\
0 & 0 & 0 & 0 & 0 & 0 & 0 & 0 & 0 & 1 & 0 \\
0 & 0 & 0 & 0 & 0 & 0 & 0 & 0 & 0 & 0 & 1 \\
\end{pmatrix}.
\tag{3.8}
$$

Additional issues arise (usually solved by interpolation) when the motion vector components are non-integer quantities (subpixel motion).

The next important step toward providing solution to the SR problem is to establish models which describe the acquisition of the LR images. Two such models are described in the next two sections, namely the warp–blur model and the blur–warp model.

3.1.1   The Warp–Blur Model

As the name implies, with this model the warping of an image is applied before it is blurred. The end-to-end system in this case is depicted in Fig. 3.3 which is a re-drawing of the system in Fig. 1.2. The LR discrete sequence is denoted by $g_l(i, j)$, with $i = 0, \ldots, M-1$, $j = 0, \ldots, N-1$. Using matrix–vector notation, each LR image is denoted by the $(M \times N) \times 1$

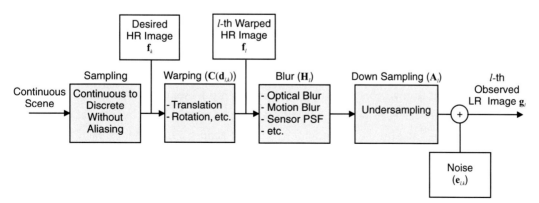

FIGURE 3.3: warp–blur model relating LR images to HR images

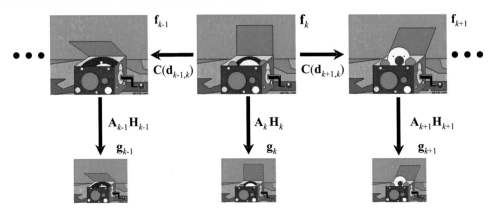

FIGURE 3.4: Graphical depiction of the relationship between the observed LR images and the HR images. See the roles played by motion compensation, blur, and downsampling

vector $\mathbf{g}_l$. The LR image $\mathbf{g}_l$ is related to the HR image $\mathbf{f}_l$ by

$$\mathbf{g}_l = \mathbf{A}_l\mathbf{H}_l\mathbf{f}_l + \eta_l \quad l = 1, 2, \ldots, L, \tag{3.9}$$

where the matrix $\mathbf{H}_l$ of size $(PM \times PN) \times (PM \times PN)$ describes the filtering of the HR image, $\mathbf{A}_l$ is the downsampling matrix of size $MN \times (PM \times PN)$, and η_l denotes the observation noise. The matrices $\mathbf{A}_l$ and $\mathbf{H}_l$ are generally assumed to be known (see, however, [7]).

Equation (3.9) expresses the relationship between the LR and HR frames $\mathbf{g}_l$ and $\mathbf{f}_l$, while Eq. (3.3) expresses the relationship between frames l and k in the HR sequence. Combining these two equations we obtain the following equation which describes the acquisition of an LR image $\mathbf{g}_l$ from the unknown HR image $\mathbf{f}_k$,

$$\mathbf{g}_l = \mathbf{A}_l\mathbf{H}_l\mathbf{C}(\mathbf{d}_{l,k})\mathbf{f}_k + \eta_l + \mu_{l,k} = \mathbf{A}_l\mathbf{H}_l\mathbf{C}(\mathbf{d}_{l,k})\mathbf{f}_k + \mathbf{e}_{l,k} , \tag{3.10}$$

where $\mu_{l,k}$ represents the registration noise and $\mathbf{e}_{l,k}$ represents the combined acquisition and registration noise. It is clear from Eq. (3.10) that $\mathbf{C}(\mathbf{d}_{l,k})$—the warp—is applied first on $\mathbf{f}_k$, followed by the application of the blur $\mathbf{H}_l$. This process is pictorially depicted in Fig. 3.4.

Note that the above equation shows the explicit dependency of $\mathbf{g}_l$ on both unknowns, the HR image $\mathbf{f}_k$ and the motion vectors $\mathbf{d}_{l,k}$. This observation model was first formulated in Irani and Peleg [80], without matrix notation, and later written in matrix form by Elad and Feuer [51]. Wang and Q_i [197] attribute this model to [51]. The acquisition model utilized in [190] for deriving frequency domain SR methods can also be written using this model (see [22, 24] for excellent reviews of frequency domain-based SR methods).

If we assume that the noise $\mathbf{e}_{l,k}$ in Eq. (3.10) is Gaussian with zero mean and variance σ^2, denoted by $N(0, \sigma^2 I)$, the above equation produces the following conditional pdf to be used within the Bayesian framework,

$$\mathbf{P}_G(\mathbf{g}_l|\mathbf{f}_k, \mathbf{d}_{l,k}) \propto \exp\left[-\frac{1}{2\sigma^2}\|\mathbf{g}_l - \mathbf{A}_l\mathbf{H}_l\mathbf{C}(\mathbf{d}_{l,k})\mathbf{f}_k\|^2\right] . \qquad (3.11)$$

Such a noise model has been used widely.

A uniform noise model is proposed by Stark and Oskoui [173] and Tekalp *et al.* [180] (see also [139] and [53]). The noise model used by these authors is oriented toward the use of the projection onto convex sets (POCS) method in SR problems. The associated conditional pdf has the form

$$\mathbf{P}_U(\mathbf{g}_l|\mathbf{f}_k, \mathbf{d}_{l,k}) \propto \begin{cases} \text{const} & \text{if } |[\mathbf{g}_l - \mathbf{A}_l\mathbf{H}_l\mathbf{C}(\mathbf{d}_{l,k})\mathbf{f}_k](i)| \leq c, \ \forall i \\ 0 & \text{elsewhere} \end{cases} \qquad (3.12)$$

where the interpretation of the index i is that it represents the ith element of the vector inside the brackets. The zero value of c can be thought of as the limit of $\mathbf{P}_G(\mathbf{g}_l|\mathbf{f}_k, \mathbf{d}_{l,k})$ in Eq. (3.11) when $\sigma = 0$.

Farsiu *et al.* [57, 58] have recently proposed the use of a generalized gaussian markov random field (GGMRF) [100] to model the noise in the image formation process for SR problems. In this case we have

$$\mathbf{P}_{GG}(\mathbf{g}_l|\mathbf{f}_k, \mathbf{d}_{l,k}) \propto \exp\left[-\frac{1}{2\sigma^p}\|\mathbf{g}_l - \mathbf{A}_l\mathbf{H}_l\mathbf{C}(\mathbf{d}_{l,k})\mathbf{f}_k\|_p^p\right] , \qquad (3.13)$$

where $\| \ . \ \|_p$ denotes the pth norm.

3.1.2 The Blur–Warp Model

Another acquisition model which has been used in the literature [94, 137, 191] first considers the blurring of the HR image, followed by warping and downsampling, as shown in Fig. 3.5.

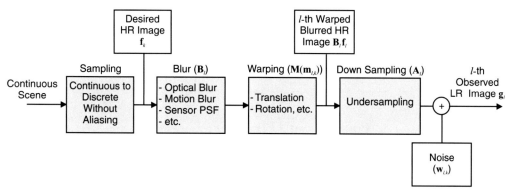

FIGURE 3.5: blur–warp model relating LR images to HR images

In this case, the observation model becomes

$$\mathbf{g}_l = \mathbf{A}_l \mathbf{M}(\mathbf{m}_{l,k}) \mathbf{B}_l \mathbf{f}_k + \mathbf{w}_{l,k} \, , \qquad (3.14)$$

where $\mathbf{w}_{l,k}$ denotes the acquisition and registration noise, $\mathbf{B}_l$ the blurring matrix for the lth HR image, $\mathbf{M}(\mathbf{m}_{l,k})$ the motion compensation operator for the blurred HR images through the use of motion vector $\mathbf{m}_{l,k}$, and $\mathbf{A}_l$ again the downsampling matrix. Different notation has been used in Eqs. (3.10) and (3.14) for the blur and warping operators in order to distinguish between these two models for the rest of the text.

The three-conditional pdfs in Eqs. (3.11)–(3.13) can be rewritten now for the blur–warp model, by substituting $\mathbf{A}_l \mathbf{H}_l \mathbf{C}(\mathbf{d}_{l,k})$ by $\mathbf{A}_l \mathbf{M}(\mathbf{m}_{l,k}) \mathbf{B}_l$ (for brevity we do not reproduce them here).

The question as to which of the two models (blur–warp or warp–blur) should be used is addressed in [197]. The authors claim that when the motion has to be estimated from the LR images, using the warp–blur model may cause systematic errors and, in this case, it is more appropriate to use the blur–warp model. They show that when the imaging blur is spatio-temporally shift invariant and the motion has only a global translational component the two models coincide. Note that in this case, the blur and motion matrices correspond to convolution matrices and thus they commute.

In the following chapter we will examine the methods used in the literature to estimate the motion vectors that relate the LR observations to the HR image we want to estimate. Another very important problem is the estimation of the blurring matrix in the image formation process. Unfortunately, not much work has been reported on how to estimate this matrix

Before concluding this section on image formation for uncompressed observations, we mention here that for both the warp–blur and the blur–warp models we have defined conditional pdfs for each LR observation $\mathbf{g}_l$ given $\mathbf{f}_k$ and $\mathbf{d}_{l,k}$. Our goal, however, is to define the conditional pdf $\mathbf{P}(\mathbf{g}|\mathbf{f}_k, \mathbf{d})$, that is, the distribution when all the observations $\mathbf{g}$ and all the motion vectors $\mathbf{d}$ for compensating the corresponding HR frames to the kth frame are taken into account. The approximation used in the literature for this joint-conditional pdf is

$$\mathbf{P}(\mathbf{g}|\mathbf{f}_k, \mathbf{d}) = \prod_{l=1}^{L} \mathbf{P}(\mathbf{g}_l|\mathbf{f}_k, \mathbf{d}_{l,k}), \qquad (3.15)$$

which implies that the LR observations are independent given the unknown HR image $\mathbf{f}_k$ and motion vectors $\mathbf{d}$.

3.2 IMAGE FORMATION MODELS FOR COMPRESSED OBSERVATIONS

In many applications, the acquired LR digital images and video are compressed. This is certainly the case when digital still or video cameras are utilized for acquisition (at least with mid- to low-end such cameras acquiring raw, i.e., uncompressed data, is not an option). Compressed data are also typically available when they are transmitted from the acquisition to the processing location.

The new scenario is graphically depicted in Fig. 2.2. The uncompressed LR sequence $\mathbf{g}_l$, $l = 1, \ldots, L$, is no longer available in this case. Instead, its compressed version, denoted by $\mathbf{y}_l$, $l = 1, \ldots, L$, is available, along with the motion vectors $\mathbf{v}_{l,m}$ provided by the encoder. The HR unknown sequence is denoted again by $\mathbf{f}$.

Let us now provide a high level description of a hybrid motion compensated video compression system, which forms the basis of existing video compression standards. The LR frames $\mathbf{g}_l$, $l = 1, \ldots, L$, are compressed with a video compression system resulting in $y_l(i, j)$, $i = 0, \ldots, M - 1$, $j = 0, \ldots, N - 1$, $l = 1, \ldots, L$. Using matrix–vector notation, each LR compressed image is denoted by the $(M \times N) \times 1$ vector $\mathbf{y}_l$. The compression system also provides the motion vectors $v(i, j, l, m)$ that predict pixel $g_l(i, j)$ from some previously coded frame $\mathbf{y}_m$. Since multiframe motion prediction is also used by some standards (e.g., H.264) vector $\mathbf{y}_m$ is expanded in this case to include more than one frame. These motion vectors that predict $\mathbf{g}_l$ from $\mathbf{y}_m$ are represented by the vector $\mathbf{v}_{l,m}$ that is formed by stacking the transmitted horizontal and vertical offsets.

During compression, frames are divided into blocks that are encoded with one of two available methods, intra-coding or inter-coding. For the first one, a linear transform such as the discrete cosine transform (DCT) is applied to each block (usually of size 8×8 pixels). The transform decorrelates the intensity data and the resulting transform coefficients are independently quantized and transmitted to the decoder. For the second method, i.e., inter-coding (see Fig. 3.6 for a graphical depiction), the blocks in a given observed frame (Fig. 3.6(b)) are first predicted by motion compensating previously compressed frames (Fig. 3.6(a)). The compensation is controlled by motion vectors (Fig. 3.6(c)) that define the spatial offset between the current uncompressed block and a reference compressed block. To improve the predicted frame (Fig. 3.6(d)), the prediction error (typically referred to as displaced frame difference (DFD)) is transformed with a linear transformation like the DCT, and the transformed coefficients are then quantized (Fig. 3.6(e) shows the quantized prediction error). At the decoder, the quantized prediction error is added to the predicted frame to generate the current compressed frame (Fig. 3.6(f)).

Using all this information, the relationship between the acquired LR frame $\mathbf{g}_l$ and its compressed observation $\mathbf{y}_l$ becomes

$$\mathbf{y}_l = \mathbf{T}^{-1} Q \left[\mathbf{T} \left(\mathbf{g}_l - MC_l(\mathbf{y}_l^P, \mathbf{v}_l) \right) \right] + MC_l(\mathbf{y}_l^P, \mathbf{v}_l), \quad l = 1, \ldots, L, \qquad (3.16)$$

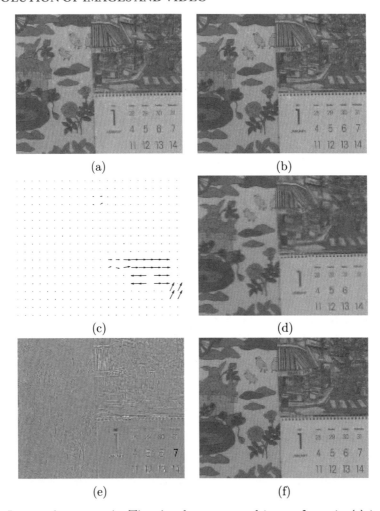

FIGURE 3.6: Inter-coding example. The already compressed image frame in (a) is used to predict through motion compensation the observed frame in (b). The process begins by finding the motion vectors between the frames shown in (a) and (b) which are shown in (c). Utilizing these motion vectors, frame (b) is predicted from (a), as shown in (d). The difference between the predicted (d) and input frame (b) is then computed, transformed using the DCT, and quantized. The resulting residual appears in (e), scaled to the range [0, 255], and is added to the prediction in (d) to generate the compressed frame in (f)

where $Q[.]$ represents the quantization procedure, $\mathbf{T}$ and $\mathbf{T}^{-1}$ are the forward and inverse transform operators, respectively, $MC_l(\mathbf{y}_l^P, \mathbf{v}_l)$ is the motion compensated prediction of $\mathbf{g}_l$ formed by motion compensating through $\mathbf{v}_l$ previously decoded frame(s) $\mathbf{y}_l^P$ as defined by the encoding method. We want to make clear here that MC_l depends on $\mathbf{v}_l$ and only a subset of $\mathbf{y}$.

To understand the structure of the compression errors, we need to model the quantization process Q in Eq. (3.16). This is a nonlinear operation that discards data in the transform domain, and it is typically realized by dividing each transform coefficient by a quantization scale factor and then rounding the result. The procedure is expressed as

$$Q[\mathbf{T}\left(\mathbf{g}_l - MC_l(\mathbf{y}_l^P, \mathbf{v}_l)\right)](i) = q(i)\,Round\left(\frac{[\mathbf{T}\left(\mathbf{g}_l - MC_l(\mathbf{y}_l^P, \mathbf{v}_l)\right)](i)}{q(i)}\right), \qquad (3.17)$$

where $[\mathbf{T}(\mathbf{g}_l - MC_l(\mathbf{y}_l^P, \mathbf{v}_l))](i)$ denotes the ith transform coefficient of the difference $\mathbf{g}_l - MC_l(\mathbf{y}_l^P, \mathbf{v}_l)$, $q(i)$ is the quantization factor for coefficient i, and $Round(.)$ is an operator that maps each value to its nearest integer.

Note that using the idempotent property of the quantization operator ($Q^2 = Q$) we obtain from Eq. (3.16)

$$Q\left[\mathbf{T}\left(\mathbf{y}_l - MC_l(\mathbf{y}_l^P, \mathbf{v}_l)\right)\right] = Q\left[\mathbf{T}\left(\mathbf{g}_l - MC_l(\mathbf{y}_l^P, \mathbf{v}_l)\right)\right]. \qquad (3.18)$$

Having a description of the compression system, we can define precisely the relationship between the HR frames and the LR-compressed observations. For the warp–blur model, using Eq. (3.10) in Eq. (3.16) we obtain

$$\mathbf{y}_l = \mathbf{T}^{-1}Q\left[\mathbf{T}\left(\mathbf{A}_l\mathbf{H}_l\mathbf{C}(\mathbf{d}_{l,k})\mathbf{f}_k + \mathbf{e}_{l,k} - MC_l(\mathbf{y}_l^P, \mathbf{v}_l)\right)\right] + MC_l(\mathbf{y}_l^P, \mathbf{v}_l). \qquad (3.19)$$

The process is pictorially depicted in Fig. 3.7.

Two prominent models for the quantization noise appear in the SR literature. The first one uses the equation

$$Q\left[\mathbf{T}\left(\mathbf{A}_l\mathbf{H}_l\mathbf{C}(\mathbf{d}_{l,k})\mathbf{f}_k + \mathbf{e}_{l,k} - MC_l(\mathbf{y}_l^P, \mathbf{v}_l)\right)\right]$$
$$= Q\left[\mathbf{T}\left(\mathbf{y}_l - MC_l(\mathbf{y}_l^P, \mathbf{v}_l)\right)\right] = Q\left[\mathbf{T}\left(\mathbf{g}_l - MC_l(\mathbf{y}_l^P, \mathbf{v}_l)\right)\right], \qquad (3.20)$$

which is obtained by utilizing Eq. (3.18) in Eq. (3.19). Since quantization errors are bounded by the quantization scale factor, it seems reasonable that the recovered HR image (when mapped to low resolution) has transform coefficients within the same interval (note that the model consequently assumes that $\mathbf{e}_{l,k}$ does not change the quantization interval). Equation (3.20) is typically referred to as the *quantization constraint* and within the Bayesian framework leads to the conditional distribution

$$\mathbf{P}_U(\mathbf{y}_l|\mathbf{f}_k, \mathbf{d}) = \begin{cases} const & \text{if } \forall i \\ & -\frac{q(i)}{2} \leq [\mathbf{T}(\mathbf{A}_l\mathbf{H}_l\mathbf{C}(\mathbf{d}_{l,k})\mathbf{f}_k - MC_l(\mathbf{y}_l^P, \mathbf{v}_l))](i) \leq \frac{q(i)}{2}. \\ 0 & \text{elsewhere} \end{cases} \qquad (3.21)$$

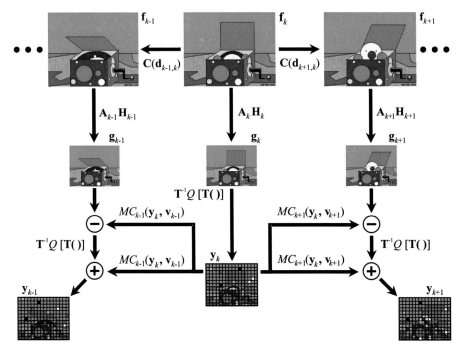

FIGURE 3.7: Graphical depiction of the relationship between the observed compressed LR images and the HR ones, for the warp–blur model. Note that in this example intra-frame compressed image $\mathbf{y}_k$ is used for the inter-frame compression of both images $\mathbf{y}_{k-1}$ and $\mathbf{y}_{k+1}$, that is, $\mathbf{y}_{k-1}^P = \mathbf{y}_{k+1}^P = \mathbf{y}_k$

This equation states that the compression, registration, and acquisition errors are uniformly distributed in the quantization interval $[-q(i)/2, q(i)/2]$ and so their means are zero and their variances are equal to $q(i)^2/12$. The quantization step size $q(i)$ is considered known, since it is provided by the encoder in the bitstream.

We note here that by defining this conditional pdf we are somewhat abusing the Bayesian notation since the appearance of $\mathbf{y}_l$ in the distribution is only through the use of Eq. (3.20).

Several authors employ the quantization constraint for SR processing. For example, it is utilized by Altunbasak *et al.* [2, 3], Gunturk *et al.* [72], Patti and Altunbasak [136], and Segall *et al.* [165, 167]. With the exception of [72], quantization is considered to be the sole source of noise within the acquisition system. This simplifies the construction of Eq. (3.21). However, since the distribution $\mathbf{P}_U(\mathbf{y}_l|\mathbf{f}_k, \mathbf{d})$ in Eq. (3.21) is not differentiable, care must still be taken when finding the HR estimate.

Modeling $\mathbf{T}^{-1}Q\mathbf{T}$ as a system that adds zero mean noise to its input leads to the second noise model. Application of this modeling to Eq. (3.19) results in

$$\mathbf{y}_l = \mathbf{A}_l\mathbf{H}_l\mathbf{C}(\mathbf{d}_{l,k})\mathbf{f}_k + \mathbf{r}_{l,k}, \qquad (3.22)$$

where $\mathbf{r}_{l,k}$ includes the errors introduced during compression, registration, and acquisition. The characteristics of this noise are not the same as the ones of the noise $\mathbf{e}_{l,k}$ in Eq. (3.10).

This second noise model is constructed in the spatial domain. This is appealing, as it justifies the use of a Gaussian distribution that is differentiable. This justification is based on the following. First, the quantization operator in Eq. (3.17) quantizes each transform coefficient independently. Thus, the quantization noise in the transform domain is not correlated between transform indices. Second, the transform operator is linear. With these two conditions, the quantization noise in the spatial domain becomes the linear sum of independent noise processes. The resulting distribution tends to be Gaussian and it is expressed, within the Bayesian framework, as [154]

$$\mathbf{P}_G(\mathbf{y}_l|\mathbf{f}_k, \mathbf{d}) \propto \exp\left[-\frac{1}{2}(\mathbf{y}_l - \mathbf{A}_l\mathbf{H}_l\mathbf{C}(\mathbf{d}_{l,k})\mathbf{f}_k)^t \mathbf{K}_Q^{-1}(\mathbf{y}_l - \mathbf{A}_l\mathbf{H}_l\mathbf{C}(\mathbf{d}_{l,k})\mathbf{f}_k)\right], \qquad (3.23)$$

where $\mathbf{K}_Q$ is the noise covariance matrix.

The Gaussian approximation for the noise appears in the work by Chen and Schultz [45], Gunturk et al. [71], Mateos et al. [108, 109], Park et al. [133, 134], and Segall et al. [165, 168, 169]. A primary difference between these efforts lies in the definition and estimation of the covariance matrix. For example, a white noise model is assumed by Chen and Schultz [45], and Mateos et al. [108, 109], while Gunturk et al. [71] develop the distribution experimentally. Segall et al. [165, 168, 169] consider a high bit-rate approximation for the quantization noise. Lower rate compression scenarios are addressed by Park et al. [133, 134], where the covariance matrix and HR frame are estimated simultaneously.

We mention here that the spatial domain noise model also incorporates errors introduced by the sensor and motion models. This is accomplished by modifying the covariance matrix $\mathbf{K}_Q$ [132]. Interestingly, since these errors are often independent of the quantization noise, incorporating the additional noise components further justifies the use of the Gaussian model.

Modeling the quantization noise is a major step in the SR process; similarly, modeling $\mathbf{v}$, the vector field provided by the encoder, also constitutes an important step in the SR process. These motion vectors introduce a departure from traditional SR techniques. In traditional approaches, the observed LR images provide the only source of information about the relationship between HR frames. When compression is introduced though, the transmitted motion vectors provide an additional observation for the SR problem.

There are several methods that exploit the motion vectors during RE. At the high level, however, each method tries to model the similarity between the transmitted motion vectors and the actual HR displacements. For example, Chen and Schultz [45] constrain the motion vectors to be within a region surrounding the actual subpixel displacements. This is accomplished with the distribution

$$\mathbf{P}_U(\mathbf{v}_l|\mathbf{f}_k, \mathbf{d}, \mathbf{y}) = \begin{cases} \text{const} & \text{if } |v_{l,i}(j) - [\mathbf{A}_D\mathbf{d}_{l,i}](j)| \leq \Delta, \ i \in PS, \forall j \\ 0 & \text{elsewhere} \end{cases}, \qquad (3.24)$$

where $\mathbf{A}_D$ is a matrix that maps the displacements to the LR grid, Δ denotes the maximum difference between the transmitted motion vectors and estimated displacements, PS represents the set of previously compressed frames employed to predict $\mathbf{f}_k$, and $[\mathbf{A}_D\mathbf{d}_{l,i}](j)$ is the jth element of the vector $\mathbf{A}_D\mathbf{d}_{l,i}$. Similarly, Mateos $et~al.$ [109] utilize the distribution

$$\mathbf{P}_G(\mathbf{v}_l|\mathbf{f}_k, \mathbf{d}, \mathbf{y}) \propto \exp\left[-\frac{\gamma_l}{2}\sum_{i \in PS}||\mathbf{v}_{l,i} - \mathbf{A}_D\mathbf{d}_{l,i}||^2\right], \qquad (3.25)$$

where γ_l specifies the similarity between the transmitted and estimated information.

Note that we are again somewhat abusing the notation since the above distributions $\mathbf{P}_U$ and $\mathbf{P}_G$ do not depend on $\mathbf{y}$. We are doing this, however, for consistency with the model for $\mathbf{v}$ to be defined next.

There are two shortcomings to modeling the transmitted motion vectors and HR displacements as similar throughout the frame. The first one is due to the fact that, in the context of compression, the optical flow is sought after and not the true motion. The second one is due to the fact that the motion vectors are determined in the rate-distortion sense and thus their accuracy depends on the underlying compression ratio, which typically varies within the frame. Segall $et~al.$ [168, 169] account for these errors by modeling the DFD within the encoder. This incorporates the motion vectors and is written as

$$\mathbf{P}_{MV}(\mathbf{v}_l|\mathbf{f}_k, \mathbf{d}, \mathbf{y}) \propto \exp\left[-\frac{1}{2}F^t(\mathbf{v}_l, \mathbf{y}, \mathbf{d}, \mathbf{f}_k)\mathbf{K}_{MV}^{-1}F(\mathbf{v}_l, \mathbf{y}, \mathbf{d}, \mathbf{f}_k)\right], \qquad (3.26)$$

where

$$F(\mathbf{v}_l, \mathbf{y}, \mathbf{d}, \mathbf{f}_k) = MC_l\left(\mathbf{y}_l^P, \mathbf{v}_l\right) - \mathbf{A}_l\mathbf{H}_l\mathbf{C}(\mathbf{d}_{l,k})\mathbf{f}_k, \qquad (3.27)$$

and $\mathbf{K}_{MV}$ is the covariance matrix of the prediction error between the original frame and its motion compensated estimate $MC_l(\mathbf{y}_l^P, \mathbf{v}_l)$. Estimates for $\mathbf{K}_{MV}$ are derived from the compressed bit-stream and therefore reflect the degree of compression.

To complete this section we indicate here that, within the Bayesian framework, we have defined probability models for each LR-compressed observation $\mathbf{y}_l$ given $\mathbf{f}_k$ and $\mathbf{d}_{l,k}$, although our goal is to define the conditional distribution $\mathbf{P}(\mathbf{y}|\mathbf{f}_k, \mathbf{d})$. The approximation used in the literature for this distribution is the following

$$P(\mathbf{y}|\mathbf{f}_k, \mathbf{d}) = \prod_{l=1}^{L} P(\mathbf{y}_l|\mathbf{f}_k, \mathbf{d}_{l,k}), \qquad (3.28)$$

which implies that the LR-compressed observations are independent given the HR image $\mathbf{f}_k$ and motion vectors $\mathbf{d}$, as was also assumed for the uncompressed observations.

Similarly, when the LR motion vectors are also used in the SR problem, we need to define $P(\mathbf{y}, \mathbf{v}|\mathbf{f}_k, \mathbf{d})$, which is approximated as

$$P(\mathbf{y}, \mathbf{v}|\mathbf{f}_k, \mathbf{d}) = \prod_{l=1}^{L} P(\mathbf{y}_l|\mathbf{f}_k, \mathbf{d}_{l,k}) \prod_{l=1}^{L} P(\mathbf{v}_l|\mathbf{f}_k, \mathbf{d}, \mathbf{y}), \qquad (3.29)$$

that is, the observed LR compressed images are statistically independent given the HR image $\mathbf{f}_k$ and motion vectors $\mathbf{d}$ and the observed LR motion vectors are statistically independent given the HR image $\mathbf{f}_k$, the motion vectors $\mathbf{d}$ and the LR compressed observations.

Finally, we mention that the presentation in this section utilized only the warp–blur model, as was mentioned earlier and as should be clear by the utilized notation. Similar distributions can be derived for the blur–warp model, although it has been utilized widely for the case of compressed observations.

3.3   LIMITS ON SUPER RESOLUTION

The techniques described so far are referred to as reconstruction-based SR techniques (to be differentiated from the recognition or learning-based SR techniques to be discussed in Chapter 5). Important elements of such techniques are the constraints imposed on the original HR image through the modeling of the observed LR images and the addition of prior information on the reconstruction.

We discuss here the limits on reconstruction-based SR methods. The first results on the topic addressed the existence of a unique HR image $\mathbf{f}_k$ resulting from the solution of the matrix equations

$$\mathbf{g}_l = \mathbf{A}_l \mathbf{H}_l \mathbf{C}(\mathbf{d}_{l,k}) \mathbf{f}_k, \quad l = 1, \dots, L. \qquad (3.30)$$

Examining the existence of a unique solution of the above equations is equivalent to studying the properties of the matrix

$$\mathbf{R} = \sum_{l} [\mathbf{A}_l \mathbf{H}_l \mathbf{C}(\mathbf{d}_{l,k})]^t [\mathbf{A}_l \mathbf{H}_l \mathbf{C}(\mathbf{d}_{l,k})]. \qquad (3.31)$$

The case corresponding to $\mathbf{H}_l = \mathbf{I}$, $l = 1 \ldots, L$, and constant translational motion in Eq. (3.1), that is,

$$d_{l,k}^x(m, n) = d_{l,k}^x \quad \text{and} \quad d_{l,k}^y(m, n) = d_{l,k}^y, \tag{3.32}$$

was first studied by Kim *et al.* [91]. The same problem but with $\mathbf{H}_l \neq \mathbf{I}$, $l = 1 \ldots, L$, was later addressed by Elad and Feuer [51]. Note that the study of these two cases is similar to the study of the limits of image denoising and image restoration respectively when all the observations may not be available.

We examine here how the magnification factor, P, the number of LR images, L, the blurring matrices $\mathbf{H}_l, l = 1, \ldots, L$, and the noise influence the quality of the HR reconstructed image.

To proceed, we need to specify a real function $\mathbf{S}(\mathbf{x})$, $\mathbf{x} \in R^2$ of the discrete image $\mathbf{f}_k(\mathbf{p})$, where $\mathbf{p} = (p, q) \in Z^2$ are the coordinates of an HR pixel. The case studied by Baker and Kanade [9] corresponds to the piecewise constant function

$$\mathbf{S}(\mathbf{x}) = \mathbf{f}_k(\mathbf{p}), \tag{3.33}$$

for all $\mathbf{x} \in (p - 0.5, p + 0.5] \times (q - 0.5, q + 0.5]$. Following [9], we assume that the registration between each pair of HR images is a global translation, that is, $\mathbf{d}_{l,k}(m, n) = (d_{l,k}^x, d_{l,k}^y)$, $\forall (m, n)$.

We now turn our attention to the blurring function. We perform our study assuming that all images are subject to the same blur (referred to as the sensor blur) which has the following form

$$a(\mathbf{x}) = \begin{cases} 1/S^2 & \text{if } |x| \leq S/2 \text{ and } |y| \leq S/2 \\ 0 & \text{otherwise} \end{cases}. \tag{3.34}$$

If there is additional blur, due for instance to the lens, the results to be described remain valid. $\mathbf{S}(\mathbf{x})$ is convolved with $a(\mathbf{x})$ before downsampling takes place for generating the LR observations.

Let us use the 1D case to better illustrate the role of P and S. As shown in Fig. 3.8, P expresses the number of HR pixels an LR unit contains, while $S/2$ represents the maximum distance the region covered by an HR pixel can be from an LR pixel, to contribute to its formation. For instance, as shown in Fig. 3.8, if $S = 1/2$, only the second and third HR pixels contribute to the formation of the first LR pixel. The same is true if $S = 1$, however, if $S > 1$ also the first and fourth HR pixels contribute to the formation of the second LR observation.

Baker and Kanade [9] proved that if PS is an integer greater than one, then regardless of the number of LR images we may obtain by translating $\mathbf{f}_k$ to produce $\mathbf{f}_l$, the image formation model does not have a unique $\mathbf{f}_k$ solution (a checkerboard pattern always produces zero values

Magnification $P=2$

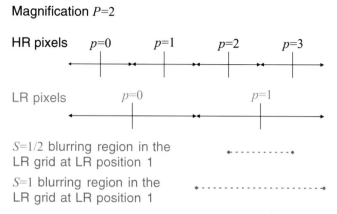

FIGURE 3.8: Graphical depiction of the role of the magnification factor P and the blurring size S for the 1D case

for the LR observations as seen in Fig. 3.9). However, if PS is not an integer or it is equal to one, with an adequate number of properly located LR observations we are able to recover the HR image (see Figs. 3.10 and 3.11).

It is reasonable to assume that given a blurring value S we can always use values P, as large as we want, as long as PS is not integer. The condition number, however, of the matrix defining the relationship between the LR images and the HR one grows at least as fast as $(PS)^2$ (see [9]). The condition number of a matrix $\mathbf{A}$, denoted by $cond(\mathbf{A})$ expresses the ill-conditionedness of the system of equations $\mathbf{a} = \mathbf{Ab}$ with $\mathbf{a}$ the known and $\mathbf{b}$ the unknown parameters. One of the definitions of the condition number is [147]

$$cond(\mathbf{A}) = \frac{w_1}{w_n},\qquad(3.35)$$

Magnification $P=2$, Blurring $S=1$

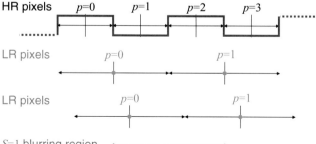

FIGURE 3.9: An HR signal always producing the same LR observations independently of the LR pixel positions, $P = 2$, $S = 1$

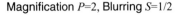

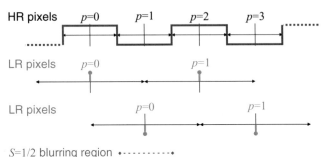

FIGURE 3.10: With these two LR observations we can recover the original HR signal

where $\mathbf{A}$ is of dimension $m \times n$ and $w_1 \geq \ldots \geq w_n \geq 0$ are its singular values. The larger the condition number, the more severe the ill-conditionedness of $\mathbf{A}$, which translates to a greater numerical sensitivity of the solution. Clearly, if $m < n$ then $cond(\mathbf{A}) = \infty$, which means that $\mathbf{A}$ is noninvertible. It is however not the noninvertibility of $\mathbf{A}$ the main concern but the number of its singular values approaching zero.

The conversion of a real-valued irradiance to an integer-valued intensity adds to the complexity of the search for the HR image. Adding as much as $(PS)^2$ to each HR value produces the same LR observation. This implies that for large magnification factors P there is a huge volume of HR values that produce the same discretized observations (see [9]). Finding the "real underlying HR image" becomes more difficult as the magnification factor increases and we have to rely on more precise knowledge of the original image than the simple smoothness constraints.

The work of Baker and Kanade [9] does not provide an explicit value for the maximum magnification factor that can effectively be achieved in real problems (even if we have a unlimited

FIGURE 3.11: With these two LR observations we cannot recover the original HR signal

number of LR observation). Lin and Shum [98] provide an upper bound on the magnification factor that depends on the image characteristics. See, however, Wang and Feng [195] for comments on [98] and the reply by Lin and Shum [99]. The above limits on SR have been established for the warp–blur model; results on the blur–warp model are presented in [198].

In the presentation so far in this section, the performance limit of SR techniques has been addressed from the algebraic perspective of solving large systems of linear equations. Robinson and Milanfar [156] analyze the problem from a statistical point of view, as outlined next.

Let us consider the image formation model

$$\mathbf{P}(\mathbf{g}; \theta) = \mathbf{P}(\mathbf{g}; \mathbf{f}_k, \mathbf{d}) \propto \exp\left[-\frac{1}{2\sigma^2} \sum_l \|\mathbf{g}_l - \mathbf{AHC}(\mathbf{d}_{l,k})\mathbf{f}_k\|^2 \right], \qquad (3.36)$$

where $\mathbf{A}$ and $\mathbf{H}$ are the downsampling and blurring operators, $\mathbf{C}(\mathbf{d}_{l,k})$ describes a translational global model and $\theta = (\mathbf{f}_k, \mathbf{d})$. The semicolon in $(\mathbf{g}; \theta)$ is used to denote that no distribution is assumed on θ.

The fundamental performance limit for SR is then analyzed in terms of the Cramer-Rao (CR) bounds (see [88] and also [157] for an introduction). Let $\theta^t = (\mathbf{f}_k^t, \mathbf{d}^t)$; then the CR bound applied to the distribution on $\mathbf{g}$ in Eq. (3.36) establishes that for any unbiased estimate $\hat{\theta}$ of θ the following matrix inequality holds

$$\mathbf{E}[(\hat{\theta} - \theta)(\hat{\theta} - \theta)^t] \geq \mathbf{J}^{-1}(\theta) \qquad (3.37)$$

where the (i, j)th element of $\mathbf{J}(\theta)$ is given by

$$\mathbf{J}_{i,j}(\theta) = -\mathbf{E}\left[\frac{\partial^2 \ln \mathbf{P}(\mathbf{g}; \theta)}{\partial \theta_i \partial \theta_j} \right] \qquad (3.38)$$

and the expected value is calculated using the pdf on $\mathbf{g}$, $\mathbf{P}(\mathbf{g}; \theta)$.

Note that the above inequality indicates that the difference between the covariance matrix of $\hat{\theta}$ and $\mathbf{J}(\theta)$ is a positive semidefinite matrix. Note also that by providing a lower bound on the covariance matrix of $\hat{\theta}$, $\mathbf{J}(\theta)$ characterizes, from an information point of view, the difficulty with which θ can be estimated.

Let us now assume that information on the HR image and motion vectors is included in the form of an *à priori* probability distribution $\mathbf{P}(\theta)$, that is, we have $\mathbf{P}(\theta, \mathbf{g}) = \mathbf{P}(\theta)\mathbf{P}(\mathbf{g}|\theta)$ where $\mathbf{P}(\mathbf{g}|\theta) = \mathbf{P}(\mathbf{g}; \theta)$ (note the difference in notation when the *à priori* pdf is introduced on θ).

Using now van Trees' inequality [67, 188] we have that for weakly biased estimators $\hat{\theta}$ (MAP estimators are weakly biased, see [67, 188] for details) the following inequality holds

$$\mathbf{E}[(\hat{\theta} - \theta)(\hat{\theta} - \theta)^t] \geq \mathbf{J}^{-1} \qquad (3.39)$$

where

$$\mathbf{J}_{i,j} = -\mathbf{E}\left[\frac{\partial^2 \ln \mathbf{P}(\theta)\mathbf{P}(\mathbf{g}|\theta)}{\partial\theta_i\partial\theta_j}\right] \qquad (3.40)$$

and the expectation is now taken with respect to the joint pdf $\mathbf{P}(\theta, \mathbf{g})$.

By studying $\mathbf{J}(\theta)$ and $\mathbf{J}$ in the above equations, Robinson and Milanfar [156] analyze the interdependency of the estimation of the HR image and motion vectors, the role played by the blur operator, and the importance of the prior information on the unknown parameters. In particular, the authors show that the performance of SR methods depends on a complex relationship between the measurement signal to noise ratio (SNR), the number of observed frames, the set of relative motions between frames, the image content, and the point spread function (PSF) of the system. A very interesting study is carried out on the uncertainty added to the SR reconstruction system when the HR translations among frames have to be estimated as well as on the *best* motion vectors to be utilized when an SR system can control and know in advance the HR motion vectors. This naturally leads to the use of the CR bounds for imaging system design under the assumption that SR processing is included as a post-processing step (see Chapter 7 for a complete description of the so-called *SR for compression* problem).

CHAPTER 4

Motion Estimation in Super Resolution

As we have already explained, SR image reconstruction requires the estimation of both the HR motion vectors and the HR image. Following the Bayesian formulation of the SR problem presented in Chapter 2, the alternate, and sequential Bayesian approaches can be developed for estimating the HR image and motion vectors. This chapter is devoted to the motion estimation problem applicable to these two SR methodologies for both compressed and uncompressed LR image sequences. Consequently, we assume here that an estimate of the original HR image $\mathbf{f}_k$ is available based on which the HR motion vectors are estimated. This motion estimation problem corresponds to finding the solution to Eq. (2.4) for the alternate approach and to the motion estimation part of the sequential methodology described in Chapter 2.

This chapter is organized as follows: In Section 4.1, the most widely used methods to estimate the HR motion vectors for uncompressed LR observations are studied. The motion estimation methods utilized when the LR sequence is compressed are studied in Section 4.2. As it was explained in Section 3.1, the estimated motion vectors may be erroneous due, for instance, to the occlusion or aperture problems. The important problem of detecting unsatisfactory motion estimates that jeopardize the quality of the reconstructed HR image is studied in Section 4.3. Consistent motion estimation over several frames is analyzed in Section 4.4. Finally, the chapter concludes with a discussion on some motion estimation-related issues which we consider relevant to the reconstruction of HR images from LR observations.

4.1 MOTION ESTIMATION FROM UNCOMPRESSED OBSERVATIONS

In this section we assume that we have access to an LR uncompressed sequence $\mathbf{g}$ consisting of the frames $\mathbf{g}_1, \ldots, \mathbf{g}_L$. Most of the HR motion estimation methods used for uncompressed LR observation, first, interpolate $\mathbf{g}$ to obtain $\mathbf{u}$, an estimate of the original HR sequence $\mathbf{f}$, and then find the displacement vectors $d_{l,k}(x, y)$ satisfying

$$u_l(x, y) = u_k(x + d_{l,k}^x(x, y), y + d_{l,k}^y(x, y)), \qquad (4.1)$$

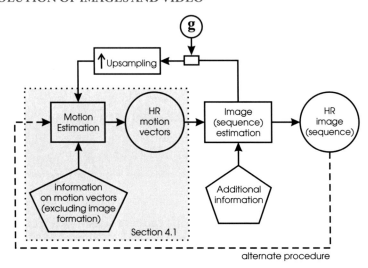

FIGURE 4.1: Graphical depiction of the process to sequentially estimate the motion vector (no information on the image formation is used) and the HR image. Note that once we have an HR image estimate we can re-estimate the HR motion vectors (alternate procedure, dashed line)

with, in some cases, the inclusion of additional constraints on the HR motion vectors (e.g., prior distributions on the HR motion vectors associated to object segmentation or smoothness constraints). This motion estimation problem corresponds to approximating the solution to Eq. (2.4) following the alternate approach and to the motion estimation part of the sequential methodology described in Chapter 2.

We must emphasize here that different interpolation methods can be used to construct **u**. In most of the methods in the literature, however, no information about the process of obtaining the LR observations from the HR image and HR motion vectors is used in estimating **d**. A graphical depiction of the process is shown in Fig. 4.1, where the topic of this section is represented by the shaded area. After an estimate of the HR motion becomes available, an HR image can be estimated using Eqs. (2.5) or (2.6), as will be discussed in detail in Chapter 5. For the alternate procedure the motion estimation step will be repeated, as indicated by the dashed line in Fig. 4.1.

Most HR motion estimation methods to estimate $\mathbf{d}_{l,k}$ from the upsampled sequence **u**, particularly those derived in the Fourier domain, are based on the assumption of purely translational image motion (see for instance the first paper on SR [190]). This *translational motion* model corresponds to

$$d_{l,k}^{x}(x, y) = p_1 \quad \text{and} \quad d_{l,k}^{y}(x, y) = p_2, \quad \forall(x, y), \qquad (4.2)$$

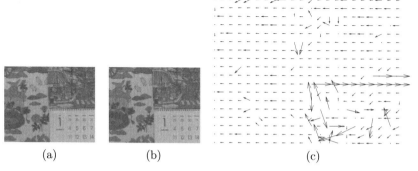

(a)   (b)   (c)

FIGURE 4.2: Two consecutive observed LR frames of the "mobile" sequence are depicted in (a) and (b). These frames are upsampled by a factor of two in each direction. Block matching is used on the upsampled images to estimate the HR motion vectors. The estimated HR motion field is plotted in (c), with each vector corresponding to an 16×16 block

either for all pixels in the lth image or for pixels within a given block or region in that image. In other words,

$$\mathbf{u}_l(x, y) = \mathbf{u}_k(x + p_1, y + p_2). \qquad (4.3)$$

Note that, for simplicity, the dependency of the motion vectors on the lth image (or its blocks) has been omitted.

Several models have been proposed in the literature for estimating translational motion. The most widely used method is Hierarchical Block Matching (see [23] for a comparison of block matching techniques on SR problems). Phase correlation methods have also been used (see [179]).

Figure 4.2 illustrates the use of block matching for HR motion estimation. Two observed 176×144 pixels LR frames are shown in Figs. 4.2(a) and 4.2(b). To estimate the HR motion field with a magnification factor of $P = 2$, first, the observed frames are upsampled by a factor of two in each direction using bicubic interpolation. Block matching is performed on these images using 16×16 blocks and a 13×13 search area. Exhaustive search and the minimum mean square error (MMSE) matching criterium are used, resulting in the HR motion vector estimates shown in Fig. 4.2(c). One motion vector is assigned to each 16×16 block, although the resolution of the motion field can be as high as one motion vector per pixel. Note that, while the algorithm is able to correctly estimate most of the motion vectors, in flat areas and areas with repeating patterns the estimated vectors do not agree with the motion of the objects. This problem may be ameliorated by the use, for instance, of a hierarchical block matching approach.

The next level of complexity in modeling the image motion is to assume that images $\mathbf{u}_l$ and $\mathbf{u}_k$ (or regions within them) are related by a horizontal shift a, a vertical shift b, and a rotation θ around the origin, that is,

$$\mathbf{u}_l(x, y) = \mathbf{u}_k(x \cos \theta - y \sin \theta + a, y \cos \theta + x \sin \theta + b). \qquad (4.4)$$

The estimation of the rotation angle and the translation is tackled in [79, 89] using the first-term of the Taylor series expansion of $\mathbf{u}_k$. Other approximations of the rotation are used in [75, 199].

Increasing the complexity of the problems, the motion is modeled using an *affine transformation* (of which translation, rotation, and scaling are particular cases), that is,

$$\mathbf{u}_l(x, y) = \mathbf{u}_k(ax + by + c, dx + ey + f), \qquad (4.5)$$

where a, b, c, d, e, and f are parameters to be estimated. This model has been used in SR problems, for instance, in [79, 80]. *Bilinear transformations*, that is,

$$\mathbf{u}_l(x, y) = \mathbf{u}_k(ax + by + cxy + d, ex + fy + gxy + h). \qquad (4.6)$$

where a, b, c, d, e, f, g, and h are parameters to be estimated, have also been used in [44].

Except when using block matching or phase correlation techniques to estimate translations, the methods used in the literature to estimate the above transformations are related to the Lucas–Kanade method as described in [10] based on [101], which is described next.

Let the warp $\mathbf{W}(\mathbf{x}; \mathbf{p})$ denote the subpixel location, a pixel $\mathbf{x} = (x, y)^t$ in frame $\mathbf{u}_l$ is mapped onto in frame $\mathbf{u}_k$ using the parameter vector

$$\mathbf{p} = (p_1, \ldots, p_n)^t, \qquad (4.7)$$

that is,

$$\mathbf{u}_l(\mathbf{x}) = \mathbf{u}_k(\mathbf{W}(\mathbf{x}; \mathbf{p})), \qquad (4.8)$$

where we have removed the dependency of the warp on the frames for simplicity. For instance, when the warp corresponds to translation, then

$$\mathbf{W}(\mathbf{x}; \mathbf{p}) = \begin{pmatrix} x + p_1 \\ y + p_2 \end{pmatrix}, \qquad (4.9)$$

with

$$\mathbf{p} = (p_1, p_2)^t. \qquad (4.10)$$

For an affine transformation we have

$$\mathbf{W}(\mathbf{x}; \mathbf{p}) = \begin{pmatrix} x + (a-1)x + by + c \\ y + dx + (e-1)y + f \end{pmatrix},$$

(4.11)

with

$$\mathbf{p} = (a-1, b, c, d, e-1, f)^t.$$

(4.12)

The vector $\mathbf{p}$ can be pixel dependent or region dependent. We denote by $\mathbf{B}$ the image region where $\mathbf{p}$ is assumed to be constant. The goal of the Lucas–Kanade algorithm is to find $\hat{\mathbf{p}}$ satisfying

$$\hat{\mathbf{p}} = \arg\min_{\mathbf{p}} \sum_{\mathbf{x} \in \mathbf{B}} \left[\mathbf{u}_l(\mathbf{x}) - \mathbf{u}_k(\mathbf{W}(\mathbf{x}; \mathbf{p})) \right]^2.$$

(4.13)

To minimize the function in the above equation, the algorithm assumes that a current estimate of $\mathbf{p}$ is available and then iteratively refines it by finding $\hat{\Delta}\mathbf{p}$ satisfying

$$\hat{\Delta}\mathbf{p} = \arg\min_{\Delta\mathbf{p}} \sum_{\mathbf{x} \in \mathbf{B}} \left[\mathbf{u}_l(\mathbf{x}) - \mathbf{u}_k(\mathbf{W}(\mathbf{x}; \mathbf{p} + \Delta\mathbf{p})) \right]^2.$$

(4.14)

This optimization problem is solved by performing a first-order Taylor expansion on $\mathbf{u}_k(\mathbf{W}(\mathbf{x}; \mathbf{p} + \Delta\mathbf{p}))$ to obtain

$$\hat{\Delta}\mathbf{p} = \arg\min_{\Delta\mathbf{p}} \sum_{\mathbf{x} \in \mathbf{B}} \left[\mathbf{u}_l(\mathbf{x}) - \mathbf{u}_k(\mathbf{W}(\mathbf{x}; \mathbf{p})) - \nabla\mathbf{u}_k^t \frac{\partial \mathbf{W}}{\partial \mathbf{p}} \Delta\mathbf{p} \right]^2,$$

(4.15)

where $\nabla\mathbf{u}_k$ is the gradient of $\mathbf{u}_k$ evaluated at $\mathbf{W}(\mathbf{x}; \mathbf{p})$ and $\partial\mathbf{W}/\partial\mathbf{p}$ is the Jacobian of the warp. If $\mathbf{W}(\mathbf{x}; \mathbf{p}) = (\mathbf{W}_x(\mathbf{x}; \mathbf{p}), \mathbf{W}_y(\mathbf{x}; \mathbf{p}))^t$ then

$$\frac{\partial \mathbf{W}}{\partial \mathbf{p}} = \begin{pmatrix} \frac{\partial \mathbf{W}_x(\mathbf{x};\mathbf{p})}{\partial p_1} & \frac{\partial \mathbf{W}_x(\mathbf{x};\mathbf{p})}{\partial p_2} & \cdots & \frac{\partial \mathbf{W}_x(\mathbf{x};\mathbf{p})}{\partial p_n} \\ \frac{\partial \mathbf{W}_y(\mathbf{x};\mathbf{p})}{\partial p_1} & \frac{\partial \mathbf{W}_y(\mathbf{x};\mathbf{p})}{\partial p_2} & \cdots & \frac{\partial \mathbf{W}_y(\mathbf{x};\mathbf{p})}{\partial p_n} \end{pmatrix}.$$

(4.16)

The least squares solution $\hat{\Delta}\mathbf{p}$ can be found from Eq. (4.15) in a straightforward way. Then, $\mathbf{p}$ is replaced by $\mathbf{p} + \hat{\Delta}\mathbf{p}$ and the new refinement $\hat{\Delta}\mathbf{p}$ is found using Eq. (4.14). The iterative procedure terminates when a convergence criterion is met. The estimation of a refinement $\Delta\mathbf{p}$ of an initial estimate of the vector $\mathbf{p}$, along with the linearization step of Eq. (4.15), is a step also taken with pel-recursive motion estimation algorithms, which solve the spatiotemporal gradient optical flow equation (see, for example, [28, 50]).

Feature matching has also been used in SR motion estimation. In computer vision it is common to estimate the parameters of a geometric transformation by automatic detection and analysis of corresponding features among the input images. Typically, in each image several hundred points of interest are automatically detected with subpixel accuracy using an algorithm such as the Harris feature detector [76]. Putative correspondences are identified by comparing the image neighborhoods around the features, using a similarity metric such as the normalized correlation. These correspondences are refined using a robust search procedure such as the random sample consensus (RANSAC) algorithm [60], which extracts only those features whose inter-image motion is consistent with the geometric transformation. Finally, those corresponding points are used in a nonlinear estimator which returns a highly accurate estimate of the transformation. These ideas have been applied by Capel and Zisserman in [31, 34] when the geometric transformation is a planar homography, also called a plane *projective transformation*, that is, when the coordinates are related by

$$\mathbf{u}_l(x, y) = \mathbf{u}_k \left(\frac{ax + by + c}{gx + hy + i}, \frac{dx + ey + f}{gx + hy + i} \right), \qquad (4.17)$$

where a, b, c, d, e, f, g, h, and i are parameters to be estimated. Note that [34] is one of the few papers that addresses the problem of photometric registration in SR.

A comparative study between the Lucas–Kanade algorithm-based methods and the ones based on the use of feature points in SR problems has been carried out in [97].

Figure 4.3 illustrates the use of a feature matching motion estimation algorithm in the SR problem. The two LR images of size 180×200 pixels displayed in Figs. 4.3(a) and 4.3(b)

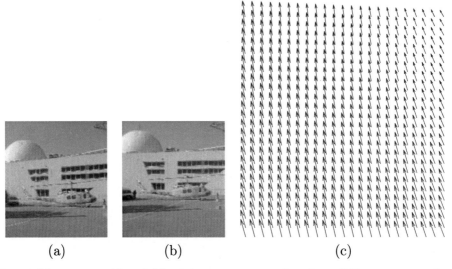

(a)   (b)   (c)

FIGURE 4.3: Two images (a) and (b) of the same scene taken from different points of view were registered using feature matching motion estimation obtaining the estimated transformation depicted in (c)

represent the same scene captured from different points of view. The images are upsampled by a factor of two so that they are registered. More than 300 corners were detected on each image using the Harris algorithm [76]. For only 112 of those corners corresponding matching points were found on the other image. Using the RANSAC algorithm [60] those 112 points were refined to obtain the 28 correspondences that were consistent with the geometric transformation. From those correspondences the transformation depicted in Fig. 4.3(c) was estimated. Note that the RANSAC algorithm provides us with a global parametric HR motion field for the whole image and cannot be used in video sequences where typically several objects undergo different transformations.

We can also use segmentation maps which can be computed automatically [40], or mark an object of interest in a key frame interactively and then track it over the rest of the frames [53]. Finally, the tracking of objects by deformable regions in SR problems is described in [14].

Before finishing this section we would like to mention an alternative approach to motion estimation in SR. Let us consider again Eq. (3.10) which relates the LR image $\mathbf{g}_l$ to the HR image $\mathbf{f}_k$, that is,

$$\mathbf{g}_l = \mathbf{A}_l \mathbf{H}_l \mathbf{C}(\mathbf{d}_{l,k}) \mathbf{f}_k + \eta_l + \mu_{l,k} = \mathbf{A}_l \mathbf{H}_l \mathbf{C}(\mathbf{d}_{l,k}) \mathbf{f}_k + \mathbf{e}_{l,k} \; . \tag{4.18}$$

Without the presence of $\mathbf{A}_l$ in the above equation, it corresponds to a blind deconvolution problem where both image and blur are unknown. So, in principle, blind deconvolution techniques can be used to solve SR imaging problems.

The presence of the downsampling matrix $\mathbf{A}_l$ complicates the problem. However, initially we can at least upsample the observed LR image $\mathbf{g}_l$ to obtain $\mathbf{g}_l^{\uparrow}$ and use the equation

$$\mathbf{g}_l^{\uparrow} = \mathbf{H}_l \mathbf{C}(\mathbf{d}_{l,k}) \mathbf{f}_k + \zeta_{l,k} \; , \tag{4.19}$$

with $\zeta_{l,k}$ denoting Gaussian independent noise, to estimate $\mathbf{H}_l \mathbf{C}(\mathbf{d}_{l,k})$, $l = 1, \ldots, L$, and $\mathbf{f}_k$ using blind deconvolution techniques (see [172]).

4.2   MOTION ESTIMATION FROM COMPRESSED OBSERVATIONS

To estimate the HR motion vectors from compressed LR observations a straightforward approach is to upsample these observed images so that any of the motion estimation methods described in the previous section are applied on them. For example, Altunbasak *et al.* [2, 3] and Patti and Altunbasak [136] follow the approach in [139] to obtain $\mathbf{d}$. However, richer models have been proposed for compressed LR observations. Such models take into account the process of obtaining the compressed LR observations from the HR images and motion vectors, and are described next. The complete process is depicted in Fig. 4.4, where the topic of this section is addressed by the system inside the shaded area. The two differences between the

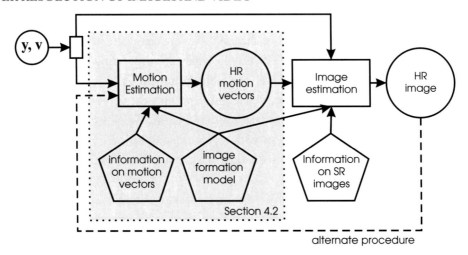

FIGURE 4.4: Graphical depiction of the process to sequentially estimate the HR motion vector (information about the image formation is used) and the HR image. Note that once we have an HR image estimate we can re-estimate the HR motion vectors (alternate procedure, dashed line)

systems in Figs. 4.1 and 4.4 are in the observed data and the system "image formation model" in Fig. 4.4, which is providing an input to both the motion and image estimation systems.

Let $\tilde{\mathbf{f}}_k$ be an estimate of the original HR image $\mathbf{f}_k$. Then, as we know, an estimate of the HR motion vectors $\mathbf{d}$ can be obtained by applying the Bayesian paradigm by either solving

$$\tilde{\mathbf{d}} = \arg\max_{\mathbf{d}} \mathbf{P}(\mathbf{d}|\tilde{\mathbf{f}}_k)\mathbf{P}(\mathbf{y}|\tilde{\mathbf{f}}_k, \mathbf{d}), \qquad (4.20)$$

where $\mathbf{P}(\mathbf{y}|\tilde{\mathbf{f}}_k, \mathbf{d})$ has been defined in Eq. (3.28) or by solving

$$\tilde{\mathbf{d}} = \arg\max_{\mathbf{d}} \mathbf{P}(\mathbf{d}|\tilde{\mathbf{f}}_k)\mathbf{P}(\mathbf{y}, \mathbf{v}|\tilde{\mathbf{f}}_k, \mathbf{d}), \qquad (4.21)$$

if the observation model also uses the LR motion vectors $\mathbf{v}$ provided by the codec (see Eq. (3.29)). In both cases $\mathbf{P}(\mathbf{d}|\tilde{\mathbf{f}}_k)$ encapsulates any prior information on the HR motion vectors to be estimated.

When

$$\mathbf{P}(\mathbf{d}|\tilde{\mathbf{f}}_k) = \mathbf{P}(\mathbf{d}) = \text{constant}, \qquad (4.22)$$

that is, when all motion vectors are equally likely and independent of the HR image (sequence), several models have been proposed to solve Eqs. (4.20) and (4.21).

Mateos *et al.* [109] combine the spatial domain model for the quantization noise in Eq. (3.23) with the distribution $\mathbf{P}(\mathbf{v}|\mathbf{f}_k, \mathbf{d}, \mathbf{y})$ defined in Eq. (3.25). The resulting block-matching cost function is given by

$$\tilde{\mathbf{d}}_{l,k} = \arg\min_{\mathbf{d}_{l,k}} \left[(\mathbf{y}_l - \mathbf{A}_l\mathbf{H}_l\mathbf{C}(\mathbf{d}_{l,k})\tilde{\mathbf{f}}_k)' \mathbf{K}_Q^{-1}(\mathbf{y}_l - \mathbf{A}_l\mathbf{H}_l\mathbf{C}(\mathbf{d}_{l,k})\tilde{\mathbf{f}}_k) + \gamma_l ||\mathbf{v}_{l,k} - \mathbf{A}_D\mathbf{d}_{l,k}||^2 \right].$$

(4.23)

Similarly, Segall *et al.* [168, 169] utilize the spatial domain noise model in Eq. (3.23) and the distribution $\mathbf{P}(\mathbf{v}|\mathbf{f}_k, \mathbf{d}, \mathbf{y})$ defined in Eq. (3.26). The cost function then becomes

$$\tilde{\mathbf{d}}_{l,k} = \arg\min_{\mathbf{d}_{l,k}} \left[(\mathbf{y}_l - \mathbf{A}_l\mathbf{H}_l\mathbf{C}(\mathbf{d}_{l,k})\tilde{\mathbf{f}}_k)' \mathbf{K}_Q^{-1}(\mathbf{y}_l - \mathbf{A}_l\mathbf{H}_l\mathbf{C}(\mathbf{d}_{l,k})\tilde{\mathbf{f}}_k) \right.$$
$$\left. + (MC_l(\mathbf{y}_l^P, \mathbf{v}_l) - \mathbf{A}_l\mathbf{H}_l\mathbf{C}(\mathbf{d}_{l,k})\tilde{\mathbf{f}}_k)' \mathbf{K}_{MV}^{-1}(MC_l(\mathbf{y}_l^P, \mathbf{v}_l) - \mathbf{A}_l\mathbf{H}_l\mathbf{C}(\mathbf{d}_{l,k})\tilde{\mathbf{f}}_k) \right]. \quad (4.24)$$

Finally, Chen and Schultz [45] use the distribution $\mathbf{P}(\mathbf{v}|\mathbf{f}_k, \mathbf{d}, \mathbf{y})$ defined in Eq. (3.24). This produces the following block-matching cost criterion

$$\tilde{\mathbf{d}}_{l,k} = \arg\min_{\mathbf{d}_{l,k} \in C_{MV}} \left[(\mathbf{y}_l - \mathbf{A}_l\mathbf{H}_l\mathbf{C}(\mathbf{d}_{l,k})\tilde{\mathbf{f}}_k)' \mathbf{K}_Q^{-1}(\mathbf{y}_l - \mathbf{A}_l\mathbf{H}_l\mathbf{C}(\mathbf{d}_{l,k})\tilde{\mathbf{f}}_k) \right], \quad (4.25)$$

where C_{MV}, following Eq. (3.24), denotes the set of displacements that satisfy the condition $|v_{l,k}(i) - [\mathbf{A}_D\mathbf{d}_{l,k}](i)| < \Delta, \forall i$.

When the quantization constraint in Eq. (3.21) is combined with the distribution $P(\mathbf{v}|\tilde{\mathbf{f}}_k, \mathbf{d}, \mathbf{y})$ we obtain as HR motion estimates

$$\tilde{\mathbf{d}}_{l,k} = \arg\max_{\mathbf{d}_{l,k} \in C_Q} P(\mathbf{v}|\tilde{\mathbf{f}}_k, \mathbf{d}, \mathbf{y})$$

(4.26)

where C_Q denotes the set of displacements that satisfy the constraint

$$-\frac{q(i)}{2} \leq [\mathbf{T}(\mathbf{A}_l\mathbf{H}_l\mathbf{C}(\mathbf{d}_{l,k})\tilde{\mathbf{f}}_k - MC_l(\mathbf{y}_l^P, \mathbf{v}_l))](i) \leq \frac{q(i)}{2}, \quad \forall i.$$

(4.27)

Together with the prior model in Eq. (4.22), explicit informative prior models for the displacements have been presented in the literature [168, 169]. There, the displacement information is assumed to be independent between frames and not dependent on the HR image so that

$$\mathbf{P}(\mathbf{d}|\tilde{\mathbf{f}}_k) = \prod_l \mathbf{P}(\mathbf{d}_{l,k}).$$

(4.28)

The displacements within each frame are then assumed to be smooth and void of coding artifacts. Note that no dependency on the HR image is assumed and, in consequence, no consistency of motion trajectories over other frames is imposed. Consistent prior motion models will be discussed in Section 4.4.

To penalize for coding errors, the displacement prior is given by

$$P(\mathbf{d}_{l,k}) \propto \exp\left[-\frac{\lambda_m}{2} \parallel \mathbf{Qd}_{l,k} \parallel^2\right], \tag{4.29}$$

where $\mathbf{Q}$ is a linear high-pass operator and λ_m the inverse of the noise variance of the normal distribution. The discrete 2D Laplacian is typically selected for $\mathbf{Q}$. With the introduction of prior information on the HR motion vectors, differential methods become common estimation methods for the displacements. These methods are based on the optical flow equation and are explored in Segall *et al.* [168, 169]. In these works, the spatial domain quantization noise model in Eq. (3.23) is combined with distributions for the motion vectors in Eq. (3.26) and displacements in Eq. (4.29). The estimation problem is then expressed as

$$\tilde{\mathbf{d}}_{l,k} = \arg\min_{\mathbf{d}_{l,k}} \left[(\mathbf{y}_l - \mathbf{A}_l\mathbf{H}_l\mathbf{C}(\mathbf{d}_{l,k})\tilde{\mathbf{f}}_k)^t \mathbf{K}_Q^{-1}(\mathbf{y}_l - \mathbf{A}_l\mathbf{H}_l\mathbf{C}(\mathbf{d}_{l,k})\tilde{\mathbf{f}}_k)\right.$$
$$+ (MC_l(\mathbf{y}_l^P, \mathbf{v}_l) - \mathbf{A}_l\mathbf{H}_l\mathbf{C}(\mathbf{d}_{l,k})\tilde{\mathbf{f}}_k)^t \mathbf{K}_{MV}^{-1}(MC_l(\mathbf{y}_l^P, \mathbf{v}_l) - \mathbf{A}_l\mathbf{H}_l\mathbf{C}(\mathbf{d}_{l,k})\tilde{\mathbf{f}}_k)$$
$$\left. + \lambda_m \mathbf{d}_{l,k}^t \mathbf{Q}^t\mathbf{Qd}_{l,k}\right]. \tag{4.30}$$

Finding the displacements is accomplished by differentiating the quantity inside the brackets in Eq. (4.30) with respect to $\mathbf{d}_{l,k}$ and setting the result equal to zero. This leads to a successive approximations algorithm [169].

Figure 4.5 shows an example of HR motion estimation from compressed observations using the algorithm in [169]. The LR-compressed frame depicted in Fig. 4.5(a), and the motion vectors transmitted from the encoder depicted in Fig. 4.5(b), are parts of the inputs to the SR algorithm in [169]. Note that the frame in Fig. 4.5(a) is the MPEG-4 compressed version

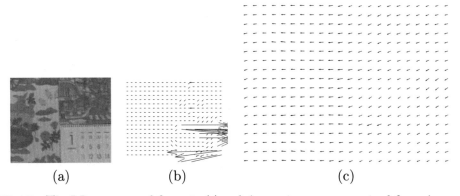

| (a) | (b) | (c) |

FIGURE 4.5: The LR-compressed frame in (a) and the motion vectors received from the encoder in (b) are used to estimate the HR motion field depicted in (c) using the algorithm in [169]

of the frame displayed in Fig. 4.2(a) and the LR motion vectors depicted in Fig. 4.5(b) were obtained by the MPEG-4 coder using the compressed frame in Fig. 4.5(a) and the original frame in Fig. 4.2(b). Also, note that the LR motion vectors in Fig. 4.5(b) are not a downsampled version of the ones displayed in Fig. 4.2(c). MPEG-4 uses the original current frame and a compressed previous frame to perform LR motion estimation in the rate-distortion sense, while the algorithm in [169] utilizes two LR-compressed frames and an algorithmic approach dictated by Eq. (4.30) to perform HR motion estimation (the result is shown in Fig. 4.5(c)). Notice that in addition to the increase in spatial resolution and the smoothness of the field, the estimation errors present in the LR motion field have been corrected in the estimated HR motion field.

An alternative differential approach is utilized by Park *et al.* [132, 133] for motion estimation. In these works, the motion between LR frames is estimated with the block-based optical flow method suggested by Lucas and Kanade [101]. Displacements are estimated for the LR frames in this case.

4.3   HOW TO DETECT UNRELIABLE MOTION ESTIMATES

As we emphasized in Chapter 3, the accuracy of the motion vectors, $\mathbf{d}$, is of the outmost importance in determining the quality of the estimated HR images. In the previous sections we have not examined the goodness or reliability of the estimated motion vectors; we will study this problem now. Accuracy of motion estimates in SR problems is a very important issue, which has not received much attention in the literature.

As is well known in motion estimation (see Fig. 3.2) there are pixels in one frame for which there are no corresponding pixels in the other frame under consideration. This is referred to as the occlusion problem, and includes covered or uncovered objects or background and objects leaving or entering the scene. Motion of nonrigid objects (i.e., elastic or deformable motion) also leads to the situation where correspondence between certain pixels in the frames cannot be established. If the rigidity of the object is a modeling assumption, the motion of nonrigid objects additionally leads to estimation errors. Similarly, other assumptions, like considering that a region, typically a block, around the pixel under consideration undergoes the same motion, lead to erroneous motion estimates.

There have been various attempts to identify the pixels in the observed LR images $\mathbf{g}_l$ or $\mathbf{y}_l$ for which no corresponding pixels can be found in the HR image $\mathbf{f}_k$ (note that the motion compensation is carried out between HR images but we only observe LR ones) and therefore they should not be taken into account when estimating $\mathbf{f}_k$. The LR pixels which can be used in the SR reconstruction problem are represented by the validity map [53], or observability map [4], or visibility map [175].

The first paper to address the use of only some areas of the observed LR images in the SR reconstruction of image $\mathbf{f}_k$ is [161]. The LR-observed images were assumed to satisfy

$$g_l(i, j) = g_k(i - v_i, j - v_j) \qquad (4.31)$$

for each LR pixel (i, j), where (v_i, v_j) is the unknown displacement vector. Resulting from the above hypothesis is that each LR pixel (i, j) in frame l can be expressed as the mean value of a block of $P \times P$ HR pixels in image $\mathbf{f}_k$ starting at the HR position $(Pi - Pv_i, Pj - Pv_j)$. The LR motion vectors (v_i, v_j) are then estimated using hierarchical block-matching. A threshold on the DFD

$$DFD^{l,k}(m, n) = |\mathbf{g}_l^\uparrow(m, n) - \mathbf{g}_k^\uparrow(m - Pv_i, n - Pv_j)|, \qquad (4.32)$$

where (m, n) denotes an HR pixel contributing to the LR pixel (i, j) in frame $\mathbf{g}_l$ and $\uparrow$ the up-sampling operation determines whether the LR observation $\mathbf{g}_l(i, j)$ is used in the reconstruction process (if the DFD is smaller than the threshold).

The above-described approach assumes white independent noise in each LR observation. As we have already explained, when instead of $\mathbf{g}$ we only have access to the LR compressed observations $\mathbf{y}$, the noise in the image formation model becomes more complex. Using Eq. (3.23) as the observation model, Alvarez et al. [4] first upsample the LR-observed images using the method proposed in [193] and then estimate the motion vectors between the upsampled images and an initial estimate of the HR image $\mathbf{f}_k$. The reconstruction method does not use the HR pixels in image l whose DFD with respect to image k is greater than a given threshold. Since those HR pixels contribute to an LR pixel in frame l which will not be taken into account when reconstructing the HR image $\mathbf{f}_k$, the covariance matrix in Eq. (3.23) has to be adapted (see [4] for details). Figure 4.6 shows two LR-observed images and the set of pixels (in black) in the first image which will not be used to reconstruct the HR image corresponding to the second image. Notice that the method discards observation which correspond to occlusions but also discards observations because of poor motion estimates. No distinction between these two very different scenarios has so far been proposed.

Eren et al. [53] (see also [54]), borrowing ideas from the reconstruction of HR images from interlaced video [138], consider the DFD between the LR image $\mathbf{g}_l$ and the corresponding motion compensated image $\mathbf{g}_k^{mc}$. Then for each LR pixel a 3×3 window is considered and the sum of the DFD is calculated. Two thresholds are used with this sum to determine if the LR pixel will be used to reconstruct the HR image $\mathbf{f}_k$. A low threshold for regions of low local variance and a high one for regions with higher local variance.

Assuming that the difference between the motion compensated upsampled LR obser-vations and the HR image $\mathbf{f}_k$ follows a normal distribution, the pdf of a given pixel $\mathbf{x}$ being visible, $\mathbf{P}_v(\mathbf{x})$, can be calculated. Note that this pdf depends on the HR image and the motion

FIGURE 4.6: From left to right, two LR-observed images and the set of pixels in the first image (in black) which will not be used to reconstruct the HR image corresponding to the second-observed image.

estimates. Furthermore, given a visibility map $\mathcal{V}$ which takes values between zero (nonvisible) and one (visible) for each pixel in the HR image, an estimate of the pdf , $\mathbf{P}_{nv}(\mathbf{x})$, of a given pixel $\mathbf{x}$ being nonvisible can also be calculated. This distribution corresponds to the histogram of the current estimate of the HR image where the contribution of each pixel is weighted by $(1 - \mathcal{V}(\mathbf{x}))$ (see [175] for details). The mean value of the distribution which takes the value one with probability $\mathbf{P}_v(\mathbf{x})$ and zero with probability $\mathbf{P}_{nv}(\mathbf{x})$ is used to update the visibility map.

4.4   CONSISTENCY OF MOTION ESTIMATES FOR SUPER RESOLUTION

In this chapter we have so far presented the HR motion estimation step of the sequential and alternate methodologies, as well as, the detection of unreliable motion estimates in SR image reconstruction. We now consider the consistency of the estimated motion vectors over several frames. While the use of validity, observability, or visibility maps, discussed in the previous section, modifies the conditional pdf of the observations given the HR image and motion

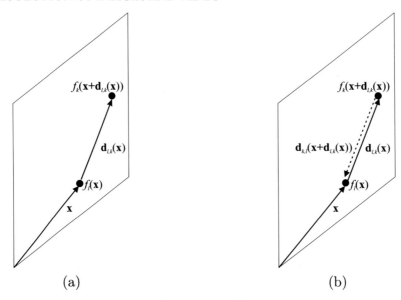

(a)   (b)

FIGURE 4.7: Graphical depiction of (a) Eq. (4.33) and (b) forward–backward consistency constraint in Eq. (4.35), also referred to as the anti-symmetry relations

vectors, consistency leads to the use of prior models on the motion vectors that depend on the HR images.

Let us consider again Eq. (3.1), which can be rewritten as

$$f_l(\mathbf{x}) = f_k(\mathbf{x} + \mathbf{d}_{l,k}(\mathbf{x})). \tag{4.33}$$

The intensity value at location $\mathbf{x} + \mathbf{d}_{l,k}(\mathbf{x})$ in the kth frame, $\mathbf{f}_k$, can also be predicted from the lth frame, $\mathbf{f}_l$, to obtain

$$f_k(\mathbf{x} + \mathbf{d}_{l,k}(\mathbf{x})) = f_l(\mathbf{x} + \mathbf{d}_{l,k}(\mathbf{x}) + \mathbf{d}_{k,l}(\mathbf{x} + \mathbf{d}_{l,k}(\mathbf{x}))). \tag{4.34}$$

The forward and backward motion estimates are expected to be consistent and so they should satisfy (see Fig. 4.7)

$$\mathbf{d}_{l,k}(\mathbf{x}) + \mathbf{d}_{k,l}(\mathbf{x} + \mathbf{d}_{l,k}(\mathbf{x})) = 0. \tag{4.35}$$

Let us now consider frames l and m to be motion compensated by frame k. Predicting frame l from frame m we obtain

$$f_l(\mathbf{x}) = f_m(\mathbf{x} + \mathbf{d}_{l,m}(\mathbf{x})), \tag{4.36}$$

and predicting frame m from frame k we have

$$f_m(\mathbf{x} + \mathbf{d}_{l,m}(\mathbf{x})) = f_k(\mathbf{x} + \mathbf{d}_{l,m}(\mathbf{x}) + \mathbf{d}_{m,k}(\mathbf{x} + \mathbf{d}_{l,m}(\mathbf{x}))). \tag{4.37}$$

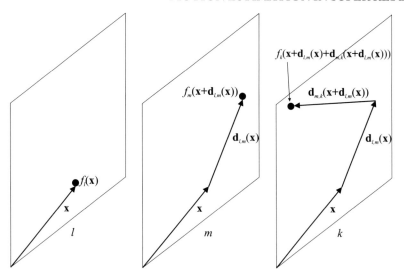

FIGURE 4.8: Motion of one pixel from frame l, to frame m, to frame k

Figure 4.8 shows a graphical depiction of Eqs. (4.36) and (4.37). Using now also the prediction of frame l by frame k finally produces the motion estimation constraint, depicted in Fig. 4.9,

$$\mathbf{d}_{l,m}(\mathbf{x}) + \mathbf{d}_{m,k}(\mathbf{x} + \mathbf{d}_{l,m}(\mathbf{x})) = \mathbf{d}_{l,k}(\mathbf{x}). \tag{4.38}$$

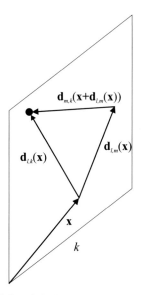

FIGURE 4.9: Graphical depiction of the global motion consistency equation (4.38), referred to as the Jacobi condition

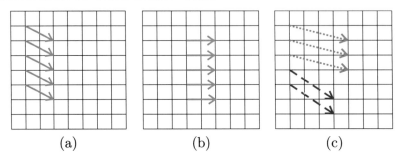

FIGURE 4.10: (a) Motion vectors relating pixels in frames $k-2$ and $k-1$. (b) Motion vectors relating pixels in frames $k-1$ and k. (c) Motion vectors relating pixels in frames $k-2$ and k. The dotted motion vectors in (c) are consistent with the vectors in (a) and (b) while the dashed ones are nonconsistent

Figure 4.10 illustrates the concept of consistency of motion vectors over three frames. Figure 4.10(a) depicts the motion vectors compensating five pixels in frame $k-2$ from five pixels in frame $k-1$. Those five pixels in frame $k-1$ are motion compensated by the corresponding pixels in frame k utilizing the motion vectors in Fig. 4.10(b). Finally, Fig. 4.10(c) depicts the motion vectors compensating the same five pixels in frame $k-2$ by pixels in frame k. In this figure, the dotted vectors in Fig. 4.10(c) are consistent with the motion vectors between frames $k-2$ and $k-1$ (Fig. 4.10(a)) and between frames $k-1$ and k (Fig. 4.10(b)), i.e., they satisfy Eq. (4.38). The dashed motion vectors in Fig. 4.10(c) are, however, nonconsistent with the vectors shown in the previous two images, i.e., they do not satisfy Eq. (4.38).

Zhao and Sawhney [203] address the forward–backward consistency of motion estimates in SR image reconstruction following the approach in [19]. Instead of using Eq. (4.33) to estimate motion vectors, a unique vector $\mathbf{d}_l(\mathbf{x})$ is calculated satisfying

$$f_l\left(\mathbf{x} - \frac{1}{2}\mathbf{d}_l(\mathbf{x})\right) = f_k\left(\mathbf{x} + \frac{1}{2}\mathbf{d}_l(\mathbf{x})\right). \qquad (4.39)$$

This approach aims at estimating the motion vectors with respect to an image located between frames l and k. For any other frame m to be registered with respect to frame k we also have

$$f_m\left(\mathbf{x} - \frac{1}{2}\mathbf{d}_m(\mathbf{x})\right) = f_k\left(\mathbf{x} + \frac{1}{2}\mathbf{d}_m(\mathbf{x})\right). \qquad (4.40)$$

To add consistency to the above motion estimates the equation

$$f_m\left(\mathbf{x} - \frac{1}{2}\mathbf{d}_m(\mathbf{x})\right) = f_l\left(\mathbf{x} - \frac{1}{2}\mathbf{d}_l(\mathbf{x})\right). \qquad (4.41)$$

is also enforced in the motion estimation process.

The above ideas are then extended to motion compensating a whole sequence with respect to a given frame (see [203]). Notice that there are some problems in this approach since the motion compensation is not performed with respect to the image to be reconstructed $\mathbf{f}_k$.

A different approach toward consistent motion estimation has been developed in [6]. The motion vectors, $d_{l,l+1}, l = 1, \ldots, k-1$, and $d_{l+1,l}, l = k, \ldots, L-1$, are first estimated using the Lucas–Kanade algorithm [101]. Then, for $l = k-2, \ldots, 1, d_{l+1,k}$ is estimated first and then $d_{l,k}$ is estimated by regularizing the Lucas–Kanade method with the term $\|d_{l,k}(\mathbf{x}_0) - d_{l+1,k}(\mathbf{x}_0 + d_{l,l+1}(\mathbf{x}_0))\|^2$. Similarly, for $l = k+2, \ldots, L, d_{l-1,k}$ is estimated first and then $d_{l,k}$ is estimated by regularizing the Lucas–Kanade method with the term $\|d_{l,k}(\mathbf{x}_0) - d_{l-1,k}(\mathbf{x}_0 + d_{l,l-1}(\mathbf{x}_0))\|^2$.

Let us consider again the forward–backward consistency equation (see Eq. (4.35)) and the global consistency equations (see Eq. (4.38)) for three frames. These consistency equations have been named as the skew anti-symmetry relations and the Jacobi condition respectively [68]. The global consistency constraint is similar to the round about constraint utilized in stereo pairs, including both disparity and motion vectors [178, 189]. Let us denote by $\mathbf{u}$ all the motion vectors between all the images in the sequence, being $\mathbf{d}$ a subset of it. All consistency conditions can be written as

$$\Psi(\mathbf{u}) = 0. \tag{4.42}$$

Farsiu *et al.* [56] investigate methods to obtain motion estimates satisfying the skew anti-symmetry relations and the Jacobi condition. Two approaches are proposed. The first consists of finding the motion vectors using the Lucas–Kanade method (either using the $l2$ or the $l1$ norms) and then project the estimated motion vectors in the space defined by Eq. (4.42). The second consists of using $\|\Psi(\mathbf{u})\|^2$ as a regularizer to the Lucas–Kanade method. The case of translational and affine motions are studied in depth in [56]. Note that the use of $\|\Psi(\mathbf{u})\|^2$ is similar to the method proposed in [6] to estimate consistent motion vectors.

4.5   SOME OPEN ISSUES IN MOTION ESTIMATION FOR SUPER RESOLUTION

In the previous sections in this chapter we have examined the currently used approaches to estimate the motion vectors for solving SR problems. We want to highlight here some motion-related issues that we believe need attention in SR problems. We exclude from this section, issues that have already been addressed in the previous sections, for instance region tracking, validity maps, and consistency of motion estimates. The list is not exhaustive and reflects our view on interesting motion estimation problems.

When several images are motion-compensated to the same HR location work needs to be done in fusing them. Figures 4.11(a), 4.11(b), and 4.11(c) represent three frames of an observed LR sequences. The first two frames in this figure together with five additional ones are used to

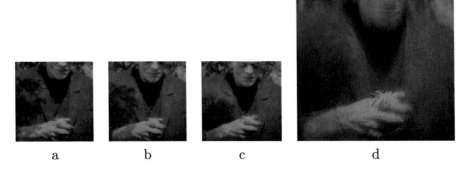

a   b   c   d

FIGURE 4.11: (a), (b), and (c) represent three frames of an observed LR sequences. Frames (a) and (b) together with five additional ones are used to improve the image in (c). The Shift and Add method in [52] applied to the observed LR sequence produces a blurred SR reconstruction shown in (d)

improve the image in Fig. 4.11c. The obtained super-resolved image using the Shift and Add method in [52] is shown in Fig. 4.11(d). It can be observed that since the additions of seven motion compensated images results in a blurred HR image.

As we already know, accurate estimates of the HR displacements are critical for the SR problem. There is work to be done in designing methods for estimating the motion utilizing blurred, subsampled, aliased, and in some cases blocky observations. Toward this goal, the use of probability distributions on optical flow as developed by Simoncelli *et al.* [171] and Simoncelli [170], as well as, coarse-to-fine estimation seem to be areas worth further exploration. Note that we could also incorporate the coherence of the distribution of the optical flow when constructing the SR image (or sequence). The use of band-pass directional filters on SR problems also appears to be an interesting area of research (see [122] and [35]). The idea of learning motion vectors seems also very promising [61, 63], even more so now that belief propagation methods used to find the motion estimates have been extended to cope with continuous variables [176]. An issue that has not received, so far, much attention (see however [155]) is the study of the accuracy limit of motion estimation.

Finally, we want to mention here an SR issue related to motion estimation which to our knowledge has not been addressed in the literature. In all the described work the magnification factor, P, has been assumed constant for the whole image or region of interest. However, due to the object motion we may be able to increase the resolution more in some parts of the images than in others. The development of a spatially adaptive (due to motion) magnifying factor has not been addressed in the literature.

CHAPTER 5

Estimation of High-Resolution Images

Following the Bayesian paradigm, in Chapter 3, we have modeled the relationship between the observed LR images and the HR image and motion vectors, with the use of conditional distributions. That is, we have modeled $\mathbf{P}(\mathbf{o}|\mathbf{f}_k, \mathbf{d})$, where $\mathbf{o}$ denotes the observation vector consisting of either the uncompressed LR observations $\mathbf{g}$ or both the LR-compressed observations $\mathbf{y}$ and the LR motion vectors $\mathbf{v}$ (plus possibly additional information provided by the bitstream, such as, the quantizer step sizes). We have then devoted Chapter 4 to the study of the distribution $\mathbf{P}(\mathbf{d}|\mathbf{f}_k)$ (which in most cases is considered independent from the HR image $\mathbf{f}_k$) and also to the estimation of the motion vectors, $\mathbf{d}$.

The next step in the application of the Bayesian paradigm to the SR problem is to study the prior models that have been proposed in the literature for the HR images, $\mathbf{P}(\mathbf{f}_k)$. We will then end up with all the elements of the Bayesian modeling of the SR problem, that is, with all the quantities in the right-hand side of the equation

$$\mathbf{P}(\mathbf{f}_k, \mathbf{d}, \mathbf{o}) = \mathbf{P}(\mathbf{f}_k)\mathbf{P}(\mathbf{d}|\mathbf{f}_k)\mathbf{P}(\mathbf{o}|\mathbf{f}_k, \mathbf{d}). \tag{5.1}$$

As we already described in Chapter 2, the work in the SR literature has primarily dealt with the following two methodologies which provide approximations to the solution

$$\hat{\mathbf{f}}_k, \hat{\mathbf{d}} = \arg\max_{\mathbf{f}_k, \mathbf{d}} \mathbf{P}(\mathbf{f}_k, \mathbf{d}|\mathbf{o}) = \arg\max_{\mathbf{f}_k, \mathbf{d}} \mathbf{P}(\mathbf{f}_k)\mathbf{P}(\mathbf{d}|\mathbf{f}_k)\mathbf{P}(\mathbf{o}|\mathbf{f}_k, \mathbf{d}). \tag{5.2}$$

Those methodologies were briefly described in Chapter 2, but they are also described next since they can be placed now in a better context.

Algorithm 1. Alternate SR methodology

1. Obtain $\hat{\mathbf{f}}^0$, an initial estimate of the HR sequence $\mathbf{f}$ (upsampling of the observed LR sequence is normally used)

2. Set $i = 0$

3. Until a convergency criterion is met

(a) Using $\hat{\mathbf{f}}^i$, find an estimate of the HR motion vectors $\mathbf{d}$, $\hat{\mathbf{d}}^{i+1}$, by any of the methods described in the previous chapter according to Eq. (5.2)

(b) Using $\hat{\mathbf{d}}^{i+1}$, find an estimate of the HR image $\mathbf{f}_k$, $\hat{\mathbf{f}}_k^{i+1}$, by any of the methods to be described in this chapter according to Eq. (5.2)

(c) Update $\hat{\mathbf{f}}^{i+1}$, the new estimated HR sequence. This step may consist of only replacing $\hat{\mathbf{f}}_k^i$ by $\hat{\mathbf{f}}_k^{i+1}$ in the sequence $\hat{\mathbf{f}}^i$

(d) Set $i = i + 1$

4. Set $\hat{\mathbf{f}}_k = \hat{\mathbf{f}}_k^i$

and

Algorithm 2. Sequential SR methodology

1. Obtain $\tilde{\mathbf{f}}$, an initial estimate of the HR sequence $\mathbf{f}$ (upsampling of the observed LR sequence is normally used)

2. Using $\tilde{\mathbf{f}}$, find an estimate of the HR motion vectors $\mathbf{d}$, $\hat{\mathbf{d}}$, by any of the methods described in the previous chapter (not necessarily utilizing Eq. (5.2))

3. Using $\hat{\mathbf{d}}$, find the estimate of the HR image $\mathbf{f}_k$, $\hat{\mathbf{f}}_k$, by any of the methods to be described in this chapter according to Eq. (5.2).

As we have described in Chapter 2, Bayesian modeling and inference provides a stronger and more flexible framework than the one needed to simply obtain estimates approximating the solution of Eq. (5.2) by the sequential or alternate methodologies. We study, in the next chapter, Bayesian inference models that go beyond these two methodologies in estimating the HR image and motion vectors.

In the following, we analyze the prior image models, $\mathbf{P}(\mathbf{f})$, used in the image estimation steps in the alternate (step 3b) and sequential (step 3) methodologies for uncompressed observations first in Section 5.1, followed by compressed observations in Section 5.2. Finally, the chapter concludes with a discussion on some image estimation-related issues which we consider relevant to the reconstruction of HR images from LR observations.

5.1   HIGH-RESOLUTION IMAGE ESTIMATION FROM UNCOMPRESSED SEQUENCES

To simplify notation, let us denote by $\mathbf{d}$, instead of $\tilde{\mathbf{d}}$, the current estimate of the HR motion vectors required to compensate any HR frame in the sequence to the kth frame. These motion vectors have been obtained by any of the motion estimation methods for uncompressed images described in Section 4.1. The intensity information has then to be estimated by solving

$$\hat{\mathbf{f}}_k = \arg \max_{\mathbf{f}_k} \mathbf{P}(\mathbf{f}_k)\mathbf{P}(\mathbf{d}|\mathbf{f}_k)\mathbf{P}(\mathbf{g}|\mathbf{f}_k, \mathbf{d}). \qquad (5.3)$$

Notice that for the alternate methodology, $\hat{\mathbf{f}}_k$ denotes the HR image estimate at a given iteration of Algorithm 1, that is, we have removed the iteration index for simplicity.

Most models proposed in the literature assume that the image intensities and motion vectors are independent, i.e., $\mathbf{P}(\mathbf{d}|\mathbf{f}_k) = \mathbf{P}(\mathbf{d})$. With such an assumption, the consistency of the motion vectors is not included in the model (see Section 4.4). Consequently, the above estimation equation becomes

$$\hat{\mathbf{f}}_k = \arg \max_{\mathbf{f}_k} \mathbf{P}(\mathbf{f}_k)\mathbf{P}(\mathbf{g}|\mathbf{f}_k, \mathbf{d}). \qquad (5.4)$$

The quality of the SR estimate greatly depends on our prior knowledge on the structure of $\mathbf{f}_k$, expressed through the model $\mathbf{P}(\mathbf{f}_k)$. There are a number of choices for it. Some HR reconstruction methods assign the same prior probability (noninformative prior) to all possible HR images $\mathbf{f}_k$ (see, for example, Stark and Oskoui [173] and Tekalp et al. [180]). This is mathematically stated in terms of prior probabilities as

$$\mathbf{P}(\mathbf{f}_k) \propto \text{constant}. \qquad (5.5)$$

With respect to the regularization framework this amounts to not utilizing a regularizer on the image. Then, in order to obtain an estimate of the HR image the error, $\mathbf{e}_{l,k}$, $l = 1, \ldots, L$, in the formation model below is minimized

$$\mathbf{g}_l = \mathbf{A}_l\mathbf{H}_l\mathbf{C}(\mathbf{d}_{l,k})\mathbf{f}_k + \mathbf{e}_{l,k} , \quad l = 1, \ldots, L , \qquad (5.6)$$

where the characteristics of the noise are defined by Eqs. (3.11), (3.12), or (3.13), and $\mathbf{A}_l$, $\mathbf{H}_l$, and $\mathbf{C}(\mathbf{d}_{l,k})$ are, respectively, the downsampling, blur, and motion compensation matrices that relate the observed LR image $\mathbf{g}_l$ to the HR image $\mathbf{f}_k$ (although the warp–blur model is used in the above equation and in the rest of the chapter, the discussion is equally applicable to the blur–warp model).

Gaussian modeling of $\mathbf{e}_{l,k}$ in combination with the noninformative prior in Eq. (5.5) has been used in the work by Irani and Peleg [80] (see also references in [24] for the so-called *simulate and correct methods*).

POCS is the method used in [173, 180] to find an estimate of the HR image assuming a flat prior model on $\mathbf{f}_k$. Most of the work on POCS for HR image estimation use the acquisition model that imposes constraints on the maximum difference between each component of $\mathbf{g}_l$ and $\mathbf{AHC}(\mathbf{d}_{l,k})\mathbf{f}_k$ (which corresponds to the use of a uniform model for $\mathbf{e}_{l,k}$ in Eq. (5.6)) with no regularization on the HR image. See however [24] for the introduction of convex

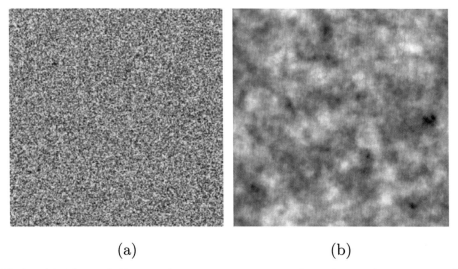

(a)   (b)

FIGURE 5.1: (a) 256×256 Image whose components are independent realizations over the range $\{0, 1, \ldots, 255\}$; no spatial structure is present in the image, (b) Sample from an SAR model for an image of size 256×256; note the spatial relationship between pixels in the image

sets as *a priori* constraints on the image using the POCS formulation, and also Tom and Katsaggelos [186, 187].

In addition to the noninformative prior mentioned so far, another model for $\mathbf{P}(\mathbf{f})$ is based on the common assumption or desirable property for $\mathbf{f}_k$ that it is smooth within homogeneous regions. A typical way to model this idea is to use the following prior distribution

$$\mathbf{P}(\mathbf{f}_k) \propto \exp\left[-\frac{\lambda}{2}\|\mathbf{Q}\,\mathbf{f}_k\|^2\right], \qquad (5.7)$$

where $\mathbf{Q}$ represents a linear high-pass operator that penalizes the estimates that are not smooth and λ controls the variance of the prior distribution (the higher the value of λ the smaller the variance of the distribution).

Figure 5.1(a) shows a realization of a random field with a uniform distribution while Fig. 5.1(b) shows a realization of a random field with the pdf in Eq. (5.7) which corresponds to a simultaneous autoregresive (SAR) model in the statistical literature [152]. Spatial structure is apparent in the latter case.

We now describe with a synthetic example the combined use of observation, HR motion, and image models in an SR problem. Let us consider the original 256×256 HR image, $\mathbf{f}_1$ (in practical applications we will not have access to it), shown in Fig. 5.2(a). This image is first convolved with a 4×4 mean filter (this convolution simulates the sensor integration procedure), and then with a Gaussian blur of variance $\sigma^2 = 2.25$. The resulting, still HR image is shown

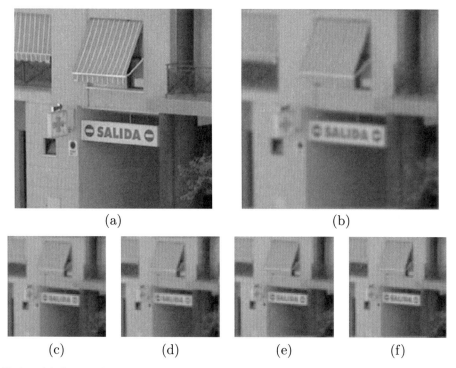

(a)   (b)

(c)   (d)   (e)   (f)

FIGURE 5.2: (a) Original 256×256 HR image, $\mathbf{f}_1$. (b) Blurred still HR image using a 4×4 mean filter plus a Gaussian blur of variance $\sigma^2 = 2.25$, (c)–(f) 64×64 Observed LR images $\mathbf{g}_1, \mathbf{g}_4, \mathbf{g}_{13}$, and $\mathbf{g}_{16}$, respectively

in Fig. 5.2(b). We proceed now to synthetically obtain the LR observations $\mathbf{g}_1, \mathbf{g}_2, \ldots, \mathbf{g}_{16}$. For $i, j = 0, \ldots, 3$, we consider the operator $\mathbf{A}_{i,j}$ that downsamples the image in Fig. 5.2(b) by a factor of four in each direction starting at the HR pixel position (i, j). The observed LR images are obtained as

$$\mathbf{g}_{l(i,j)} = \mathbf{A}_{i,j}\mathbf{H}\mathbf{f}_1 + \mathbf{e}_{l(i,j),1}, \tag{5.8}$$

where $\mathbf{H}$ represents the concatenation of the mean and Gaussian filters, $l(i, j) = i + j * 4 + 1$, and $\mathbf{e}_{l(i,j),k}$ denotes Gaussian noise with SNR $= 30$ dB.

Observe that Eq. (5.8) can also be written as

$$\mathbf{g}_{l(i,j)} = \mathbf{A}_{0,0}\mathbf{H}\mathbf{C}(\mathbf{d}_{l(i,j),1})\mathbf{f}_1 + \mathbf{e}_{l(i,j),1}, \tag{5.9}$$

where $\mathbf{C}(\mathbf{d}_{l(i,j),1})$ shifts $\mathbf{f}_1$ by (i, j) to create the HR image $\mathbf{f}_{l(i,j)}$. Notice that while Eq. (5.8) corresponds to the blur–warp model, Eq. (5.9) represents the warp–blur modeling. As we explained in Section 3.1, the two models coincide for global translational motion. This LR observation model corresponds to the LR image formation model in Eq. (3.11). Figures 5.2(c) to 5.2(f) show the 64×64 observed LR images $\mathbf{g}_1$, $\mathbf{g}_4$, $\mathbf{g}_{13}$, and $\mathbf{g}_{16}$, respectively.

With access to only $\mathbf{g} = (\mathbf{g}_1, \ldots, \mathbf{g}_{16})$ or a subset of it we proceed to estimate $\mathbf{f}_1$. First, we upsample the images in $\mathbf{g}$ by a factor $P = 4$ and register all of them with respect to the upsampled version of $\mathbf{g}_1$. The registration being performed using block matching, with the estimated motion vector is denoted by $\hat{\mathbf{d}}$. Note that $\hat{\mathbf{d}}$ contains sixteen 2D vectors, each one corresponding to an estimate of the HR motion vector between $\mathbf{f}_l$ and $\mathbf{f}_1$, $l = 1, \ldots, 16$.

Using the HR image model in Eq. (5.7) with $\mathbf{Q}$ the Laplacian operator (SAR model), the goal of the sequential methodology is to find $\hat{\mathbf{f}}_1$ satisfying

$$\hat{\mathbf{f}}_1 = \arg\min_{\mathbf{f}_1} \left\{ \frac{\lambda}{2} \|\mathbf{Q}\,\mathbf{f}_k\|^2 + \sum_{l(i,j)\in\mathcal{L}} \frac{\beta}{2} \|\mathbf{g}_{l(i,j)} - \mathbf{A}_{0,0}\mathbf{HC}(\hat{\mathbf{d}}_{l(i,j),1})\mathbf{f}_1\|^2 \right\} , \qquad (5.10)$$

where λ and β are the inverses of the HR image and LR observation model variances and $\mathcal{L}$ denotes a subset of the 16 LR-observed images (we will examine how the SR reconstruction improves with the number of observations). Notice that in real problems, the blur operator will not be known, usually the HR motion will be more complex than a simple translation and λ and β will have to be estimated. We will analyze the estimation of these parameters in the next chapter.

The optimization in Eq. (5.10) generates the results shown in Fig. 5.3. Figure 5.3(a) shows the reconstructed HR image using only the observation $\mathbf{g}_1$. Figures 5.3(b)–5.3(d) show three reconstructions using two LR images ($\mathbf{g}_1$ and $\mathbf{g}_3$), four LR images ($\mathbf{g}_1, \mathbf{g}_3, \mathbf{g}_7, \mathbf{g}_{13}$), and the 16 LR observations, respectively. Observe that with only four LR observation the "SALIDA" sign can be well distinguished.

Two other prior models are based on the Huber functional and were proposed by Schultz and Stevenson [161] and Hardie et al. [75] (see Park et al. [135] for additional references on prior models). More recently, total variation [32], anisotropic diffusion [90], and compound models [149] have all been applied to the SR problem.

Farsiu et al. [58] suggest the use of the so-called bilateral total variation prior model (some authors do not consider the sum of the $l1$-norm of the first derivatives a good approximation of the total variation) defined by

$$\mathbf{P}(\mathbf{f}_k) \propto \exp\left[-\lambda \sum_{l=-P,}^{P} \sum_{\substack{m=0, \\ l+m \geq 0}}^{P} \alpha^{|l|+|m|} \|\mathbf{f}_k - S_x^l S_y^m \mathbf{f}_k\|_1 \right], \qquad (5.11)$$

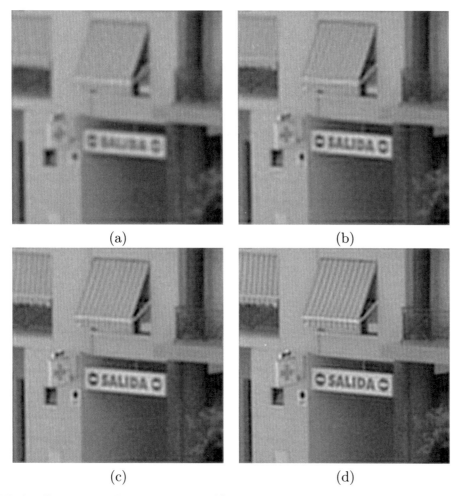

(a)

(b)

(c)

(d)

FIGURE 5.3: Reconstructed HR image using (a) only observation $\mathbf{g}_1$; (b) two LR images ($\mathbf{g}_1$ and $\mathbf{g}_3$); (c) four LR images ($\mathbf{g}_1, \mathbf{g}_3, \mathbf{g}_7, \mathbf{g}_{13}$); and (d) all 16 LR observations

where $\|\theta\|_1 = \sum |\theta_i|$ denotes the l1-norm, λ is a scale parameter, the operators S_x^l and S_y^m shift $\mathbf{f}_k$ by l and m pixels in the horizontal and vertical directions, respectively, and $0 < \alpha < 1$ is applied to provide a spatially decaying effect to the summation of the l1-norm terms.

An example of SR reconstruction using the bilateral total variation prior model [58] is presented in Fig. 5.4. The method uses a GGMRF to model the noise in the image formation process (see Eq. (3.13)), and utilizes the sequential methodology to estimate the HR motion vectors using the hierarchical motion estimation method described in [15]. A sequence consisting of 16 LR frames captured with a commercial video camera was used. The (unknown) camera PSF was assumed to be an 11×11 Gaussian kernel with variance $\sigma^2 = 2.25$ and a

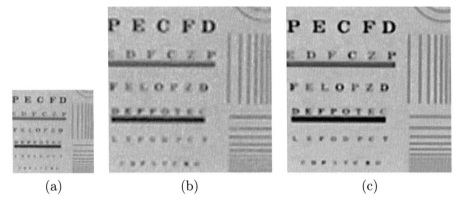

(a)   (b)   (c)

FIGURE 5.4: SR image reconstruction applied to a sequence of 16 LR frames. (a) An 80 × 80 part of the first LR frame in the sequence; (b) *Corel® Paint Shop Pro® X* smart size filtering of the image in (a); (c) super resolved image using the method in [58]

4 × 4 mean was used as the (unknown) sensor integration operator. Figure 5.4(a) shows an 80 × 80 part of the first frame in the sequence. Figure 5.4(b) displays the result of applying the smart size filter in *Corel® Paint Shop Pro® X* to the frame in Fig. 5.4(a) to increment its resolution by a factor of 4 in each direction. The reconstructed HR image, using the MDSP resolution enhancement software [55], is shown in Fig. 5.4(c).

In the previous paragraphs we have presented certain image prior models (regularizers) that can be used in combination with the observation model to find an estimate of the HR image once the motion vectors are known. Obviously, many more prior models can be used since the problem is formally the same with the one that appears in Bayesian image restoration (reconstruction), where a large number of prior models have been used. It still remains an open problem determining which image prior model is the *best* for a particular HR reconstruction problem.

Most of the prior models described above require a careful definition of it at the boundaries of the HR image (this is also true in most cases for the LR image formation model). Ng and Yip [125] examine the use of the Neumann boundary condition on the image (the scene outside the observed area is a reflection of the original scene with respect to the boundary) for quadratic observation and image models. They also examine the use of cosine transform preconditioners for solving linear systems of equations arising from the HR image reconstruction with multisensors problem (see also [124]).

The prior image models mentioned so far were defined in the intensity domain, where also the processing (using Algorithms 1 and 2) is taking place. A departure from this is the work by Chan *et al.* [37] in which the degradation model by Bose and Boo [25] is used.

The observation model in [25] can be viewed as passing the HR image through a blurring kernel built from the tensor product of a univariate low-pass filter of the form $[\frac{1}{2} + \epsilon, 1, \ldots, 1, \frac{1}{2} - \epsilon]$, where ϵ is the displacement error, followed by downsampling of the blurred image and noise addition. If the one-dimensional downsampling by a factor K starting at position $l = 0, \ldots, K - 1$ is denoted by D_l^K, then observing the whole set of LR images corresponds to the application of the tensor product $D_{l1}^K \otimes D_{l2}^K$, $l1, l2 = 0, \ldots, K - 1$, to the blurred HR image, which is formally the same with a classical image restoration problem.

Using the LR image formation model in [25], Chan *et al.* [37] propose the use of biorthogonal wavelets whose scaling function corresponds to the blurring function associated with the blurring kernel. The use of wavelets is combined with the assumption that the prior HR image distribution is given by

$$\mathbf{P}(\mathbf{f}_k) \propto \exp\left[-\frac{\lambda}{2}\|\mathbf{f}_k\|^2\right]. \tag{5.12}$$

When the displacement error ϵ in the blurring kernel is LR image dependent, alternative wavelet decompositions have been proposed in [38, 39].

As another example Willett *et al.* [200] formulate the LR observation model in terms of the discrete wavelet transform (DWT) coefficients θ of the HR image $\mathbf{f}_k$ [104] and use

$$\mathbf{P}(\theta) \propto \exp[-\tau\|\theta\|_1] \tag{5.13}$$

as the image prior model, where τ is a scale parameter.

We finally describe learning techniques applied to the modeling of the HR image prior as pertaining to the SR problem. Capel and Zisserman [33] propose the modeling of HR images using principal component analysis (PCA). PCA applied to SR imaging is basically a prior distribution learning technique from a dataset of HR images. A linear transformation chooses a new coordinate system for the images such that the greatest variance by any projection of the dataset comes to lie on the first axis (called the first principal component), the second greatest variance on the second axis, and so on. PCA is then used for reducing the dimensionality of a dataset while retaining those characteristics of it that contribute the most to its variance, by keeping lower-order principal components and ignoring higher-order ones. Once the principal components have been found, the goal in SR imaging is to calculate the coefficients of $\mathbf{f}_k$ in the new (reduced) coordinate system.

As a PCA example, Fig. 5.5(a) shows a 32×84 part of a registration plate. In Fig. 5.5(b), the image represented by its 33 (out of 10 000 total components) is shown. The information is clearly very well preserved by these 33 components.

Mathematically, given a set of HR training images, the HR image $\mathbf{f}_k$ is modeled as

$$\mathbf{f}_k = V\mathbf{p} + \mu, \tag{5.14}$$

(a)   (b)

FIGURE 5.5: (a) Image of a part of car registration plate of size 32×84; (b) the same plate represented using the first 33 principal components obtained from the whole set of possible four digit plates

where V represents the set of the principal components basis vectors and μ the average of the training images.

Given V and μ the SR problem can be formulated in terms of estimating $\mathbf{f}_k$ directly or estimating $\mathbf{p}$ instead. In the latter case the following prior model can be used on $\mathbf{p}$,

$$\mathbf{P}(\mathbf{p}) \propto \left\{ -\frac{1}{2} \mathbf{p}^t \Sigma^{-1} \mathbf{p} \right\}, \tag{5.15}$$

where Σ is the diagonal matrix of component variances obtained from the PCA.

In the former case we can proceed as follows to define a prior model on $\mathbf{f}_k$. Let

$$\hat{\mathbf{p}} = \arg \min_{\mathbf{p}} \| \mathbf{f}_k - V\mathbf{p} - \mu \|^2. \tag{5.16}$$

Then we have (note that $V^t V = I$)

$$\hat{\mathbf{p}} = V^t(\mathbf{f}_k - \mu), \tag{5.17}$$

and in consequence

$$\| \mathbf{f}_k - V\hat{\mathbf{p}} - \mu \|^2 = \| (I - VV^t)(\mathbf{f}_k - \mu) \|^2. \tag{5.18}$$

We can then use the following HR image prior model in the intensity domain

$$\mathbf{P}(\mathbf{f}_k) \propto \exp \left[-\frac{\lambda}{2} \| (I - VV^t)(\mathbf{f}_k - \mu) \|^2 \right], \tag{5.19}$$

where λ is again the scale parameter.

The above PCA methodology has also been applied to obtain HR face images utilizing eigenfaces [73]. As described in [42], eigen-image decomposition for a class of similar objects is currently a very popular and active area of research in pattern recognition with, we believe, some potential in SR imaging. Terms like eigen-faces, eigen-shape, eigen-car, etc., are increasingly being used in the literature to specify the domain of recognition. The concept derives its origin from the above-described task of finding principal components of a signal from an ensemble of its observations. See [42] for the application of eigen-image concepts to SR from only one LR observation.

The works by Capel and Zisserman [33] on the application of PCA to SR and Baker and Kanade [9] on learning priors from databases of images similar to the one to be reconstructed can also be considered as a recognition-based SR imaging approach. These works do not aim at only reconstructing the gray levels in the image but also identifying the image as one in the database.

In [9], image priors are learnt using the following approach. Let us assume that we know the process to obtain the LR observation $\mathbf{g}_k$ from its corresponding HR image $\mathbf{f}_k$. Let us also assume that we have access to a dataset of images $\mathcal{D} = \{\mathbf{u}_1, \ldots, \mathbf{u}_m\}$ which share common characteristics with the HR image we want to reconstruct; for example, a dataset of face images when the goal is the reconstruction of an HR face.

Since the process of obtaining the LR observation is known we can apply it to the HR images in $\mathcal{D}$ to obtain their corresponding LR observations $\mathcal{D}^\downarrow = \{\mathbf{u}_1^\downarrow, \ldots, \mathbf{u}_m^\downarrow\}$. Let us now define a set of operators $\mathcal{O} = \{O_1, \ldots, O_R\}$ on the LR images. These operators consist of first- and second-order derivatives applied at different image resolutions.

For each pixel i in the LR observation $\mathbf{g}_k$, the image $\mathbf{u}_k^\downarrow$ in $\mathcal{D}^\downarrow$ which contains the pixel $n(i)$ with most similar characteristics in the least square sense is found, that is,

$$[\hat{m}(i), \hat{n}(i)] = \arg\min_{m,n} \sum_{r=1}^{R} \lambda_r \|(O_r \mathbf{g}_k)(i) - (O_r \mathbf{u}_m^\downarrow)(n)\|^2 \tag{5.20}$$

where λ_r weights the contribution of each operator O_r to the total sum of squares. Note that $m(i)$ denotes an image in the dataset $\mathcal{D}^\downarrow$ and $n(i)$ corresponds to the pixel in such image most similar to pixel i in the LR observation $\mathbf{g}_k$.

The LR observation $\mathbf{g}_k(i)$ is obtained from a set of HR image values in image $\mathbf{f}_k$ whose pixel positions are generically denoted by $i^\uparrow$. Each pixel $i^\uparrow$ in image $\mathbf{f}_k$ is associated with the pixel $\hat{n}^\uparrow(i^\uparrow)$ in image $\mathbf{u}_{\hat{m}(i)}$ which is obtained from pixel $\hat{n}(i)$ in $\mathbf{u}_{\hat{m}(i)}^\downarrow$ the same way pixel $i^\uparrow$ in $\mathbf{f}_k$ is obtained from pixel i in $\mathbf{g}_k$. Similarity of characteristics between $\mathbf{f}_k$ and the images in the dataset is enforced with the use of the following Gaussian prior on the HR image

$$\mathbf{P}(\mathbf{f}_k) = \prod_{i^\uparrow} \exp\left[-\frac{1}{2} \sum_{s=1}^{S} \mu_s \|(Q_s \mathbf{f}_k)(i^\uparrow) - (Q_s \mathbf{u}_{\hat{m}(i)})(\hat{n}^\uparrow(i^\uparrow))\|^2 \right]. \tag{5.21}$$

The set of operators $\mathcal{Q} = \{Q_1, \ldots, Q_S\}$ also consists of first- and second-order derivatives applied at different image resolutions. Note that $\mathcal{Q}$ consists of operator which are applied to HR images while those on $\mathcal{O}$ are applied to LR images; furthermore, the same set of first- and second-order operators we do not have to apply during the learning and reconstruction–

recognition phases. Each μ_s weights the contribution of operator Q_s to the total sum of squares.

To illustrate the use of the method in [9], let us assume that we have access to the original HR image depicted in Fig. 5.6(a). This HR image, of size 96×128, was first convolved using a 2×2 mean filter simulating sensor integration and then downsampled by a factor of two in each direction. The observed 48×64 LR image is displayed in Fig. 5.6(b). This image is shown upsampled by pixel replication in Fig. 5.6(c) for a better visual comparison.

A subset of the Color FERET database [142, 143] was used to learn the image prior. Only 200 grayscale frontal images were included in the training database. All the images in the training database were resized to the size of the HR image and were aligned with an 96×128 upsampled version of the observed LR image. This alignment was performed by hand marking the location of the center of the eyes and the tip of the nose. Using these three points, an HR affine warp was estimated and applied to the images in the database to guarantee that all faces appear in approximately the same position.

The images in the database were then downsampled to the size of the LR-observed image using the procedure carried out on the original HR image to obtain the LR observation (notice that this process is assumed to be known in [9]). For the observed LR image and each image in the downsampled training database, a four-level Gaussian pyramid was created by recursively convolving the images with a 5×5 Gaussian filter with $\sigma = 0.5$ and downsampling them by a factor of two. The first- and second-order derivatives in the horizontal and vertical directions, as well as, the Laplacian was calculated for all images in the pyramids (201 pyramids). These first- and second-order derivatives and the Laplacian applied to the image pyramids constitute the set $\mathcal{O}$. The best match (image, pixel) for each pixel in the observed LR is found in the database using Eq. (5.20). The search for the minimum is carried out in a 7×7 window centered at each image pixel. Similarity of characteristics between the HR image $\mathbf{f}_k$ and the images in the database is then enforced by using the prior in Eq. (5.21) where $\mathcal{Q}$ consists of the first-order derivatives applied at four different resolutions. Note that in this equation the quantities $(Q_s \mathbf{u}_{\hat{m}(i)})(\hat{n}^\uparrow(i^\uparrow))$ are numbers calculated from the registered HR images in the database while $(Q_s \mathbf{f}_k)(i^\uparrow)$ is written in terms of the derivatives of the unknown image $\mathbf{f}_k$. The resulting estimated HR image is displayed in Fig. 5.6(d).

The works by Capel and Zisserman [33] and Baker and Kanade [9] fall, in terms of prior modeling, between the use of generic prior models and the use of sampling exemplars from training images in SR imaging problems. The sampling of exemplars has been pursued by Candocia and Principe [30] and Freeman et al. [62]. See also the work by Pickup et al. [144] and Wang et al. [196] which addresses the presence of an unknown blur.

Freeman et al. [62] consider the problem of obtaining an HR image from one LR observation. The method has so far not been extended to video sequences. Their approach

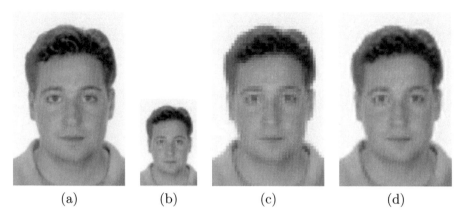

(a)   (b)   (c)   (d)

FIGURE 5.6: (a) Real HR image. (b) Observed LR image. (c) Upsampling of the observed LR image in (b) by pixel replication. (d) Reconstructed HR image using the method in [9]

starts from a collection of HR images which are degraded using the same process that was applied to the original HR image $\mathbf{f}_k$ in order to obtain the observed LR image $\mathbf{g}_k$. An initial analytic interpolation, using for example cubic splines, is then applied to the LR images. We now have two collections of images of the same size; collection $\mathcal{A}$, consisting of the HR images, and collection $\mathcal{B}$ consisting of the upsampled degraded images.

Freeman *et al.* [62] assume that the highest spatial frequency components of the upsampled LR images are the most important in predicting the extra details in their corresponding HR images. They therefore filter out the lowest frequency components in all images in $\mathcal{B}$. They also assume that the relationship between high- and low-resolution patches of pixels is essentially independent of local image contrast, and thus, they apply a local contrast normalization. Images in $\mathcal{A}$ are also high-pass filtered and locally contrast normalized. $\mathcal{A}$ and $\mathcal{B}$ no longer contain images but contrast normalized filtered version of the images originally in these collections. An example is shown in Fig. 5.7. An LR image is shown in Fig. 5.7(a) and its interpolated version in Fig. 5.7(c). The HR image is shown in Fig. 5.7(b) while the contrast normalized high-pass and band-pass filtered versions of the images in Figs. 5.7(b) and 5.7(c) are shown respectively in Figs. 5.7(d) and 5.7(e). The image in Fig. 5.7(d) belongs to $\mathcal{A}$ while the image in Fig. 5.7(e) belongs to $\mathcal{B}$. Each 5×5 patch, P, in the images in $\mathcal{A}$, is now associated with a patch, Q, of size 7×7 in the corresponding image in $\mathcal{B}$ that contains P in its center.

The HR image to be estimated is then divided into overlapping 5×5 patches. To each of these patches P corresponds a patch of size 7×7 Q in the upsampled version of the image $\mathbf{g}_k$. The collection $\mathcal{B}$ is searched to find the most similar patches (typically 16) to Q. These 7×7 patches in $\mathcal{B}$ and the corresponding 5×5 patches in $\mathcal{A}$ are used in the reconstruction process. The prior model of the SR problem is defined as follows: for overlapping 5×5 patches

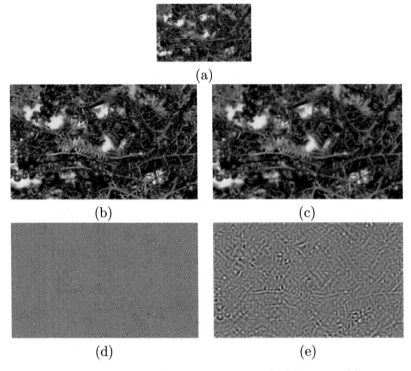

FIGURE 5.7: Image pre-processing steps for training images. (a) LR image; (b) its corresponding HR source; (c) initial interpolation of the LR image to the higher pixel sampling resolution; (d) and (e) contrast normalized high-pass and band-pass filtered versions of the images in (b) and (c), respectively. Images (d) and (e) are now utilized in the training set.

the shared pixels should have similar values, and therefore the differences between neighboring patches in their overlapping regions is calculated and assumed to have a Gaussian distribution. The prior model consists of the product of all these Gaussian distributions over overlapping patches. The observation model uses again a quadratic penalty function on differences between the filtered upsampled observed LR image patch and the candidate filtered upsampled LR patch (as we have explained earlier, typically 16) found in $\mathcal{B}$.

The MAP estimate represents the HR reconstructed image. Finding the MAP estimate can be computationally intractable; the authors in [62] use the belief propagation algorithm in [140] for graphs without loops. Note that the graph that links neighboring patches in the HR image and each 5×5 HR patch to its corresponding 7×7 LR patch has loops and so the algorithm in [140], which is designed for graphs with no loops, does not converge to the global MAP (see [61]). An extension of the belief propagation algorithm in [61, 62] to continuous variables can be found in [176].

In this book we are studying SR methods which in most cases take the form of an iterative spatial procedure. There are, however, other interesting methods which we would like to briefly mention here.

Interpolation–restoration SR methods consider the SR problem as an interpolation problem with nonuniformly sampled data followed, in some cases, by restoration techniques. The paper by Sauer and Allebach [159] can be considered as the first paper in this area (see Nguyen and Milanfar [127] for a short but very interesting summary of interpolation–restoration methods and more recently [22]). Among others, the following papers could be classified as using interpolation–restoration techniques. Chiang and Boult [46] couple warping and resampling in their SR method. Rajan and Chaudhuri [149] propose a generalized interpolation method by dividing the SR image into subspaces. Fruchter and Hook [64] present a local SR interpolation method ("drizzle") for astronomical images. Zakharov *et al.* [202] use the Karhunen–Loeve expansion of the objects to be super resolved and Vandewalle *et al.* [192] use frequency domain-based methods and bicubic interpolation to obtain SR images. Some SR methods performed in a transform domain can also be classified as interpolation–restoration techniques (see for instance [127], and also Daubos and Murtagh [48] where the registration is performed in the wavelet domain). An SR method which could be considered to be as an interpolation method is based in dynamic trees and is presented in [174]. Finally, learning interpolation kernels in SR problems is proposed in [30].

5.2   HIGH-RESOLUTION IMAGE ESTIMATION FROM COMPRESSED SEQUENCES

To simplify notation, let us denote again by $\mathbf{d}$, instead of $\tilde{\mathbf{d}}$, the current estimate of the HR motion vector required to compensate any HR frame in the sequence to the kth frame. This motion vector has been obtained by any of the motion estimation methods for compressed images described in Section 4.2. The intensity information has then to be estimated by solving

$$\hat{\mathbf{f}}_k = \arg\max_{\mathbf{f}_k} \mathbf{P}(\mathbf{f}_k)\mathbf{P}(\mathbf{d}|\mathbf{f}_k)\mathbf{P}(\mathbf{y}, \mathbf{v}|\mathbf{f}_k, \mathbf{d}), \tag{5.22}$$

where $\mathbf{y}$ and $\mathbf{v}$ denote, respectively, the LR compressed observation and LR motion vectors provided by the encoder. The various observation models $\mathbf{P}(\mathbf{y}, \mathbf{v}|\mathbf{f}_k, \mathbf{d})$ used in the literature have been examined in Section 3.2 and the Bayesian modeling of the SR problem has been described in Chapter 2. Notice that for the alternate methodology, $\hat{\mathbf{f}}_k$ in Eq. (5.22) denotes the HR image estimate at a given iteration of Algorithm 1; we have removed the iteration index for simplicity.

As in the case of uncompressed observations, most of the models proposed in the literature to reconstruct HR images from LR compressed observations assume that the HR image $\mathbf{f}_k$ and motion vectors **d** are independent, that is, $\mathbf{P}(\mathbf{d}|\mathbf{f}_k) = \mathbf{P}(\mathbf{d})$. With such an assumption, the consistency of the motion vectors is not included in the model (see Section 4.4). Consequently, the above estimation equation becomes:

$$\hat{\mathbf{f}}_k = \arg\max_{\mathbf{f}_k} \mathbf{P}(\mathbf{f}_k)\mathbf{P}(\mathbf{y}, \mathbf{v}|\mathbf{f}_k, \mathbf{d}). \qquad (5.23)$$

All the prior HR image models $\mathbf{P}(\mathbf{f}_k)$ proposed in the previous section for uncompressed observations can also be used to reconstruct HR images from LR-compressed observations. Note, however, that compression introduces artifacts in the observed LR images. In general, these coding artifacts are taken into account by introducing in the prior image model the requirement that the original HR image should be free of them. Let us briefly describe these compression artifacts.

The first type of artifacts are blocking artifacts (an example is shown in Fig. 5.8). They are objectionable and annoying at all bitrates of practical interest, and their severity increases as the

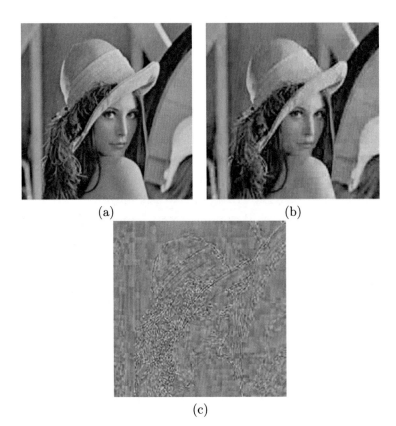

(a)

(b)

(c)

FIGURE 5.8: (a) Original 8 bpp image; (b) DCT-based compressed image at 0.46 bpp (compression ratio 53:1); (c) difference between original and compressed images making visible the blocking artifacts

bitrate decreases (or the compression ratio increases). They are due to the fact that with most codecs and all the image and video compression standards, the image is artificially segmented into equally sized blocks which are processed independently. The first step of the processing is the application of a de-correlation operator, typically the DCT. Since the transform considers each block independently, pixels outside the block region are ignored and the continuity across block boundaries is not captured. This is perceived as a synthetic, grid-like error, and sharp discontinuities appear between blocks in the reconstructed image varying regions. Note that DCT is applied to images and also to differences between image frames and their predictions by motion compensation (see Section 3.2 for details).

The transformed intensities within each block are then independently quantized. Block artifacts can be further enhanced by poor quantization decisions. Compression standards do not define the strategy for allocating the available bits among the various components the original source has been decomposed into (i.e., motion, texture, and in some cases segmentation). Instead, the system designer has complete control of providing solution to this so-called rate control problem. This allows for the development of encoders for a wide variety of applications. Very often heuristic allocation strategies are utilized, which may assign considerably different quantizers to neighboring regions even though they have similar visual content. Embedded in this comment is the fact that it is in general difficult to assess the visual quality of an image or video by objective metrics. The commonly used mean squared error (MSE) or peak signal-to-noise ratio (PSNR) metrics do not always reflect the visual quality of a reconstructed image.

Other artifacts are also attributed to the improper allocation of bits. In satisfying delay constraints, encoders operate without knowledge of future sequence activity. Thus, bits are allocated based on a prediction of such future activity. When this prediction is invalid, the encoder must quickly adjust the degree of quantization to satisfy a given rate constraint. The encoded video sequence possesses a temporally varying image quality, which manifests itself as a temporal flicker.

High spatial frequency features, such as, edges and impulse-like features introduce another type of coding error. In still image compression, the transformed intensities are typically quantized by a spectrally shaped or perceptually weighted quantizer, which performs uniform fine quantization of the low-frequency coefficients and uniform coarse quantization of the high-frequency coefficients. This leads to oscillations or ringing in the vicinity of strong edges, as demonstrated by the example of Fig. 5.9. In video compression, it is the transformed motion-based prediction error (DFD) which is quantized. Typically a flat across frequencies quantizer is used in this case. Large DCT values of the DFD which indicate a serious failure of the motion estimation and compensation step, are eliminated in this case. This failure can be due to a number of factors, like the particular implementation of motion estimation (the simple

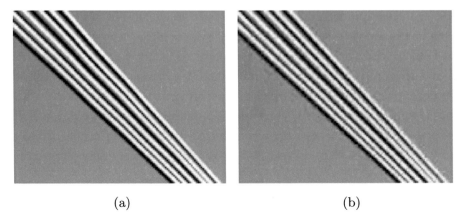

(a)   (b)

FIGURE 5.9: (a) Original image; (b) DCT-based compressed image at a compression rate 0.42 bpp. Notice the ringing artifacts around the edges

fact that block-based motion estimation is performed, the fact that the size of the search region may not be adequate, etc.) and the available bitrate. Boundaries of moving objects are typically present in the DFD, representing the high spatial frequency information which is reduced or eliminated during quantization, thus giving rise to ringing or mosquito artifacts.

To encapsulate the statement that images are correlated and absent of blocking and ringing artifacts, the prior distribution

$$\mathbf{P}(\mathbf{f}_k) \propto \exp\left[-\frac{1}{2}\left(\lambda_1\|\mathbf{Q}_1\mathbf{f}_k\|^2 + \lambda_2\|\mathbf{Q}_h\mathbf{f}_k\|^2 + \lambda_3\|\mathbf{Q}_v\mathbf{f}_k\|^2\right)\right] \qquad (5.24)$$

is utilized in [165, 166] (and references therein). Here, $\mathbf{Q}_1$ represents a linear high-pass operation that penalizes super-resolution estimates that are not smooth, $\mathbf{Q}_h$ and $\mathbf{Q}_v$ represent respectively linear horizontal and vertical high-pass operators that penalize estimates with blocking artifacts, and λ_1, λ_2, and λ_3 control the influence of the norms. A common choice for $\mathbf{Q}_1$ is the discrete 2D Laplacian while common choices for $\mathbf{Q}_h$ and $\mathbf{Q}_v$ are simple horizontal and vertical difference operators applied at the block boundary locations. Other distributions could also be incorporated into the estimation procedure. For example, Huber's function could replace the quadratic norm, as discussed in [45].

Figure 5.10 shows the application of the SR method in [108] to reconstruction an HR image from a sequence of compressed LR images. The method uses a sequential methodology in which the HR motion vectors are first estimated using a linear combination of an upsampled version of the LR motion vectors provided by the coder and the HR motion estimates obtained by utilizing block matching techniques on an initial estimate of the HR image sequence. The reconstruction method makes use of the Gaussian observation model in Eq. (3.23) and the

(a)   (b)   (c)

FIGURE 5.10: (a) Frame 3 of the "mobile" LR video sequence compressed at 256 kpbs. (b) Reconstruction by *Corel® Paint Shop Pro® X* smart size filtering. (c) Reconstruction by the SR algorithm in [108]. The method in [108] clearly increases the resolution of the compressed image and reduces the blocking and ringing artifacts

prior model in Eq. (5.24). Additionally, the proposed algorithm enforces temporal smoothness between the HR frame $\mathbf{f}_k$ and estimates of the remaining HR images in the HR video sequence $\mathbf{f}$.

The color "mobile" sequence was used in this experiment. Each frame, of size 720×576 pixels, was subsampled to obtain a common intermediate format (CIF) frame (progressively scanned format with 360×288 pixels/frame at 30 frames/sec). These frames represent the original HR sequence. They were further subsampled horizontally and vertically by a factor of two, to obtain the quarter common intermediate format (QCIF) (180×144 pixels/frame at 30 frames/sec.) LR frames. The first 40 frames of the sequence were compressed at 256 kbps using the baseline mode MPEG-4 video coding standard. Figure 5.10(a) shows the central part of the third compressed LR frame of the sequence. Its HR reconstruction using the smart size filter in *Corel® Paint Shop Pro® X* is depicted in Fig. 5.10(b). The HR reconstruction obtained with the SR algorithm in [108] is displayed in Fig. 5.10(c). The method in [108] clearly increases the resolution of the compressed image and reduces the blocking and ringing artifacts.

5.3   SOME OPEN ISSUES IN IMAGE ESTIMATION FOR SUPER RESOLUTION

In the previous sections in this chapter we have examined the currently used approaches for estimating the HR image in SR problems. We want to highlight here some issues on the topic that we believe deserve additional attention. We exclude from this section, issues that have been already addressed in the previous sections.

While the modeling of the HR image in the spatial domain may still be of some interest, we believe that its modeling in a transformed domain has a greater potential. Among others, the use of scale mixtures of Gaussians in the wavelet domain [146] and the works by Figueiredo and Nowak [59], and Bioucas-Dias [18] are of interest in SR problems. Sparse priors [92, 182] could also be used in SR problems. However, there are, in some cases, probably conflicts between the prior on the HR image and the measurements in the LR observations. In all possible domains, spatial and transformed, we believe that learning the characteristics of the elements providing the representation from a database is also of interest. This approach would include, for instance, the learning of derivatives in the spatial domain, coefficients (or their derivatives) in the wavelet domain and also the coefficients in a sparse representation. The learning approach will have to address the problem of selecting the database based on which the various characteristics should be learnt from. This is a very interesting problem, that could probably be addressed with the use of model selection techniques [87, 119, 162].

CHAPTER 6

Bayesian Inference Models in Super Resolution

Throughout this book we have been studying the Bayesian modeling and inference for SR image reconstruction problems. This study has provided us with a sound justification of the process of finding the estimates of HR image and motion vectors using the alternate and sequential methodologies (see Chapter 2). These methodologies have been shown to provide approximations to the solution of the Bayesian estimation problem

$$\hat{\mathbf{f}}_k, \hat{\mathbf{d}} = \arg \max_{\mathbf{f}_k, \mathbf{d}} \mathbf{P}(\mathbf{f}_k)\mathbf{P}(\mathbf{d}|\mathbf{f}_k)\mathbf{P}(\mathbf{o}|\mathbf{f}_k, \mathbf{d}), \qquad (6.1)$$

where $\mathbf{o}$ denotes the observation vector consisting of either the uncompressed LR observations $\mathbf{g}$ or both the LR-compressed observations $\mathbf{y}$ and the motion vectors $\mathbf{v}$, and $\mathbf{f}_k$, and $\mathbf{d}$ denote the HR image and motion vectors, respectively.

Bayesian inference provides us with a stronger and more flexible framework than the one needed to simply obtain estimates approximating the solution of Eq. (6.1). Together with the simultaneous estimation of the HR images and motion vectors, we examine in this chapter how the Bayesian paradigm can handle three very important SR issues which have not received much interest in the literature; namely, (i) the estimation of the unknown parameters involved in the prior and conditional distributions, (ii) the marginalization in $\mathbf{f}_k$ or $\mathbf{d}$ to obtain $\mathbf{P}(\mathbf{d}|\mathbf{o})$ and $\mathbf{P}(\mathbf{f}_k|\mathbf{o})$, respectively, and (iii) the calculus in closed form, the approximation, or the simulation of the posterior distribution $\mathbf{P}(\mathbf{f}_k, \mathbf{d}|\mathbf{o})$.

The chapter is organized as follows. In Section 6.1, we formulate the HR reconstruction problem within the Bayesian framework. This formulation will include as unknowns in addition to the HR image and motion vectors all the parameters in the prior and observation models. A more realistic and practical formulation of the problem thus results. In Section 6.2, we discuss solution approaches to the HR reconstruction problem different from the sequential and alternate ones. These solutions will be studied as inference models under the Bayesian framework. Finally, the chapter concludes with a discussion on some SR Bayesian modeling and inference issues which we consider relevant to the reconstruction of HR images from LR observations.

6.1   HIERARCHICAL BAYESIAN FRAMEWORK FOR SUPER RESOLUTION

A fundamental principle of the Bayesian philosophy is to regard all observable and unobservable variables as unknown stochastic quantities, assigning probability distributions to them based on subjective beliefs. Thus in our SR problem the original image $\mathbf{f}_k$, the motion vectors $\mathbf{d}$, and the observations $\mathbf{o}$ are all treated as samples of random fields, with corresponding probability distributions that model our knowledge about the imaging process and the nature of the HR image and motion vectors. These distributions depend on parameters which will be denoted by Ω. The parameters of the prior distributions are termed *hyperparameters*.

Often Ω is assumed known (or is first estimated separately from $\mathbf{f}_k$ and $\mathbf{d}$). Alternatively, we may adopt the *hierarchical* Bayesian framework whereby Ω is also assumed unknown, in which case we also model our prior knowledge of their values. The probability distributions of the hyperparameters are termed *hyperprior* distributions. This *hierarchical* modeling allows us to write the joint global distribution

$$\mathbf{P}(\Omega, \mathbf{f}_k, \mathbf{d}, \mathbf{o}) = \mathbf{P}(\Omega)\mathbf{P}(\mathbf{f}_k, \mathbf{d}|\Omega)\mathbf{P}(\mathbf{o}|\Omega, \mathbf{f}_k, \mathbf{d}), \tag{6.2}$$

where $\mathbf{P}(\mathbf{o}|\Omega, \mathbf{f}_k, \mathbf{d})$ is termed the *likelihood* of the observations.

Typically, as we have noted in earlier chapters, we assume that $\mathbf{f}_k$ and $\mathbf{d}$ are *à priori* conditionally independent, given Ω, that is,

$$\mathbf{P}(\mathbf{f}_k, \mathbf{d}|\Omega) = \mathbf{P}(\mathbf{f}_k|\Omega)\mathbf{P}(\mathbf{d}|\Omega). \tag{6.3}$$

The task is to perform inference using the posterior

$$\mathbf{P}(\mathbf{f}_k, \mathbf{d}, \Omega|\mathbf{o}) = \frac{\mathbf{P}(\Omega)\mathbf{P}(\mathbf{f}_k, \mathbf{d}|\Omega)\mathbf{P}(\mathbf{o}|\Omega, \mathbf{f}_k, \mathbf{d})}{\mathbf{P}(\mathbf{o})}. \tag{6.4}$$

We have studied in the previous chapters the distributions $\mathbf{P}(\mathbf{f}_k, \mathbf{d}|\Omega)$ and $\mathbf{P}(\mathbf{o}|\Omega, \mathbf{f}_k, \mathbf{d})$ that appear in the Bayesian modeling of the SR problem in Eq. (6.2). We complete this modeling by studying now the distribution $\mathbf{P}(\Omega)$. In the nomenclature of hierarchical Bayesian modeling the formulation of $\mathbf{P}(\Omega)$ constitutes the second stage, the first stage being the formulation of $\mathbf{P}(\mathbf{f}_k, \mathbf{d}|\Omega)$ and $\mathbf{P}(\mathbf{o}|\Omega, \mathbf{f}_k, \mathbf{d})$.

As with the prior and conditional distributions studied in previous chapters, there have been a number of model for $\mathbf{P}(\Omega)$. For the cases studied so far of known values for the hyperparameters they correspond to using degenerate distributions (delta functions) for the hyperpriors. A degenerate distribution on Ω is defined as

$$\mathbf{P}(\Omega) = \delta(\Omega, \Omega_0) = \begin{cases} 1, & \text{if } \Omega = \Omega_0 \\ 0, & \text{otherwise} \end{cases}. \tag{6.5}$$

When Ω is unknown most of the methods proposed in the literature use the model

$$P(\Omega) = \text{constant.} \qquad (6.6)$$

However, a large part of the Bayesian literature is devoted to finding hyperprior distributions $P(\Omega)$ for which $P(\Omega, \mathbf{f}_k, \mathbf{d}|\mathbf{o})$ can be calculated or approximated in a straightforward way. These are the so-called *conjugate priors* [16], which were developed extensively in Raiffa and Schlaifer [148]. Besides providing for easy calculation or approximations of $P(\Omega, \mathbf{f}_k, \mathbf{d}|\mathbf{o})$, conjugate priors have the intuitive feature of allowing one to begin with a certain functional form for the prior and end up with a posterior of the same functional form, but with the parameters updated by the sample information.

Taking the above considerations about conjugate priors into account, the literature in SR offers different *à priori* models for the parameters depending on their types. Unfortunately, those parameter distributions are almost exclusively of use for Gaussian modeling of the first stage of the hierarchical SR problem formulation. For parameters corresponding to inverses of variances, the gamma distribution $\Gamma(\omega|a_\omega^o, b_\omega^o)$ has been used. It is defined by

$$P(\omega) = \Gamma\left(\omega|a_\omega^o, b_\omega^o\right) = \frac{(b_\omega^o)^{a_\omega^o}}{\Gamma(a_\omega^o)} \omega^{a_\omega^o - 1} \exp[-b_\omega^o\, \omega], \qquad (6.7)$$

where $\omega > 0$ denotes a hyperparameter in Ω, and $b_\omega^o > 0$ is the scale parameter and $a_\omega^o > 0$ the shape parameter which are assumed to be known. The gamma distribution has the following mean, variance, and mode

$$E[\omega] = \frac{a_\omega^o}{b_\omega^o} \quad \text{Var}[\omega] = \frac{a_\omega^o}{(b_\omega^o)^2}, \quad \text{Mode}[\omega] = \frac{a_\omega^o - 1}{b_\omega^o}. \qquad (6.8)$$

Note that the mode does not exist when $a_\omega^o \leq 1$ and that mean and mode do not coincide.

For mean vectors, the corresponding conjugate prior is a normal distribution. Additionally, for covariance matrices Σ, the hyperprior is given by an inverse Wishart distribution (see [65]). These two hyperpriors can be of use, for instance, when the HR image prior is a multidimensional Gaussian distribution with unknown mean and covariance matrices.

To our knowledge these hyperprior models have only been used by Humblot and Mohammad-Djafari [78] in SR problems. Note that this paper also utilizes a modeling of the original HR image that includes segmentation.

6.2 INFERENCE MODELS FOR SUPER-RESOLUTION RECONSTRUCTION PROBLEMS

There are a number of different ways we may follow in estimating the hyperparameters and the HR image and motion vectors using Eq. (6.4). Depending on the chosen distributions on the right-hand side of Eq. (6.2) finding $\mathbf{P}(\mathbf{f}_k, \mathbf{d}, \Omega|\mathbf{o})$ analytically may be difficult, so approximations are often needed. The estimation methods discussed in the two previous chapters seek point estimates of the parameters $\mathbf{f}_k$ and $\mathbf{d}$ assuming that the values of the hyperparameters are known, that is, $\Omega = \Omega_0$. This reduces the SR problem to one of optimization, i.e., the maximization of $\mathbf{P}(\mathbf{f}_k, \mathbf{d}|\Omega_0, \mathbf{o})$. However, the Bayesian framework provides other methodologies [65, 84, 120] for estimating the distributions of $\mathbf{f}_k$, $\mathbf{d}$, and Ω which deal better with uncertainty; approximating or simulating the posterior distribution are two such options. These different inference strategies and examples of their use will now be presented, proceeding from the simplest to the more complex.

One possible point estimate is provided by the MAP solution, which is represented by the values of $\mathbf{f}_k$, $\mathbf{d}$, and Ω that maximize the posterior probability density:

$$\{\hat{\mathbf{f}}_k, \hat{\mathbf{d}}, \hat{\Omega}\}_{\mathrm{MAP}} = \arg\max_{\mathbf{f}_k, \mathbf{d}, \Omega} \mathbf{P}(\mathbf{f}_k, \mathbf{d}, \Omega|\mathbf{o}). \qquad (6.9)$$

Note that we have defined in Chapter 2 the MAP solution assuming that Ω is known.

The maximum likelihood (ML) estimate attempts instead to maximize the likelihood $\mathbf{P}(\mathbf{o}|\mathbf{f}_k, \mathbf{d}, \Omega)$ with respect to all the unknown parameters, that is,

$$\{\hat{\mathbf{f}}_k, \hat{\mathbf{d}}, \hat{\Omega}\}_{\mathrm{ML}} = \arg\max_{\mathbf{f}_k, \mathbf{d}, \Omega} \mathbf{P}(\mathbf{o}|\Omega, \mathbf{f}_k, \mathbf{d}). \qquad (6.10)$$

Note, however, that in this case we can only estimate the parameters in Ω that are present in the conditional distribution $\mathbf{P}(\mathbf{o}|\Omega, \mathbf{f}_k, \mathbf{d})$.

Assuming known values for Ω, that is, $\Omega = \Omega_0$, the MAP and ML HR image and motion vector estimates become respectively

$$\{\hat{\mathbf{f}}_k, \hat{\mathbf{d}}\}_{\mathrm{MAP}} = \arg\max_{\mathbf{f}_k, \mathbf{d}} \mathbf{P}(\mathbf{o}|\mathbf{f}_k, \mathbf{d}, \Omega_0)\mathbf{P}(\mathbf{f}_k, \mathbf{d}|\Omega_0) \qquad (6.11)$$

and

$$\{\hat{\mathbf{f}}_k, \hat{\mathbf{d}}\}_{\mathrm{ML}} = \arg\max_{\mathbf{f}_k, \mathbf{d}} \mathbf{P}(\mathbf{o}|\mathbf{f}_k, \mathbf{d}, \Omega_0). \qquad (6.12)$$

Let us consider again the solution of Eq. (6.11). A major problem in the optimization is the simultaneous estimation of the unknown $\mathbf{f}_k$ and $\mathbf{d}$. A widely used approach is to alternate the steps of maximizing the function $\mathbf{P}(\mathbf{o}|\mathbf{f}_k, \mathbf{d}, \Omega_0)\mathbf{P}(\mathbf{f}_k, \mathbf{d}|\Omega_0)$ with respect to one unknown while fixing the other unknown. This optimization procedure corresponds to the iterated conditional modes (ICM) approach proposed by Besag [17] and justifies the use of our sequential and alternate methodologies discussed in Chapters 4 and 5 as ICM estimation procedures.

One of the attempts to the simultaneous estimation of the HR image and motion parameters in Eq. (6.11) is the work by Hardie $et\ al.$ [74]. For uncompressed observations the authors use as HR image model

$$P(\mathbf{f}_k|\lambda) \propto \exp\left[-\frac{\lambda}{2}\parallel \mathbf{Q}\mathbf{f}_k \parallel^2\right], \tag{6.13}$$

where $\mathbf{Q}$ denotes the Laplacian operator and λ is the inverse of the known variance of the prior model (see Eq. (5.7)). The motion vectors are assumed to represent global HR image translations and their prior probability is assumed to be flat (all translations have the same probability). Finally the observation model corresponds to a global translation at the HR level followed by blur and downsampling. The noise of the LR observations is assumed to be white Gaussian.

Given an HR image estimate, the motion vectors are estimated using block-matching techniques but, in contrast to most of the methods proposed in the literature, the motion estimation is performed using the real observation model of the HR images given the HR image and motion vectors. Once the HR motion vectors are calculated the update of the HR image is performed using gradient descend methods.

Note that the method in [74] is still an alternate method and, in consequence, an ICM procedure. The same methodology has been applied by Mateos *et al.* [109] and Segall *et al.* [169] to LR compressed sequences.

Figure 6.1 shows an example of the use of the alternate methodology (ICM estimate of the HR image and motion vectors) as presented in [169]. The method starts with the distribution for the spatial domain quantization noise model defined in Eq. (3.23), and the distribution for the motion vectors in Eq. (3.26), which are combined with the prior distributions for the displacements in Eq. (4.29) and HR image in Eq. (5.24). Then, the estimation of the HR image and motion vectors corresponds to the optimization problem

$$\hat{\mathbf{f}}_k, \hat{\mathbf{d}}_{l,k} = \arg\min_{\mathbf{f}_k, \mathbf{d}_{l,k}} \big[(\mathbf{y}_l - \mathbf{A}_l\mathbf{H}_l\mathbf{C}(\mathbf{d}_{l,k})\mathbf{f}_k)^t\mathbf{K}_Q^{-1}(\mathbf{y}_l - \mathbf{A}_l\mathbf{H}_l\mathbf{C}(\mathbf{d}_{l,k})\mathbf{f}_k)$$
$$+ (MC_l(\mathbf{y}_l^P, \mathbf{v}_l) - \mathbf{A}_l\mathbf{H}_l\mathbf{C}(\mathbf{d}_{l,k})\mathbf{f}_k)^t\mathbf{K}_{MV}^{-1}(MC_l(\mathbf{y}_l^P, \mathbf{v}_l) - \mathbf{A}_l\mathbf{H}_l\mathbf{C}(\mathbf{d}_{l,k})\mathbf{f}_k)$$
$$+ \lambda_m \parallel \mathbf{Q}\mathbf{d}_{l,k} \parallel 2 + \lambda_1 \parallel \mathbf{Q}_1\mathbf{f}_k \parallel^2 + \lambda_2 \parallel \mathbf{Q}_b\mathbf{f}_k \parallel^2 + \lambda_3 \parallel \mathbf{Q}_v\mathbf{f}_k \parallel^2 \big], \tag{6.14}$$

which is solved using the alternate methodology.

The original "mobile" sequence was downsampled to CIF format to form the LR sequence. This LR sequence was compressed utilizing the MPEG-4 bitstream syntax. Besides the first image in the sequence, which is intra-coded, each frame is compressed as a P-frame. This restricts the reference frame for the motion vectors to be the temporally preceding frame. The VM5+ rate control mechanism is utilized for bit allocation to obtain a bitrate of 1 Mbps, and all frames are encoded. Figure 6.1(a) shows the 176×144 central part of a frame of the compressed LR sequence.

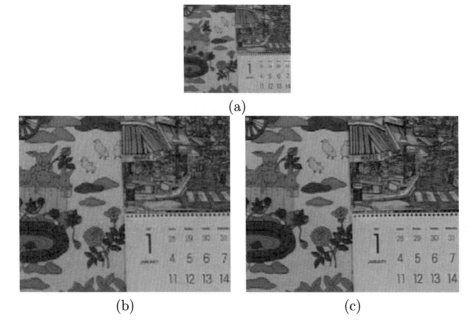

FIGURE 6.1: (a) A frame of the "mobile" LR sequence compressed at 1 Mbps. Result obtained by the application of (b) *Corel® Paint Shop Pro® X* smart size filtering (c) the SR method in [169]. The method in [169] enhances the resolution throughout the image and attenuates coding noise. Notice the sharpness of the numbers and text as well as the vertical features in the central-upper part of the frame

The algorithm is initialized using the following procedure. First, the encoded LR image $\mathbf{y}_k$ is bilinearly interpolated. This provides the initial estimate of the HR intensities, $\mathbf{f}_k^i$, $i = 0$. High-resolution motion estimate $\mathbf{d}_{l,k}^i$, $l = 1, \ldots, L$, is then calculated. Equation (6.14) with $\mathbf{f}_k = \mathbf{f}_k^i$ is employed within a multiscale framework (see [169] for the details) for this task. The procedure continues alternating between the estimation of the HR image and the HR motion vectors.

Nine frames, four before and four after the target frame and the target frame itself, were utilized to estimate the HR image $\mathbf{f}_5$ using Eq. (6.14). The 176×144 central part of the estimated HR image is shown in Fig. 6.1(c). The same central part of the reconstruction using *Corel® Paint Shop Pro® X* smart size filter is displayed in Fig. 6.1(b) for comparison purposes. The SR method in [169] produces a much sharper image (see, for instance, the numbers in the calendar or the vertical and diagonal features in the central-upper part of the frame), enhancing the resolution throughout the image and attenuating coding noise.

In the discussion so far none of the three unknowns, $\mathbf{f}_k$, $\mathbf{d}$, and Ω have been integrated first to perform inference on only a subset of them. As Tipping and Bishop [181] point out, the MAP approach and its alternate maximization approximations do not take into account the uncertainty in the HR image when estimating the registration parameters and the consequential effect on the estimation of the registration parameters.

We can therefore approach the SR inference problem by first calculating

$$\hat{\mathbf{d}}, \hat{\Omega} = \arg\max_{\mathbf{d},\Omega} \int_{\mathbf{f}_k} \mathbf{P}(\Omega)\mathbf{P}(\mathbf{f}_k, \mathbf{d}|\Omega)\mathbf{P}(\mathbf{o}|\Omega, \mathbf{f}_k, \mathbf{d})\mathrm{d}\mathbf{f}_k , \qquad (6.15)$$

and then selecting as HR reconstruction the image

$$\hat{\mathbf{f}}_k[\hat{\mathbf{d}}, \hat{\Omega}] = \arg\max_{\mathbf{f}_k} \mathbf{P}(\mathbf{f}_k|\hat{\Omega})\mathbf{P}(\mathbf{o}|\hat{\Omega}, \mathbf{f}_k, \hat{\mathbf{d}}). \qquad (6.16)$$

We can also marginalize $\mathbf{d}$ and Ω first to obtain

$$\bar{\mathbf{f}}_k = \arg\max_{\mathbf{f}_k} \int_{\mathbf{d},\Omega} \mathbf{P}(\Omega)\mathbf{P}(\mathbf{f}_k, \mathbf{d}|\Omega)\mathbf{P}(\mathbf{o}|\Omega, \mathbf{f}_k, \mathbf{d})\mathrm{d}\mathbf{d} \cdot \mathrm{d}\Omega . \qquad (6.17)$$

The above two inference models are named *Evidence-* and *Empirical*-based analysis [111], respectively. The marginalized variables are called *hidden variables.*

The expectation–maximization (EM) algorithm, first described in [49], is a very popular technique in signal processing for iteratively solving ML and MAP problems that can be regarded as having hidden variables. Its properties are well studied, that is, it is guaranteed to converge to a *local* maximum of the likelihood or the posterior distribution of the nonhidden variables given the observations. It is particularly well-suited to providing solutions to inverse problems in image restoration, blind deconvolution, and super resolution problems, since it is obvious that the unobserved image, $\mathbf{f}_k$, represents a natural choice for the hidden variables and in consequence for utilizing Eq. (6.15).

Tom and Katsaggelos [184, 185, 187] have used the EM algorithm in SR image reconstruction assuming an auto regression (AR) prior HR image model with unknown parameters and unknown global translations between HR images. The prior model used on such translation was a uniform distribution (see also [201]).

Assuming that the global translation between the HR images were known and using the SAR HR image model and the Gaussian noise image formation model, the Evidence analysis has been applied to SR in [117] to first estimate the variances of the HR image prior distribution and the image formation model and then the HR image. Additionally, a study on the quality of the reconstruction as the number of available LR observations increases is carried out in [117, 110]. We use again the original HR image in Fig. 5.2 to illustrate the use of the Evidence analysis in estimating the above-mentioned unknown variances and evaluate the effect of increasing the number of available LR observations. The 16 LR-observed images were obtained from the original HR image in Fig. 6.2(a) following the procedure used in the example corresponding to Fig. 5.2, in this case without the inclusion of a Gaussian blur. One of the LR observations is shown in Fig. 6.2(b), upsampled by pixel replication of the size of the HR image for displaying purposes. The SR algorithm in [117] which estimates the unknown variance parameters and the HR image using the Evidence analysis is applied to $L = \{1, 2, 4, 6, 8, 12, 16\}$ randomly chosen LR images. The corresponding HR image estimates are shown in Figs. 6.2(d)–6.2(j). Clearly, as the number of input images increases, the reconstructed image is closer to the original one. It is interesting to notice that after a certain point the quality of the reconstruction does not dramatically increase with the number of LR observations, since the prior distribution is able to effectively recover the missing information. Note, however, that if we only have one LR image, bicubic interpolation (Fig. 6.2(c)) is almost as effective as SR with one observation (Fig. 6.2(d)).

We have also tested this method with LR text images. In Fig. 6.3(a), we show an HR image of size 256×64 pixels. We obtained a set of 16 LR images following the procedure used in the example corresponding to Fig. 6.2, in this case, adding Gaussian noise with 40 dB SNR. Then we ran the reconstruction algorithm in [117] on different sets of randomly chosen LR images.

Figure 6.3(b) depicts the zero-order hold upsampled image of one of the LR images and Figs. 6.3(c)–6.3(f) depict respectively the estimated HR images using 1, 8, 12, and 16 LR images. Visual inspection of the resulting images shows that the HR-estimated images are considerably better than the zero-order hold upsampled image in Fig. 6.3(b). We needed, however, a greater number of LR input images than in the previous experiments to obtain an HR image with good visual quality. As can be seen, the improvement is not significant when the number of LR images increases from 12 to 16.

We want to note here that the estimation of the variances of the image and observation distribution has also been approached from the regularization point of view as is represented, for example, by the work of Bose *et al.* [27], Nguyen [126], Nguyen *et al.* [128, 129, 130], and to some extent [36] and [77].

The algorithm in [132] is derived following a regularization approach toward SR. The quantization noise (being the dominant noise) is modeled in the spatial domain as colored noise with a parameterized spatially-adaptive covariance matrix. The parameters of the covariance matrix as well as the regularization parameter are evaluated at each iteration step based on the available partially reconstructed HR image. The method builds on the results obtained in [86]. Within the Bayesian framework the algorithm in [132] can be obtained by treating the parameters of the covariance matrix and the regularization parameter (treated as the ratio of two variables) as hyperparameters and following a hierarchical Bayesian approach. This was exactly done in [113] in deriving the algorithm in [86] following a hierarchical Bayesian approach.

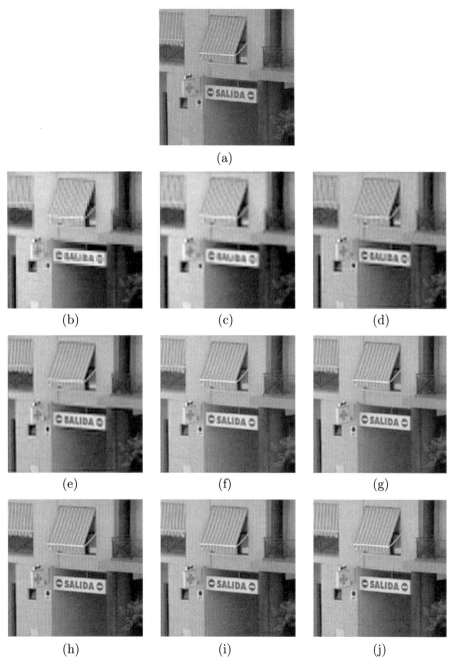

FIGURE 6.2: (a) Original HR image; (b) n LR image (out of a set of 16). It has been upsampled by pixel replication to the size of the high-resolution image for displaying purposes; (c) bicubic interpolation of the image in (b); (d) – (j) super-resolved images using the algorithm in [117] using $L = \{1, 2, 4, 6, 8, 12, 16\}$ randomly chosen LR images as input

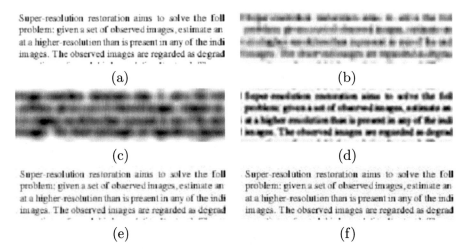

FIGURE 6.3: (a) HR image; (b) upsampled LR image; (c)–(f) reconstructions by the algorithm described in [117] using $L = \{1, 8, 12, 16\}$ LR images

Figure 6.4 depicts the experimental results. Four LR images are synthetically generated from the HR 256×256 pixels "shop" image. The HR image is translated with one of the subpixel shifts $\{(0, 0), (0, 0.5), (0.5, 0), (0.5, 0.5)\}$, blurred, decimated by a factor of two in both the horizontal and vertical directions, and compressed at 1.5 bpp with a JPEG coder–decoder (codec) to generate the 128×128 compressed LR observations. The blurring function is a 2×2 moving average filter that simulates the LR sensor point spread function. One of the LR images before compression is shown in Fig. 6.4(a) and its compressed version in Fig. 6.4(b). The subpixel shifts are estimated to be $\{(0, 0), (0, 0.46), (0.47, 0), (0.48, 0.47)\}$ with the method described in [73]. Figure 6.4(c) shows the enlarged image of one of the LR observations by bicubic interpolation, and Fig. 6.4(d) shows the HR image reconstructed by the algorithm described in [73]. For comparison, the result of the application of algorithm in [132] without the estimation of the quantization noise (the noise is treated as white, a model probably applicable at low compression ratios) is shown in Fig. 6.4(e). Finally, the result of the algorithm in [132] that includes the simultaneous estimate for the quantization noise parameters and the regularization parameter is presented in Fig. 6.4(f).

Assuming that the motion between HR images consists of shifts and rotations, Tipping and Bishop [181] use the Evidence analysis to integrate out the HR image and estimate the unknown parameters (shifts and rotations included). They compare the use of the EM and conjugate gradient (CG) algorithms to find the parameter estimates. Humblot and Mohammad-Djafari [78] also use the Evidence analysis to marginalize $\mathbf{f}_k$ and estimate all the unknown hyperparameters assuming known translational motion. Note that this work uses the proper priors defined at the end of Section 6.1 to model prior knowledge on the hyperparameters.

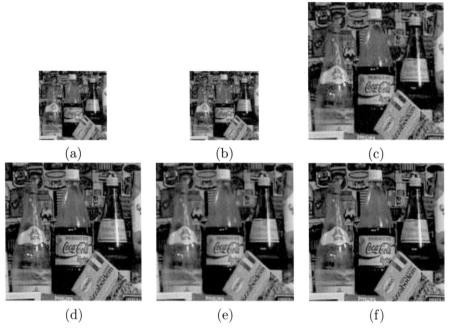

FIGURE 6.4: (a) One of the LR images before compression; (b) compressed version of the LR image in (a); (c) bicubic interpolation of the LR image in (b); HR reconstruction by (d) the algorithm in [73]; (e) the algorithm in [132] without the estimation of the quantization noise; (f) the algorithm in [132] estimating the quantization noise and regularization parameters.

The Evidence-based analysis can also be used to marginalize the image $\mathbf{f}_k$ as well as the unknown parameters Ω to obtain $\mathbf{P}(\mathbf{d}|\mathbf{o})$, and then calculate the mode of this posterior distribution. The observation model can then be used to obtain an estimate of the original HR image $\mathbf{f}_k$. This approach has been applied, for instance, in blind deconvolution (BD) problems but not in SR image reconstruction (see [20]).

It is rarely possible to calculate in closed form the integrals involved in the Evidence- and Empirical-based Bayesian inference. To address this problem we can use approximations of the integrands. Let us consider the integral in Eq. (6.15); then for each value of $\mathbf{d}$ and Ω we can calculate

$$\hat{\mathbf{f}}_k[\mathbf{d}, \Omega] = \arg \max_{\mathbf{f}_k} \mathbf{P}(\mathbf{f}_k|\Omega)\mathbf{P}(\mathbf{o}|\mathbf{f}_k, \mathbf{d}, \Omega), \qquad (6.18)$$

and perform a second-order Taylor's expansion of $\log \mathbf{P}(\mathbf{o}|\mathbf{f}_k, \mathbf{d}, \Omega)$ around $\hat{\mathbf{f}}_k[\mathbf{d}, \Omega]$. Note that we have assumed that the HR image and motion vectors are independent. As a consequence of the approximation, the integral in Eq. (6.15) is performed over a distribution on $\mathbf{f}_k$ that is Gaussian and usually easy to calculate. This methodology is called Laplace distribution approximation [87, 103], but it has not been applied to SR problems. We believe it has great potential in estimating global registration parameters in SR problems.

Variational Bayesian methods are generalizations of the EM algorithm. The EM algorithm has proven to be very useful in a wide range of applications; however, in many problems its application is not possible because the posterior distribution cannot be specified. The variational methods overcome this shortcoming by approximating $\mathbf{P}(\mathbf{f}_k, \mathbf{d}, \Omega | \mathbf{o})$ by a simpler distribution $\mathbf{Q}(\mathbf{f}_k, \mathbf{d}, \Omega)$ obtained by minimizing the Kullback-Leibler (KL) divergence between the variational approximation and the exact distribution. In addition to providing solution to the SR problem, the study of the distribution $\mathbf{Q}(\mathbf{f}_k, \mathbf{d}, \Omega)$ allows us to examine the quality of the solution.

The variational approximation applied to SR aims at approximating the intractable posterior distribution $\mathbf{P}(\mathbf{f}_k, \mathbf{d}, \Omega | \mathbf{o})$ by a tractable one denoted by $\mathbf{Q}(\mathbf{f}_k, \mathbf{d}, \Omega)$. For an arbitrary joint distribution $\mathbf{Q}(\mathbf{f}_k, \mathbf{d}, \Omega)$, the goal is to minimize the KL divergence, given by [93]

$$
\begin{aligned}
& KL(\mathbf{Q}(\mathbf{f}_k, \mathbf{d}, \Omega) \parallel \mathbf{P}(\mathbf{f}_k, \mathbf{d}, \Omega | \mathbf{o})) \\
&= \int \mathbf{Q}(\mathbf{f}_k, \mathbf{d}, \Omega) \log\left(\frac{\mathbf{Q}(\mathbf{f}_k, \mathbf{d}, \Omega)}{\mathbf{P}(\mathbf{f}_k, \mathbf{d}, \Omega | \mathbf{o})}\right) d\mathbf{f}_k \cdot d\mathbf{d} \cdot d\Omega \\
&= \int \mathbf{Q}(\mathbf{f}_k, \mathbf{d}, \Omega) \log\left(\frac{\mathbf{Q}(\mathbf{f}_k, \mathbf{d}, \Omega)}{\mathbf{P}(\mathbf{f}_k, \mathbf{d}, \Omega, \mathbf{o})}\right) d\mathbf{f}_k \cdot d\mathbf{d} \cdot d\Omega + \text{const}, \quad (6.19)
\end{aligned}
$$

which is always nonnegative and equal to zero only when $\mathbf{Q}(\mathbf{f}_k, \mathbf{d}, \Omega) = \mathbf{P}(\mathbf{f}_k, \mathbf{d}, \Omega | \mathbf{o})$, which corresponds to the EM estimation procedure.

To avoid the computational complexity caused by the dependency of the posterior distribution of $(\mathbf{f}_k, \mathbf{d}, \Omega)$ given the data, the distribution $\mathbf{Q}(\mathbf{f}_k, \mathbf{d}, \Omega)$ is factorized using

$$
\mathbf{Q}(\mathbf{f}_k, \mathbf{d}, \Omega) = \mathbf{Q}(\mathbf{f}_k)\mathbf{Q}(\mathbf{d})\mathbf{Q}(\Omega). \quad (6.20)
$$

For a vector parameter $\theta \in \{\mathbf{f}_k, \mathbf{d}, \Omega\}$, we denote by Θ_θ the subset of Θ with θ removed; for example, for $\theta = \mathbf{f}_k$, $\Theta_{\mathbf{f}_k} = (\mathbf{d}, \Omega)$ and $\mathbf{Q}(\Theta_{\mathbf{f}_k}) = \mathbf{Q}(\Omega)\mathbf{Q}(\mathbf{d})$. An iterative procedure can be developed to estimate the distributions of the parameters $(\mathbf{f}_k, \mathbf{d}, \Omega)$. At each iteration, the distribution of the parameter θ is estimated using the current estimates of the distribution of Θ_θ, that is,

$$
\mathbf{Q}^k(\theta) = \arg\min_{\mathbf{Q}(\theta)} KL(\mathbf{Q}^k(\Theta_\theta)\mathbf{Q}(\theta) \parallel \mathbf{P}(\Theta \mid \mathbf{o})). \quad (6.21)
$$

Variational distribution approximation has been applied to SR problems in [176]. This approach has also been applied recently to the SR-related problem of blind deconvolution (see Likas and Galatsanos [96] and Molina *et al.* [114]).

The most general approach to perform inference for SR problem is to simulate the posterior distribution in Eq. (6.4). This in theory allows us to perform inference on arbitrarily complex models in high-dimensional spaces, when no analytic solution is available. Markov chain Monte Carlo (MCMC) methods (see e.g. [8, 120, 158]) attempt to approximate the posterior distribution by the statistics of samples generated from a Markov chain (MC). The most simple example of MCMC is the Gibbs sampler which has been used in classical image restoration in conjunction with Markov random field (MRF) image models [66]. Once we have accumulated samples, these samples can be used to find point estimates and other statistics of their distributions. Humblot and Mohammad-Djafari [78] propose the use of estimation procedures that combine MAP and Gibbs sampling approaches for SR problems.

6.3   SOME OPEN ISSUES IN SUPER RESOLUTION BAYESIAN INFERENCE

Throughout this book we have been studying the Bayesian modeling and inference for SR image reconstruction problems. The modeling has included the definition of the observation and the HR image and motion models. We have introduced in this chapter the use of hyperpriors on the unknown hyperparameters of the above models.

The observation and prior models discussed in the three previous chapters have provided the building blocks for the application of the Bayesian alternate and sequential methodologies and have also been used to justify most of the so far proposed SR methods.

Having described in the previous two chapters some open issues on motion and image modeling, we want to indicate here some unchartered inference problem in SR.

As we mentioned at the end of Chapter 2, the book is being presented following a kind of historical order that reflects the way the various results have appeared in the literature over time. This in some sense results in presenting the various methods from the less to the more complicated ones. The alternate and sequential methodologies constitute the first and less complicated inference models. The inference models presented in this chapter constitute the next natural step in SR Bayesian inference.

In our opinion, the evidence and empirical analysis, the Laplacian and variational distribution approximations and the simulation of the posterior distributions hold great potential in SR image reconstruction. Two main reasons have probably prevented the application of such analysis (inference models) in SR problems. The first one is the fact that by using distributions different from the Gaussian one, the use of conjugate priors and the calculation of the needed integrals become much more complex. The second one is related to the fact that the use of motion models different from global translations makes again the calculation of the needed integrals a very challenging problem. Nonetheless, none of the above hard problem should prevent the application of inference models which take the uncertainty in the HR image and motion vectors into account in a more direct and effective way than the sequential and alternate methodologies.

CHAPTER 7

Super-Resolution for Compression[1]

In the previous chapters we have discussed methods to improve the spatial resolution of (possibly compressed) image and video sequences in order to obtain a level of detail that was not available in the original input due to limits on the imaging system. In this chapter, we will describe methods that use intentional downsampling of the video sequence as part of the pre-processing step before being compressed, and the application of SR techniques as a post-processing of the compressed sequence. This scheme can be utilized as a bitrate control mechanism to optimize the overall quality of the HR reconstructed video.

The main difference between the SR methods described in previous chapters and the SR for compression methods described here is that, in the latter case, the original sequence is available and, hence, the performance of the reconstruction can be evaluated, thus creating an interesting set of possibilities.

Two quite different (in terms of bits) scenarios are well suited for SR for compression. As Segall *et al.* [164] describe, although the bitrates of current high-definition systems ensure fidelity in representing an original video sequence, they preclude widespread availability of high-definition programming. For example, satellite- and Internet-based distribution systems are poorly suited to deliver a number of high-rate channels, and video on demand applications must absorb a significant increase in storage costs. Furthermore, prerecorded DVD-9 stores less than an hour of high-definition video. Additionally, with the current proliferation of HDTV sets, there is a growing number of DVD players and DVD recorders that can now upscale standard DVD playback output to match the pixel count of HDTV in order to display the signal on an HDTV set with higher definition. In this context, SR can be applied at the output of the DVD player to increase not only the resolution of the video sequence but, also, to increase its quality.

Within the same high-bitrate scenario, DVD recorders cannot record in HDTV standards. This is due to the spatial resolution used in DVD format and the limited space of current DVDs for the storage needs of HDTV signals. An alternative for storing HDTV programs

---

[1] This chapter is based on [13] and has been written in collaboration with D. Barreto, Research Institute for Applied Microelectronics, IUMA, University of Las Palmas de Gran Canaria 35017, Spain.

with DVD recorders is to pre-process and downsample the HDTV signal to a resolution supported by the standard DVD recorder prior to the compression and storing process. Then, when playing back the compressed video, an SR process can be applied to recover the signal at the original HDTV resolution. Note that the pre-processing and downsampling operations can be different depending on the characteristics of the frame being processed. In this case, it will be highly desirable that both the coder and the SR algorithm share the information about how the LR DVD video sequence has been obtained from the HDTV signal.

Another scenario for SR for compression is described by Bruckstein *et al.* [29]. The use of the DCT at low bitrates introduces disturbing blocking artifacts. It appears that at such bitrates an appropriately downsampled image compressed using the DCT and later interpolated (or filtered using SR techniques), can be visually better than the HR image compressed directly with a DCT scheme at the same number of bits.

Downsampling and upsampling play a key role in designing video codecs which support spatial scalability (e.g., MPEG-4, H.264). Such codecs split a single video source (or video object plane) into a base layer (lower spatial resolution) and continuous or discrete enhancement layers (higher spatial resolution). SR concepts and techniques can be applied to existing video compression standards that support scalability without altering the compliance to the standard. More importantly, however, is that they can be applied to the design of new compression codecs and standards.

In this chapter we present a framework of how to apply SR concepts and techniques in designing new video compression codecs. We address along the way the plethora of challenges and opportunities in such an effort. The material in this chapter is not meant to provide mature and well-studied solutions but instead to show some preliminary results of the application of this paradigm and to generate interest in it.

The chapter is organized as follows. In Section 7.1, we mathematically formulate the video compression problem and define pre- and post-processing filters that, when applied to the video sequence, provide a means to control the bitrate and image quality. Note that pre- and post-processing are not part of any of the video compression standards and in at this poing of the analysis pre-processing does not include downsampling. In Section 7.2, we present some initial results in controlling the bitrate and the image quality by downsampling the original video sequence, coding it and then, at the decoder, bringing it back to its original spatial resolution. Such an approach might require the modification of the codec structure or it might be implementable by sending side information to the decoder along with the video data.

7.1   PRE- AND POST-PROCESSING OF VIDEO SEQUENCES

When a hybrid motion-compensated and transform-based coder is used to compress the original sequence $\mathbf{f} = \{\mathbf{f}_1, \ldots, \mathbf{f}_L\}$, the reconstructed *l*th frame is given by (see Section 3.2 and

Eq. (3.16))

$$\hat{\mathbf{f}}_l = \mathbf{T}^{-1} Q \left[\mathbf{T} \left(\mathbf{f}_l - MC_l(\hat{\mathbf{f}}_l^P, \mathbf{v}_l) \right) \right] + MC_l(\hat{\mathbf{f}}_l^P, \mathbf{v}_l), \quad l = 1, \ldots, L, \qquad (7.1)$$

where $Q[.]$ represents the quantizer, $\mathbf{T}$ and $\mathbf{T}^{-1}$ the forward and inverse transforms, respectively, and $MC_l(\hat{\mathbf{f}}_l^P, \mathbf{v}_l)$ the motion-compensated prediction of $\mathbf{f}_l$ formed by motion compensating previously decoded frame(s) $\hat{\mathbf{f}}_l^P$ with the use of the motion vectors $\mathbf{v}_l$.

If the bitrate is high enough, the reconstructed image sequence, $\hat{\mathbf{f}}$ will be almost indistinguishable from the original one $\mathbf{f}$. However, in situations where the bit budget is limited, some artifacts such as blocking or ringing appear (see Section 5.2). In controlling and in achieving, in general, a better rate-distortion tradeoff, the original HR sequence can be pre-processed to obtain

$$\mathbf{b}_l = \mathbf{H}_l \mathbf{f}_l, \quad l = 1, \ldots, L, \qquad (7.2)$$

where $\mathbf{H}_l$ is a $(PM \times PN) \times (PM \times PN)$ pre-processing matrix that does not include downsampling. The sequence $\mathbf{b} = \{\mathbf{b}_1, \ldots, \mathbf{b}_L\}$ is then compressed using the compression scheme of Eq. (7.1) to obtain $\mathbf{b}^c = \{\mathbf{b}_1^c, \ldots, \mathbf{b}_L^c\}$, the compressed sequence. By taking into account the frame relation in Eq. (3.3), that is, by replacing $\mathbf{f}_l$ by $\mathbf{C}(\mathbf{d}_{l,k})\mathbf{f}_k$ we have that the lth compressed frame, $\mathbf{b}_l^c$, is given by

$$\mathbf{b}_l^c = \mathbf{T}^{-1} Q \left[\mathbf{T} \left(\mathbf{H}_l \mathbf{C}(\mathbf{d}_{l,k})\mathbf{f}_k - MC_l(\mathbf{b}_l^{c\,P}, \mathbf{v}_l) \right) \right] + MC_l(\mathbf{b}_l^{c\,P}, \mathbf{v}_l), \qquad (7.3)$$

where $MC_l(\mathbf{b}_l^{c\,P}, \mathbf{v}_l)$ is the motion-compensated prediction of $\mathbf{b}_l$ formed by motion compensating previously decoded frame(s) $\mathbf{b}_l^{c\,P}$ with the use of the motion vectors $\mathbf{v}_l$.

This sequence can now be post-processed, by a filter $\mathbf{O}_l$, to obtain another estimate of the original lth frame, that is,

$$\hat{\mathbf{f}}_l = \mathbf{O}_l[\mathbf{b}^c], \quad l = 1, \ldots, L. \qquad (7.4)$$

Note that the pre- and post-processing filters may be image dependent, the whole sequence $\mathbf{b}^c$ or a subset of it (not just $\mathbf{b}_l^c$) may be used to obtain $\hat{\mathbf{f}}_l$, and $\mathbf{O}_l$ may be a linear or a nonlinear filter on $\mathbf{b}^c$.

Molina *et al.* [112] suggest that the best post-processed frame, $\hat{\mathbf{f}}$, could, for instance, be calculated as

$$\hat{\mathbf{f}}_k = \arg \min_{\mathbf{f}_k} \sum_{l=1}^{L} \| \mathbf{T}^{-1} Q \left[\mathbf{T} \left(\mathbf{H}_l \mathbf{C}(\mathbf{d}_{l,k})\mathbf{f}_k - MC_l(\mathbf{b}_l^{c\,P}, \mathbf{v}_l) \right) \right] + MC_l(\mathbf{b}_l^{c\,P}, \mathbf{v}_l) - \mathbf{b}_l^c \|^2,$$
$$k = 1, \ldots, L, \qquad (7.5)$$

where the motion vectors $\mathbf{d}_{l,k}$, $l = 1, \ldots, L$, are assumed to be known. Another approach leads to the following minimization problem, by simply modeling $\mathbf{T}^{-1}Q\mathbf{T}$ as introducing Gaussian noise with constant variance,

$$\hat{\mathbf{f}}_k = \arg \min_{\mathbf{f}_k} \sum_{l=1}^{L} \| \mathbf{H}_l \mathbf{C}(\mathbf{d}_{l,k})\mathbf{f}_k - \mathbf{b}_l^c \|^2, \tag{7.6}$$

which is an ill-posed problem. Note that most of the SR algorithms proposed in Chapters 5 and 6 can be adapted to solve this problem by simply setting the downsampling matrix $\mathbf{A}_l$ as the identity operator. For instance, Segall [163] proposes the simultaneous estimation of the kth post-processed image and the corresponding motion vectors using the methodology developed by Segall *et al.* [169] for SR image reconstruction.

The reciprocal problem, that of designing the optimal pre-processor given knowledge of the post-processor ($\mathbf{O}_l$ in Eq. (7.4)) can also be addressed. In this case, the problem is to find the best reconstructed sequence by designing the best pre-processing matrix $\mathbf{H}_l$ for each frame l such that

$$\hat{\mathbf{H}}_l = \arg \min_{\mathbf{H}_l} \| \hat{\mathbf{f}}_l - \mathbf{f}_l \|^2 . \tag{7.7}$$

Clearly $\hat{\mathbf{f}}_l$ depends on $\mathbf{H}_l$. By simply modeling the compression process as an independent Gaussian noise process for each frame we have that

$$\hat{\mathbf{H}}_l = \arg \min_{\mathbf{H}_l} \| \mathbf{O}_l \mathbf{H}_l \mathbf{f}_l - \mathbf{f}_l \|^2, \tag{7.8}$$

which is again an ill-posed problem. A more realistic modeling of the compression process that also takes into account the previously estimated $\mathbf{H}_n$, $n < l$ is studied in Segall [163] and Segall *et al.* [169].

7.2   INCLUDING SUPER RESOLUTION INTO THE COMPRESSION SCHEME

In this section we present a framework according to which the original video sequence is intentionally downsampled in addition to pre-processing (or the pre-processing now includes downsampling), then compressed, and finally post-processed using an SR technique to bring the reconstructed sequence to the original spatial resolution [29, 112]. Clearly, since the downsampling is intentional and controlled (as opposed to an undesirable effect as was the case in the presentation so far) and, in addition, information regarding the downsampling process could be transmitted from the encoder to the decoder, this becomes a problem with new challenges and opportunities. In addition, the quality of the reconstructed HR image needs to be evaluated for a given bitrate (a rate-distortion consideration or a rate-control problem).

More specifically, an original frame $\mathbf{f}_l$ in the sequence $\mathbf{f}$ is pre-processed by a system represented by a matrix $\mathbf{H}_l$ of size $(PM \times PN) \times (PM \times PN)$ and then downsampled by a factor P using the $(M \times N) \times (PM \times PN)$ matrix $\mathbf{A}_l$, producing the LR frame

$$\mathbf{g}_l = \mathbf{A}_l \mathbf{H}_l \mathbf{f}_l, \quad l = 1, \ldots, L. \tag{7.9}$$

Note that this process is the same with the one described by Eq. (3.9) for the warp–blur model, again the only difference being that here intentional filtering and downsampling is applied in order to keep the bitrate and the distortion under control.

The sequence $\mathbf{g} = \{\mathbf{g}_1, \ldots, \mathbf{g}_L\}$ is compressed using Eq. (3.16) to obtain $\mathbf{y} = \{\mathbf{y}_1, \ldots, \mathbf{y}_L\}$ and, at the decoder, it is post-processed and upsampled using a system represented by $\mathbf{U}_l$ to obtain an estimate of the original lth frame in the sequence, that is,

$$\hat{\mathbf{f}}_l = \mathbf{U}_l[\mathbf{y}], \quad l = 1, \ldots, L. \tag{7.10}$$

Note again that the pre- and post-processing may be image dependent, the whole sequence $\mathbf{y}$ or a subset of it (not just $\mathbf{y}_l$) may be used to obtain $\hat{\mathbf{f}}_l$, and $\mathbf{U}_l$ does not have to be a linear operator.

Usually the upsampling is performed as a post-processing step with little or no information from the coder. Furthermore, as is justifiable in most cases, an open-loop system is considered, that is, the pre-processor does not have knowledge of the post-processor, and vice versa. However, it would be desirable that the pre-processor and coder, and the decoder and post-processor share more information, hence allowing for better reconstructions. Figure 7.1 depicts the structure of the coder and decoder and highlights in red the parts defined by the standards (see, for instance, [82, 83]) and in yellow the parts that are suitable for proprietary algorithms. Most of the standards allow for the inclusion of user information either in the bitstream or through the use of auxiliary channels, that is, extra information about the video sequence that can be decoded and used to increase the picture quality or to provide some add-on value to the codec. Therefore, extra functionalities, such as SR for compression, could be incorporated in existing standards.

A scalable video coder (like the scalable addition to MPEG-4—scalable video coding (SVC) or Annex F—) also includes the steps of the intentional downsampling (spatially and temporally) of the original video during encoding and upsampling during decoding. Spatial scalability, for example, is achieved by first encoding a lower-resolution base layer, and then creating the enhancement layers which contain the coded differences between the reconstructed base layer upsampled to the desired resolution and the video sequence at the resolution of the current layer. This enhancement layer information from the coder is used to increase the quality of the HR video sequence. In this scenario, the upsampling operation is normative and

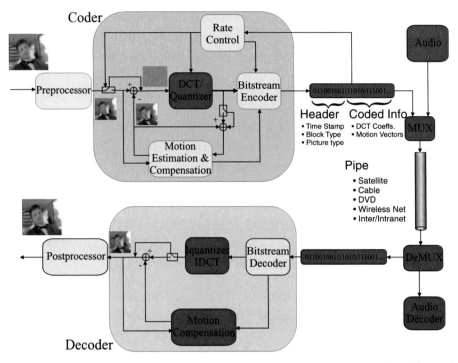

FIGURE 7.1: Standard hybrid video codec showing the parts that are standardized (in red) and the parts that represent proprietary algorithms (in yellow)

is defined to be an interpolation process (see Reichel *et al.* [151]), while the downsampling is nonnormative. However, only a small number of downsampling filters are currently utilized for SVC testing and development. These filters are found in the MPEG-4 verification model (see Li *et al.* [95]). They possibly originate from Bjontegaard and Lillevold [21]. The specific selection of these filters is described by Sullivan and Sun [177].

Clearly, the proposed framework of SR for compression goes beyond scalable video coders. In its simpler form, it can be applied to existing SVC standards (by optimally designing, for example, the nonnormative downsampling filter), or to both scalable and nonscalable video compression standards by appropriately designing the nonnormative pre- and post-processing parts and possibly utilizing the capabilities offered by the standards in conveying information from the encoder to the decoder. In its general form, however, the proposed framework can lead to the design of new video compression algorithms.

As an example of the application of the proposed framework, we assume that the filtering and downsampling steps performed by the encoder (i.e., operators $\mathbf{A}_l$ and $\mathbf{H}_l$ in Eq. (7.9)) are known to the decoder. Then, an SR algorithm can be used at the decoder to reconstruct the original HR image. In [112] an adaptation of the method proposed by Segall *et al.* [169] was

FIGURE 7.2: (a) A part of frame 17 of the original "rush-hour" sequence; (b) corresponding part of the original HR sequence compressed at 2 Mbps; (c) the same part of the reconstructed sequence using the method in [112]

used. It was tested on the "rush-hour" sequence, taken at rush-hour in Munich, available at the Technische Universität München (http://www.ldv.ei.tum.de/liquid.php?page=70). The scene is being observed with a fixed camera and has a large depth of field. The sequence has 500 frames of size 1920×1080 pixels and a frame rate of 25 fps stored in progressive format. A part of frame 17 of the sequence is displayed in Fig. 7.2(a).

For compressing the video sequence, we utilize the MPEG-4 bitstream syntax, which describes a hybrid motion-compensated and block DCT compression system. Apart from the first image in the sequence, which is intra-coded, each frame is compressed as a P-frame. This restricts the reference frame for the motion vectors to be the temporally previous frame. The rate control VM5+ mechanism is utilized for bit allocation. We used different bitrates ranging from 0.5 Mbps to 8 Mbps. A part of frame 17 of the original sequence compressed at 2 Mbps is displayed in Fig. 7.2(b).

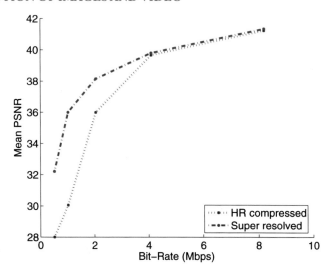

FIGURE 7.3: Rate distortion curves for the "rush-hour" sequence. The SR for video compression approach in [112] was used, which leads to significant quality improvement at low bitrates

Each frame of the original sequence was downsampled to a size of 960×540 pixels by first processing it with the filter $0.25\mathbf{I}_{2\times 2}$ and then discarding every-other pixel in the horizontal direction and every-other line in the vertical direction. The resulting LR sequence was compressed using the same bitrate with the original one and then an SR algorithm was applied to reconstruct the sequence. Note that exact knowledge of the filtering and downsampling procedures is assumed by the method. A part of frame 17 of the reconstructed sequence compressed at 2 Mbps is displayed in Fig. 7.2(c).

Figure 7.3 depicts the evolution of the mean PSNR for the whole sequence as we increase the bitrate. Note that the proposed method results in an increase of as much as 4.2 dB in mean PSNR over the conventional method. This increase is obtained at lower bitrates, since at higher bitrates enough bits are available to obtain a high quality reconstructed video by the conventional method. Notice, however, that the proposed method performs equally well at high bitrates.

Note also that using this approach it is possible for certain bitrates to obtain similar PSNR quality as with the direct compression of the original sequence but at about half the bitrate. This is, for example, the case when compressing the original sequence at 2 Mbps and the downsampled sequence at 1 Mbps, resulting in both cases to a reconstructed video of approximately 36 dB in mean PSNR, as can be seen in Fig. 7.3.

7.2.1   Region-Based Super Resolution for Compression

In this section we present a more sophisticated manifestation of the proposed framework. According to it, side information (denoted by ζ) is conveyed by the encoder to the decoder, within the bounds of existing video compression standards. The side information ζ is the result of the application of the function SI on the HR sequence, that is,

$$\zeta = \mathrm{SI}(\mathbf{f}). \tag{7.11}$$

Utilizing this side information each original frame $\mathbf{f}_l$ in the sequence $\mathbf{f}$ is pre-processed and downsampled by a factor P using the $(M \times N) \times (PM \times PN)$ matrix $\mathbf{A}_l^\zeta$ to produce the LR frame

$$\mathbf{g}_l = \mathbf{A}_l^\zeta \mathbf{f}_l, \quad l = 1, \ldots, L. \tag{7.12}$$

Equation (7.12) is a generalization of Eq. (7.9), where $\mathbf{A}_l^\zeta$ includes both $\mathbf{A}_l$ and $\mathbf{H}_l$ in Eq. (7.9) but the pre-processing now is a function of the actual data $\mathbf{f}$ through the use of the side information. It is also noted that $\mathbf{A}^\zeta$ represents not only spatial processing but also temporal processing, such as temporal downsampling (notice that SI in Eq. (7.11) is a function of the whole sequence $\mathbf{f}$, not just the current frame $\mathbf{f}_l$). The LR sequence $\mathbf{g} = \{\mathbf{g}_1, \ldots, \mathbf{g}_L\}$ is lossy compressed producing the LR-compressed sequence $\mathbf{y} = \{\mathbf{y}_1, \ldots, \mathbf{y}_L\}$ while the side information ζ can be sent to the decoder in a compressed or uncompressed fashion.

   A specific application of the above-described version of the proposed paradigm is described next in detail, based on the work by Barreto *et al.* [13]. The specific type of side information generated is segmentation information. Segmentation is performed at the encoder on groups of images to classify them into three types of regions or blocks according to motion and texture. The resulting segmentation map defines the downsampling process at the encoder, and it is also provided to the decoder as side information in order to guide the SR process.

   For compatibility with common compression standards [81–83], image division into blocks is considered in pre-processing the HR image and post-processing the decoded sequence. Therefore, it is assumed that the blocks used for the compression of the LR images are of size $q \times r$, while each HR image $\mathbf{f}_l$ is of size $PM \times PN$ and it is divided into blocks of size $Pq \times Pr$, where P represents the downsampling factor. This way, there are the same number of blocks, namely $\frac{M}{q} \times \frac{N}{r}$, in both the HR and LR images.

   A generic block $f_l^B[bm, bn]$ in $\mathbf{f}_l$ is defined as

$$f_l^B[bm, bn] = \Big(f_l(Pq \cdot (bm - 1) + i, \ Pr \cdot (bn - 1) + j)$$

$$\mid \ i = 1, \ldots, Pq, \ \ j = 1, \ldots, Pr \Big). \tag{7.13}$$

FIGURE 7.4: Example of a particular GOP

Similarly, $\mathbf{g}_l$ is divided into blocks of size $q \times r$ (an operation performed by the coder). One block in $\mathbf{g}_l$ is denoted by $g_l^B[bm, bn]$, and is defined as

$$g_l^B[bm, bn] = \Big(g_l(q \cdot (bm - 1) + i, \; r \cdot (bn - 1) + j)$$

$$| \;\; i = 1, \ldots, q \;\; j = 1, \ldots, r \Big). \tag{7.14}$$

Block coordinates in Eqs. (7.13) and (7.14) are represented by $bm = 1, \ldots, \frac{M}{q}$ and $bn = 1, \ldots, \frac{N}{r}$, the components of the block are lexicographically ordered by rows, and the superscript B denotes that block notation is used.

The compression system also provides useful motion information for the SR process. It calculates the motion vectors $v^B(bm, bn, l, k)$ that predict block $g_l^B[bm, bn]$ from some previously coded frame $\mathbf{y}_k$. These motion vectors, which can be obtained from the compressed bitstream, are represented by the $(2 \times \frac{M}{q} \times \frac{N}{r}) \times 1$ vector $\mathbf{v}_{l,k}^B$ that is formed by lexicographically stacking the transmitted horizontal and vertical offsets.

Besides the motion estimation and the block-based structure of typical codecs, this approach also considers their frame organization into groups of pictures. Three types of images are usually present in each group of pictures (GOP): intra-coded frames (I-frames), predictively-coded frames (P-frames), and bidirectionally predictively-coded frames (B-frames). Intra-coded frames are coded independently of previous or future frames. Predictively-coded frames are predicted from temporally preceding I- or P-frames, and bidirectionally predictively-coded frames are predicted from the previous and the subsequent I or P-frames for bidirectional pictures. For an example of a GOP, see Fig. 7.4.

Figure 7.5 shows a pictorial description of the overall region-based SR for compression model, which is divided into two main blocks: the transmitter and the receiver. At transmission, the HR video is pre-processed (segmented and downsampled) before compression. At reception, the bitstream is first decompressed and then post-processed according to the segmentation map in order to provide an estimation of the original sequence. The segmentation, downsampling, and upsampling procedures are explained next in more detail.

7.2.1.1 Motion and texture segmentation

The segmentation algorithm described in [13] provides one of the possible solutions for defining the side information ζ in Eq. (7.11). It classifies regions in the HR video sequence according to

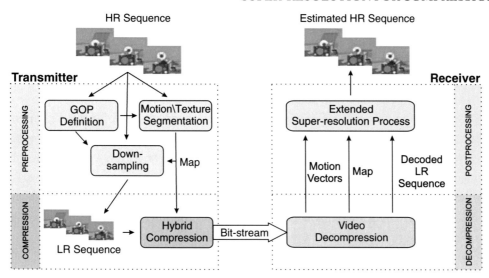

FIGURE 7.5: Graphical description of the region-based SR for the compression problem. At the transmitter pre-processing and downsampling are performed before compression. The bitstream is decompressed at the receiver and an extended SR method is applied for post-processing and upsampling the LR sequence

their amount of motion and texture. The motivation to perform this segmentation is given as a conclusion in [12], where the performance of the SR algorithm is studied with respect to the motion estimation techniques used (block matching, commonly used by compression standards, among them). As expected, the motion estimates are usually poorer in flat areas than in areas with texture. When using SR techniques to obtain HR video sequences from LR-compressed video sequences, a distinction between these two types of regions can help to improve HR estimates as well as, to reduce the computational load. Flat areas can be well recovered using simple interpolation techniques while textured areas with motion can be reconstructed using motion-based SR. Unfortunately, motionless textured regions are the ones most affected by the downsampling and compression processes and in those regions motion-based SR techniques do not produce the desired results. Therefore, texture classification (which will lead to a frame-dependent downsampling process) must be combined with motion segmentation to produce better estimates of the HR sequences.

The final segmentation map contains up to three labels: "motion (M)," "no motion and flat (F)," and "no motion and textured (T)." These segmentation labels are added to the compressed bitstream as user data information, thus the video compression method to be used has to allow for the transmission of this segmentation as side information (see, for instance, MPEG-2 [82], MPEG-4 [83], and H.264 [81] standards).

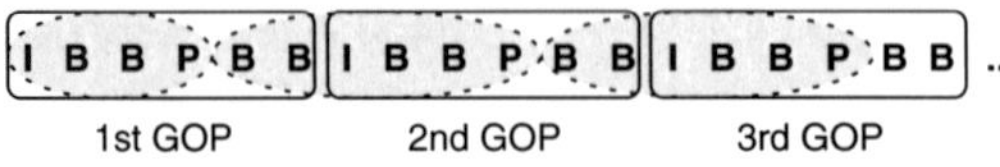

FIGURE 7.6: Each ellipse in shade denotes a GOI used for motion and texture segmentation

To reduce the amount of side information to be transmitted, instead of using a segmentation map for each HR image, such a map is obtained for groups of images. In the following description, a segmentation mask is produced for each set of images corresponding to the I-frame up to its nearest P-picture in the current GOP and down to but not including the nearest P-picture in the previous GOP (see Fig. 7.6 for an example of a GOI). Note that although the description will be carried out for this group of images, the method can also be applied to other partitions of the original HR sequence.

We now proceed with the description of the segmentation procedure. Since PCA [105] is a powerful tool for analyzing data of high-dimensionality, the first principal component is used as the basis to build the segmentation map. Note that PCA has already been used in SR problems as a way to summarize prior information on images (see [33], [73], [42]).

The resulting mean PCA image μ is taken as an input to the texture classification process and the first principal component **pc1**, which explains most of the variance of the data, is used for the motion segmentation. To assign a label to each block, a threshold τ is established experimentally over the results of applying the *Intra Sum of Absolute Differences* ($Intra\_SAD$) to both PCA images μ and **pc1** . The classification is then performed for each block $B[bm, bn]$ of size $Pq \times Pr$ in HR by first evaluating

$$Intra\_SAD[bm, bn] = \frac{1}{Pq \cdot Pr} \sum_{(i,j) \in B[bm,bn]} |x(i, j) - \bar{x}^B[bm, bn]|, \qquad (7.15)$$

where $x(i, j)$ represents either μ or **pc1** and $\bar{x}^B[bm, bn]$ its corresponding the mean value in the (bm, bn) block. PCA image μ, and the first principal component **pc1**. In the first case, the left-hand side of Eq. (7.15) will be renamed M_{SAD}, while in the second case it will be renamed T_{SAD}. Following the ideas in [160] the threshold

$$\tau = mean(Intra\_SAD) - \frac{std(Intra\_SAD)}{2} \qquad (7.16)$$

is used to detect motion (τ_M) and texture (τ_T).

The quantities in Eqs. (7.15) and (7.16) are used for the classification of each block of pixels $B(bm, bn)$ into motion, M, textured, T, and non-textured or flat, F, regions

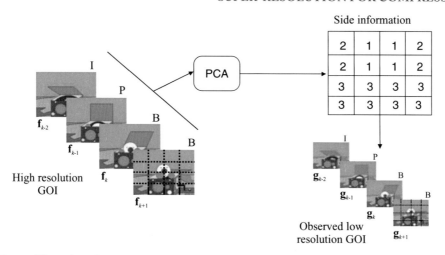

FIGURE 7.7: The side information is a matrix containing per block the labels 1, 2, or 3, corresponding respectively to block types M, F, or T, calculated from the application of PCA on a GOI in HR

according to

$$
\zeta(bm, bn) = \begin{cases} M & \text{if } M_{SAD}[bm, bn] > \tau_M \\ T & \text{if } M_{SAD}[bm, bn] \leq \tau_M \text{ and } T_{SAD}[bm, bn] > \tau_T \\ F & \text{if } M_{SAD}[bm, bn] \leq \tau_M \text{ and } T_{SAD}[bm, bn] \leq \tau_T \end{cases} \qquad (7.17)
$$

The side information (segmentation), ζ, is a matrix containing per block the labels 1, 2, or 3, corresponding respectively to the M, F, or T block types, and calculated from the application of PCA on a group of images (GOI) in HR. Figure 7.7 depicts this process graphically. As the number of blocks in LR and HR remain constant, as already explained, the map is the same for both resolutions.

7.2.1.2 Downsampling process
There are several filters that can be used for downsampling (A_l^ζ in Eq. (7.12)). Some of them can be found, for instance, in [177], but a different method is implemented in [13] for the particular case of the region-based SR. This downsampling process depends on the type of block and the picture prediction type. Without loss of generality we will assume that $P = 2$ and that the GOI is of the type IBBP. Clearly, the proposed downsampling approach can be applied to other types of GOIs.

For blocks belonging to M and F classes the downsampling is carried out according to

$$
\mathbf{g}_l^B[bm, bn] = \Big(\mathbf{f}_l((bm-1)q + 2i - 1, (bn-1)r + 2j - 1)
$$

$$
\mid i = 1, \ldots, q \;\; j = 1, \ldots, r\Big), \qquad (7.18)
$$

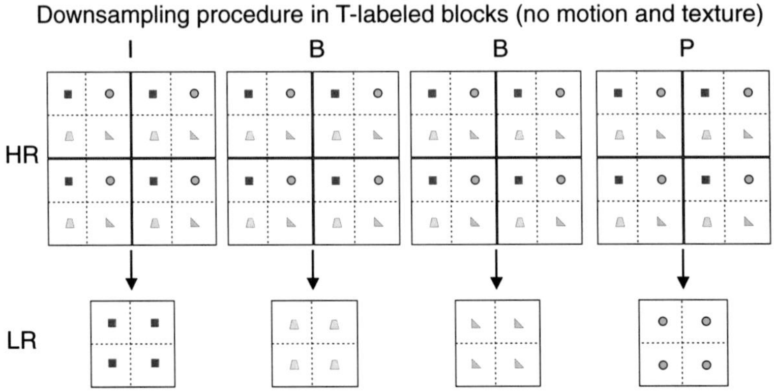

FIGURE 7.8: Example of the downsampling procedure applied to a 4 × 4 block classified as textured (*T*). The downsampling pattern depends on the prediction image type. For I-frames and P-frames, the selection of the pixel is performed according to Eqs. (7.18) and (7.19), respectively. For B-frames, the downsampling pattern is mathematically described by Eqs. (7.20) and (7.21)

where $\mathbf{g}_l^B$ is the *l*th LR block (bm, bn), and $bm = 1, \ldots, \frac{M}{q}$, $bn = 1, \ldots, \frac{N}{r}$. Note that this downsampling works under the hypothesis that when motion is present, SR techniques can be used to increase the resolution of images, and for flat regions without motion, simple interpolation techniques can be used to increase the size of an image without loss of details.

For blocks in the *T* class the downsampling pattern depends on the prediction type of the block as depicted in Fig. 7.8. I-frames are also downsampled according to Eq. (7.18), like *M* and *F* regions.

For P-pictures belonging to the *T* class, the downsampling pattern changes to:

$$\mathbf{g}_l^B[bm, bn] = \Big(\mathbf{f}_l((bm - 1)q + 2i - 1, (bn - 1)r + 2j)$$
$$\mid i = 1, \ldots, q, \quad j = 1, \ldots, r\Big). \qquad (7.19)$$

For blocks in B-pictures belonging to the *T* class, the downsampling pattern alternates between these two models:

$$\mathbf{g}_l^B[bm, bn] = \Big(\mathbf{f}_l((bm - 1)q + 2i, (bn - 1)r + 2j - 1)$$
$$\mid i = 1, \ldots, q, j = 1, \ldots, r\Big), \qquad (7.20)$$

and

$$\mathbf{g}_l^B[bm, bn] = \Big(\mathbf{f}_l((bm - 1)q + 2i, (bn - 1)r + 2j)$$
$$\mid i = 1, \ldots, q, \quad j = 1, \ldots, r\Big). \qquad (7.21)$$

7.2.1.3 Upsampling procedure

The segmentation map ζ is included in the bitstream, in this case, as part of the user data field provided by the compression standard and transmitted along with the compressed video information $\mathbf{y}$ to the decoder. There, the reconstructed LR sequence $\mathbf{y} = \{\mathbf{y}_1, \ldots, \mathbf{y}_L\}$ is post-processed using the upsampling operator $\mathbf{U}^\zeta$ to obtain an estimate of the original HR sequence $\hat{\mathbf{f}}$, that is,

$$\hat{\mathbf{f}} = \mathbf{U}^\zeta[\mathbf{y}]. \qquad (7.22)$$

This upsampling operator over one block of pixels $y_l^B[bm, bn]$ becomes

$$\mathbf{U}^\zeta(y_l^B[bm, bn]) = \begin{cases} BLI(y_l^B[bm, bn]) & \text{if } \zeta(bm, bn) = F \\ SR(y_l^B[bm, bn]) & \text{if } \zeta(bm, bn) = M \, , \\ AMR(y_l^B[bm, bn]) & \text{if } \zeta(bm, bn) = T \end{cases} \qquad (7.23)$$

where BLI stands for bilinear interpolation, AMR stands for artificial motion reconstruction, and SR stands for the application of SR techniques. The SR and AMR procedures are briefly described below (see [13] for more details).

The SR upsampling procedure for the M-labeled blocks follows an interpolation-based SR approach [58, 106]. This SR method is advantageous from a computational point of view, providing solutions close to real time. An HR grid, associated to the lth decompressed image in the sequence $\mathbf{y}$, is filled not only with values from $\mathbf{y}_l$ but also from adjacent frames for P- and B-pictures when there is subpixel motion. The motion vectors $\mathbf{v}_{lk}^B$ are provided by the encoder and are used as initial estimates. An example is shown in Fig. 7.9.

For the AMR process the upsampling operation is applied according to the downsampling process previously described, which depends on the frame prediction type. The final

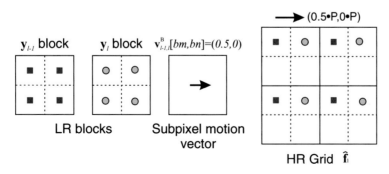

FIGURE 7.9: An example of the SR technique applied to an M-type block of a P-frame. Projection of LR samples onto an HR grid according to the refined transmitted motion vectors

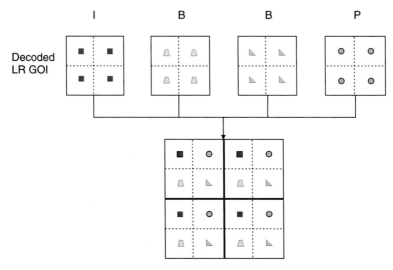

FIGURE 7.10: Reconstruction of a block classified as no motion and textured (T) for an LR block of size 2×2 and a downsampling factor of 2

HR-estimated block, for a downsampling factor of $P = 2$, will be the combination of four upsampled LR blocks, as shown in the example of Fig. 7.10.

The above described region-based SR method was tested on the "Deadline" sequence in CIF format (352×288 pixels) at a frame rate of 30 fps stored in progressive format and the results were compared with the traditional coding of the HR sequence. Frame 3 of the "Deadline" sequence is shown in Fig. 7.11(a).

The MPEG-4 codec was used for both algorithms to compress 16 frames of the image sequence (from frame 3 to frame 18) using a GOP of size 6 with structure IBBPBB at bitrates ranging from 32 kbps to 4 Mbps. Frame 3 of the sequence compressed at 512 kbps is shown in Fig. 7.11(b).

Using the region-based SR approach, each macroblock of the original sequence was classified into motion (M), texture (T), and flat (F) classes using Eq. (7.17) to generate the side information file. Figure 7.11(c) depicts the classification for the first GOI of the "Deadline" sequence in "motion" (black), "no motion and flat" (gray) and "no motion and textured" (white). Then each macroblock was downsampled by a factor of two as described in Section 7.2.1.2. The MPEG-4 coder was applied to this downsampled sequence including the side information (which is not compressed) into the bitstream in the User Data field of the group of videoobjectplane (GOV) header. At the decoder, the received sequence was upsampled using Eq. (7.23).

FIGURE 7.11: (a) Frame 3 of the "Deadline" sequence; (b) frame 3 of the sequence compressed at 512 kbps. (c) macroblock classification for the first GOI of the sequence; (d) reconstruction using the method in [13] of frame 3 of the LR sequence compressed at 512 kbps

The rate-distortion curve for the sequence is depicted in Fig. 7.12. The mean PSNR for the HR-compressed image sequence ranges from 27.8 dB to 41 dB while using the SR approach the mean PSNR ranges from 31.8 dB to 37 dB. As was the case with the SR for compression method in [112], the method in [13] outperforms compressing the HR sequence at low and

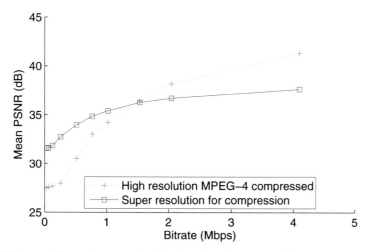

FIGURE 7.12: The rate-distortion curve for the "Deadline" sequence. The SR for compression method in [13] outperforms compressing the HR sequence at low and medium bitrates

medium bitrates but not at high bitrates where the HR-compressed sequence already has very good quality.

Frame 3 of the reconstructed LR sequence compressed at 512 kbps is displayed in Fig. 7.11(d). This frame corresponds to an I-frame in the GOP sequence. We observe that the blocking artifacts that appear in Fig. 7.11(b), are not present in the image in Fig. 7.11(d), due to the interpolations performed over flat areas. In motion classified regions, the quality of the SR reconstruction highly depends on the accuracy of the motion vectors provided by the encoder and their posterior refinement and the type of frame. In I-frames only spatial interpolation is used (this is why the regions with motion in Fig. 7.11(d) do not present the level of detail of the HR image). Motion areas are better resolved in B-frames, since information from the current and two reference frames are used for the reconstruction. Alternative SR methods could clear this limitations. Finally, textured areas with no motion are very well recovered as can be observed from the detail of Fig. 7.11(d).

We believe that with the specific implementations of the proposed framework of SR for compression presented in this chapter, we have only touched the "tip of the iceberg." There is a plethora of issues to be addressed before efficient coders can be developed, to really challenge the state of the art in video compression. For example, approaches that keep the decoder simple, possibly increasing the complexity of the coder, can be developed. These approaches may borrow ideas from the pre-processing and post-processing techniques described in Section 7.1. Since the original HR sequence is available at the coder, the downsampling, coding, and reconstruction processes can define critical information which may be efficiently represented in the rate-distortion sense and transmitted to the decoder. There is, therefore, a number of information types that can be sent and the way to sent the information, as well as, the control of the number of bits used to transmit it, are also critical problems to be considered.

In the previous discussion, we have assumed that the downsampling factor is constant and always the same for each frame in the sequence. However, in the SR for compression framework, downsampling may vary depending on the frame (for instance, we may use a higher downsampling factor for B-frames than for I-frames), or depending on the region (for instance, flat regions could be downsampled with a higher factor than detailed regions). Also, it is possible to use different downsampling schemes as well as different filters, including temporal filters, depending on the local characteristics of the frame. This is still a very open topic where a number of optimization problems can be formulated depending on what is considered to be known or unknown.

Epilogue

In this monograph we have presented the super-resolution problem for images and video from a Bayesian point of view. In doing so we dealt with both cases of compressed and uncompressed data. We first defined the problem and pointed to some current and potential applications where the problem is encountered. We proceeded, in Chapter 2, by providing a Bayesian formulation of the SR problem and defining the elements required to carry out Bayesian inference. We then followed a systematic approach in the next three chapters in describing the various ways these elements have been specified in the literature.

Thus, in Chapter 3, we described the models for obtaining the LR observations from the HR original data. Any one of such models specifies the form of the likelihood function, i.e., the conditional probability of the observations given the unknown quantities (the HR image and the motion or global displacement vectors). This is a critical step in the application of SR to a specific application. It usually requires detailed and deep understanding of the underlying physics and the (opto-electronic) devices involved in acquiring the LR images. Typically, a number of unknown parameters defining such models are involved and therefore they need to be estimated utilizing the available data. This is a topic of the outmost important, i.e., finding the most appropriate degradation model generating the observation in a specific application. It certainly requires good scientific and engineering judgment but also calls for approaches which will handle the estimation of the unknown parameters (most published results consider such critical parameters to be known).

In Chapter 4, we dealt with the general and important topic of motion estimation or registration (when global motion is present). This is a topic with various problem formulations and numerous applications in video processing, robotics, and computer vision, among others. We treated the topic within the context of super resolution (following the Bayesian paradigm). The same way however that a number of techniques and results were brought into the SR problem from other areas, the developments in motion estimation for SR can be exported to other areas and problems. The accuracy of the motion vectors is of the outmost importance for the SR problem. Work is still required on this topic in order for SR to reach its full potential. Some of the open issues and ideas for future developments on this topic are outlined at the end of Chapter 4.

Modeling of the unknown (HR) image is a step encountered in all estimation problems (when a probabilistic approach is followed a prior probability density function needs to be specified). Through this modeling, all our prior knowledge about the original image is incorporated into the solution approach. Typically, we have no way of knowing the exact properties of the original image, so then our "wish-list" of such properties are incorporated into the solution process. We, for example, expect the image to be (locally) smooth or void of compression artifacts (this is a reasonable expectation). The challenge is to device spatially and temporally adaptive models which will result in tractable solutions. We discuss the topic of prior image modeling in Chapter 5, where we also discuss ways for solving for the unknown quantities, now that all the elements of the Bayesian inference have been specified. At the end of this chapter we also mention some of the challenges and open problems on the topic.

The full potential of Bayesian inference is described in Chapter 6. It allows us, for example, to model and estimate all the unknown hyperparameters. If computationally efficient algorithms can be obtained in this case, then this is a major step in greatly increasing the usefulness and applicability of SR techniques. The user wants to have the "comforting feeling" that the parameters estimated are the "best" ones based on the available data and that they are not chosen based on the experience (or lack thereof) of the user. In addition, the Bayesian framework allows us to marginalize or integrate out one or more of the unknowns and then solve for the remaining ones. The general tradeoff is that the more complicated the models or the more complicated the problem formulation the harder to obtain a solution. One therefore has to resort to either approximations or simulations of the posterior distribution. Capitalizing on the full capabilities of the Bayesian framework certainly represents a fertile ground of research and development. Some of these open issues on Bayesian inference are mentioned at the end of the book.

Finally in Chapter 7, compression enters the picture again. Now, however, we are not interested in super-resolving compressed images and videos but in designing new image and video coders in which SR techniques and ideas will play a central role. We now intentionally introduce the degradations (e.g., blurring and down-sampling) so that the "degraded" signal is more "suitable" for compression. Of course, we now have full control of these degradations (they can be, for example, spatiotemporally adaptive) and equally importantly we can access their effectiveness simply because the original HR image is available. An SR technique now simply predicts an HR image, and the prediction error can, for example, be encoded and sent to the decoder. All these decisions (e.g., types of filters and spatially adaptive down-sampling factors to be used, choice of the appropriate SR technique per region, and types of auxiliary information to be send to the decoder) need to be made in a rate-distortion or resource-utility framework. This represents a challenging rate control problem. The topic of Chapter 7 certainly represents "unchartered waters" and has the potential to greatly impact future image and video compression standards.

A key question for the scientist is if SR techniques have reached their full potential or if there are still some important open problems on the topic. We believe that a lot can still be done and that the best is still to come! The practitioner in the field would probably like to know, for a given system, what is the limit of the performance of the state of the art SR algorithm, or what is the maximum magnification factor she/he could obtain. This is certainly a hard question to answer; as mentioned at the end of Chapter 3, it is only recently that such a question has been attempted to be answered.

As it should have become clear by now in this book we concentrated in providing and explaining and analyzing the mathematical tools required to solve an SR problem. We did not focus on any specific application, although we mentioned a number of the active or potential application areas of the technology. We also did not provide any code for any of the algorithms mentioned, although for some of the examples, additional implementation details were provided. As is usually the case, the difficulty in implementing an algorithm can range widely. For some of the algorithms a number of parameters need to be set, which is an art in itself. We believe that with a good understanding of the mathematics and the algorithms we describe, one could in principle provide solutions to any SR or RE problem encountered in practice. This is indeed our hope and expectation that this book will prove to be useful to a number of you, its future readers out there and this of course would make our efforts worthwhile.

Bibliography

[1] T. Akgun, Y. Altunbasak, and R. M. Mersereau, "Super-resolution reconstruction of hyperspectral images," *IEEE Trans. Image Process.*, vol. 14, pp. 1860–1875, 2005. doi:10.1109/TIP.2005.854479

[2] Y. Altunbasak and A. J. Patti, "A maximum a posteriori estimator for high resolution video reconstruction from MPEG video," in *IEEE Int. Conf. Image Process.*, Vancouver, BC, Canada, vol. 2, pp. 649–652, 2000.

[3] Y. Altunbasak, A. J. Patti, and R. M. Mersereau, "Super-resolution still and video reconstruction from MPEG-coded video," *IEEE Trans. Circuits Syst. Video Technol.*, vol. 12, pp. 217–226, 2002. doi:10.1109/76.999200

[4] L. D. Alvarez, J. Mateos, R. Molina, and A. K. Katsaggelos, "High resolution images from compressed low resolution video: Motion estimation and observable pixels," *Int. J. Imaging Syst. Technol.*, vol. 14, pp. 58–66, 2004. doi:10.1002/ima.20008

[5] L. D. Alvarez, R. Molina, and A. K. Katsaggelos, "High resolution images from a sequence of low resolution observations," in T. R. Reed, Ed. *Digital Image Sequence Processing, Compression and Analysis*, Florida, USA, Chap. 9, CRC Press, 2004, pp. 233–259.

[6] L. D. Alvarez, R. Molina, and A. K. Katsaggelos, "Motion estimation in high resolution image reconstruction from compressed video sequences," in *IEEE Int. Conf. Image Process.*, Singapore, vol. I, pp. 1795–1798, 2004.

[7] H. A. Aly and E. Dubois, "Specification of the observation model for regularized image up-sampling," *IEEE Trans. Image Process.*, vol. 14, pp. 567–576, 2005.

[8] C. Andrieu, N. de Freitras, A. Doucet, and M. I. Jordan, "An introduction to MCMC for machine learning," *Machine Learn.*, vol. 50, pp. 5–43, 2003.

[9] S. Baker and T. Kanade, "Limits on super-resolution and how to break them," *IEEE Trans. Pattern Anal. Machine Intell.*, vol. 24, pp. 1167–1183, 2002. doi:10.1109/TPAMI.2002.1033210

[10] S. Baker and I. Matthews, "Lucas-Kanade 20 years on: A unifying framework: Part 1," *Int. J. Comput. Vis.*, vol. 56, pp. 221–255, 2004. doi:10.1023/B:VISI.0000011205.11775.fd

[11] M. K. Banham and A. K. Katsaggelos, "Digital image restoration," *IEEE Signal Process. Mag.*, vol. 3, pp. 24–41, 1997.

[12] D. Barreto, L. D. Alvarez, and J. Abad, "Motion estimation techniques in super-resolution image reconstruction. A performance evaluation," in *Virtual Observatory.*

Plate Content Digitalization, Archive Mining and Image Sequence Processing, M. Tsvetok, V. Golev, F. Murtagh, and R. Molina, Eds. 2005, pp. 254–268, Heron Press, Sofia (Bulgaria).

[13] D. Barreto, L. D. Alvarez, R. Molina, A. K. Katsaggelos, and G. M. Callicó, "Region based super-resolution for compression," *Multidimensional Systems and Signal Processing, special issue on Super Resolution*, March 2007, DOI 10.1007/s11045-007-0019-y.

[14] B. Bascle, A. Blake, and A. Zisserman, "Motion deblurring and super resolution," in *Proc. Eur. Conf. Comput. Vis.*, Cambridge, UK, pp. 312–320, 1996.

[15] J. R. Bergen, P. Anandan, K. J. Hanna, and R. Hingorani, "Hierarchical model-based motion estimation," in *ECCV'92: Proc. Second Eur. Conf. Comput. Vis.*, Santa Margherita Ligure, Italy, pp. 237–252, 1992.

[16] J. O. Berger, *Statistical Decision Theory and Bayesian Analysis*, Chaps. 3 and 4. New York: Springer Verlag, 1985.

[17] J. Besag, "On the statistical analysis of dirty pictures," *J. R. Stat. Soc. Ser. B*, vol. 48, pp. 259–302, 1986.

[18] J. M. Bioucas-Dias, "Bayesian wavelet-based image deconvolution: A GEM algorithm exploiting a class of heavy-tailed priors," *IEEE Trans. Image Process.*, vol. 15, pp. 937–951, 2006. doi:10.1109/TIP.2005.863972

[19] S. Birchfiled, "Derivation of the Kanade-Lucas-Tomasi tracking equation," http://vision.stanford.edu/~birch/klt, 1996.

[20] T. E. Bishop and J. R. Hopgood, "Blind image restoration using a block-stationary signal model," in *IEEE Int. Conf. Acoustics, Speech Signal Process.*, Toulouse, France, May 2006.

[21] G. Bjontegaard and K. O. Lillevold, "H.263 anchors—technical description," in *ISO/IEC JTC1/SC29/WG11 MPEG95/322, Dallas meeting*, Nov. 1995.

[22] S. Borman, *"Topics in multiframe superresolution restoration,"* Ph.D. thesis, University of Notre Dame, Notre Dame, IN, 2004.

[23] S. Borman, M. A. Robertson, and R. L. Stevenson, "Block-matching sub-pixel motion estimation from noisy, under-sampled frames —An empirical performance evaluation," in *Vis. Commun. Image Process. '99*, San Jose, USA, vol. 3653, pp. 1442–1451, 1999.

[24] S. Borman and R. Stevenson, "Spatial resolution enhancement of low-resolution image sequences. A comprehensive review with directions for future research," Technical report, Laboratory for Image and Signal Analysis (LISA), University of Notre Dame, Notre Dame, IN, USA, July 1998.

[25] N. K. Bose and K. J. Boo, "High-resolution image reconstruction with multisensors," *Int. J. Imaging Syst. Technol.*, vol. 9, pp. 141–163, 1998.

[26] N. K. Bose, R. H. Chan, and M. K. Ng, Eds., "Special issue on high-resolution image reconstruction," *Int. J. Imaging Syst. Technol.*, vol. 14, 2004.

[27]   N. K. Bose, S. Lertrattanapanich, and J. Koo, "Advances in superresolution using L-curve," *IEEE Int. Symp. Circuits Syst.*, vol. 2, pp. 433–436, 2001.

[28]   J. C. Brailean and A. K. Katsaggelos, "Simultaneous recursive motion estimation and restoration of noisy and blurred image sequences," *IEEE Trans. Image Process.*, vol. 4, pp. 1236–1251, 1995.

[29]   A. Bruckstein, M. Elad, and R. Kimmel, "Down scaling for better transform compression," *IEEE Trans. Image Process.*, vol. 12, pp. 1132–1144, 2003. doi:10.1109/TIP.2003.816023

[30]   F. M. Candocia and J. C. Principe, "Superresolution of images with learned multiple reconstruction kernels," in *Multimedia Image and Video Processing*, L. Guan, S.-Y. Kung, and J. Larsen, Eds. CRC Press, 2000, Florida, USA, pp. 67–95.

[31]   D. P. Capel, *Image Mosaicing and Superresolution*. Springer, London, UK, 2004.

[32]   D. P. Capel and A. Zisserman, "Super-resolution enhancement of text image sequence," in *Int. Conf. Pattern Recognit.*, Barcelona, Spain, pp. 600–605, 2000.

[33]   D. P. Capel and A. Zisserman, "Super-resolution from multiple views using learnt image models," in *Proc. IEEE Comput. Soc. Conf. Comput. Vis. Pattern Recognit.*, Hawaii, USA, vol. 2, pp. 627–634, 2001.

[34]   D. P. Capel and A. Zisserman, "Computer vision applied to super resolution," *IEEE Signal Process. Mag.*, vol. 20, pp. 75–86, 2003. doi:10.1109/MSP.2003.1203211

[35]   J. Chamorro-Martínez, "Desarrollo de modelos computacionales de representación de secuencias de imágenes y su aplicación a la estimación de movimiento (in Spanish)", Ph.D. thesis, University of Granada, 2001.

[36]   R. H. Chan, T. F. Chan, L. X. Shen, and Z. W. Shen, "Wavelet algorithms for high-resolution image reconstruction," Technical report, Department of Mathematics, Chinese University of Hong Kong, 2001.

[37]   R. H. Chan, T. F. Chan, L. X. Shen, and Z. W. Shen, "Wavelet algorithms for high-resolution image reconstruction", *SIAM J. Sci. Comput.*, vol. 24, pp. 1408–1432, 2003. doi:10.1137/S1064827500383123

[38]   R. H. Chan, T. F. Chan, L. X. Shen, and Z. W. Shen, "Wavelet deblurring algorithms for spatially varying blur from high-resolution image reconstruction," *Linear Algebra Appl.*, vol. 366, pp. 139–155, 2003. doi:10.1016/S0024-3795(02)00497-4

[39]   R. H. Chan, S. D. Riemenschneider, L. X. Shen, and Z. W. Shen, "Tight frame: An efficient way for high-resolution image reconstruction," *Appl. Comput. Harmonic Anal.*, vol. 17, pp. 91–115, 2004. doi:10.1016/j.acha.2004.02.003

[40]   M. M. Chang, M. I. Sezan, and M. Tekalp, "Simultaneous motion estimation and segmentation," *IEEE Trans. Image Process.*, vol. 6, pp. 1326–1332, 1997.

[41]   S. Chaudhuri, Ed., *Super-Resolution from Compressed Video*. Kluwer Academic Publishers, Boston/Dordrecht/London, 2001.

[42] S. Chaudhuri and J. Manjunath, *Motion-free Super-Resolution*. Springer, New York, USA, 2005.

[43] S. Chaudhuri and D. R. Taur, "High-resolution slow-motion sequencing: how to generate a slow-motion sequence from a bit stream," *IEEE Signal Process. Mag.*, vol. 22, pp. 16–24, 2005. doi:10.1109/MSP.2005.1406471

[44] P. Cheeseman, B. Kanefsky, R. Kraft, J. Stutz, and R. Hanson, "Super-resolved surface reconstruction from multiple images," in *Maximum Entropy and Bayesian Methods*, G. R. Heidbreder, Ed. The Netherlands: Kluwer, 1996, pp. 293–308.

[45] D. Chen and R. R. Schultz, "Extraction of high-resolution video stills from MPEG image sequences," in *Proc. IEEE Int. Conf. Image Process.*, Chicago, Illinois, USA, vol. 2, pp. 465–469, 1998.

[46] MC. Chiang and T. E. Boult, "Efficient super-resolution via image warping," *Image Vis. Comput.*, vol. 18, pp. 761–771, 2000. doi:10.1016/S0262-8856(99)00044-X

[47] J. Cui, Y. Wang, J. Huang, T. Tan, and Z. Sun, "An iris image synthesis method based on PCA and super-resolution," in *Proc. 17th Int. Conf. Pattern Recognit.*, Cambridge, UK, vol. IV, pp. 471–474, 2004.

[48] T. Daubos and F. Murtagh, "High-quality still images from video frame sequences," in *Investigative Image Processing II*, vol. 4709, Zeno J. Geradts and Lenny I. Rudin, Eds. SPIE, Bellingham, WA, USA, 2002, pp. 49–59.

[49] A. D. Dempster, N. M. Laird, and D. B. Rubin. "Maximum likelihood from incomplete data via the E-M algorithm," *J. R. Stat. Soc. Ser. B*, vol. 39, pp. 1–37, 1977.

[50] S. N. Efstratiadis and A. K. Katsaggelos, "Nonstationary ar modeling and constrained recursive estimation of the displacement field," *IEEE Trans. Circuits Syst. Video Technol.*, vol. 2, pp. 334–346, 1992. doi:10.1109/76.168901

[51] M. Elad and A. Feuer, "Restoration of a single superresolution image from several blurred, noisy, and undersampled measured images," *IEEE Trans. Image Process.*, vol. 6, pp. 1646–1658, 1997. doi:10.1109/83.650118

[52] M. Elad and Y. Hel-Or, "A fast super-resolution reconstruction algorithm for pure translational motion and common space invariant blur," *IEEE Trans. Image Process.*, vol. 10, pp. 1187–1193, 2001. doi:10.1109/83.935034

[53] P. E. Eren, M. I. Sezan, and A. M. Tekalp, "Robust, object-based high-resolution image reconstruction from low-resolution video," *IEEE Trans. Image Process.*, vol. 6, pp. 1446–1451, 1997. doi:10.1109/83.624970

[54] P. E. Eren and A. M. Tekalp, "Bi-directional 2-D mesh representation for video object rendering, editing and superesolution in the presence of occlusions," *Signal Process. Image Commun.*, vol. 18, pp. 321–336, 2003. doi:10.1016/S0923-5965(02)00129-7

[55]   S. Farsiu, *MDSP Resolution Enhancement Software*. Santa Cruz: University of California, 2004.

[56]   S. Farsiu, M. Elad, and P. Milanfar, "Constrained, globally optimal, multi-frame motion estimation," in *IEEE Workshop Stat. Signal Process.*, Burdeaux, France, 2005.

[57]   S. Farsiu, D. Robinson, M. Elad, and P. Milanfar, "Robust shift and add approach to super-resolution," in *Proc. SPIE Appl. Digital Image Process. XXVI*, San Diego, California, vol. 5203, pp. 121–130, 2003.

[58]   S. Farsiu, D. Robinson, M. Elad, and P. Milanfar, "Fast and robust multi-frame super-resolution," *IEEE Trans. Image Process.*, vol. 13, pp. 1327–1344, 2004. doi:10.1109/TIP.2004.834669

[59]   M. A. T. Figueiredo and R. D. Nowak, "An EM algorithm for wavelet-based image restoration," *IEEE Trans. Image Process.*, vol. 12, pp. 906–916, 2003.

[60]   M. A. Fischler and R. C. Bolles, "Random sample consensus: A paradigm for model fitting with applications to image analysis and automatic cartography," *Commun. ACM*, vol. 24, pp. 381–395, 1981. doi:10.1145/358669.358692

[61]   W. T. Freeman, J. A. Haddon, and E. C. Pasztor, "Learning motion analysis," in *Statistical Theories of the Brain*, R. Rao, B. Olshausen, and M. Lewicki, Eds. Cambridge, MA: MIT Press, 2001.

[62]   W. T. Freeman, T. R. Jones, and E. C. Pasztor, "Example based super-resolution," *IEEE Comput. Graphics Appl.*, vol. 22, pp. 56–65, 2002. doi:10.1109/38.988747

[63]   W. T. Freeman, E. C. Pasztor, and O. T. Carmichael, "Learning low-level vision," *Int. J. Comput. Vis.*, vol. 40, pp. 25–47, 2000. doi:10.1023/A:1026501619075

[64]   A. S. Fruchter and R. N. Hook, "Drizzle: A method for the linear reconstruction of undersampled images," *Publ. Astron. Soc. Pacific*, vol. 114, pp. 144–152, 2002. doi:10.1086/338393

[65]   A. Gelman, J. B. Carlin, H. S. Stern, and D. R. Rubin. *Bayesian Data Analysis*. Chapman & Hall, Florida, USA, 2003.

[66]   S. Geman and D. Geman, "Stochastic relaxation, Gibbs distributions, and the Bayesian restoration of images," *IEEE Trans. Pattern Anal. Machine Intell.*, vol. PAMI-6, pp. 721–741, 1984.

[67]   R. D. Gill and B. Y. Levit, "Application of the Van Trees inequality: A Bayesian Cramer-Rao bound," *Bernoulli*, vol. 1, pp. 59–79, 1995. doi:10.2307/3318681

[68]   V. M. Govindu, "Lie-algebraic averaging for globally consistent motion estimation," in *Int. Conf. Comput. Vis. Patern Recognit.*, Washington, DC, vol. 1, pp. 684–691, 2004.

[69]   H. Greenspan, G. Oz, N. Kiryati, and S. Peled, "MRI inter-slice reconstruction using super resolution," *Magn. Reson. Imaging*, vol. 20, pp. 437–446, 2002. doi:10.1016/S0730-725X(02)00511-8

[70] B. K. Gunturk, Y. Altunbasak, and R. M. Mersereau, "Color plane interpolation using alternating projections," *IEEE Trans. Image Process.*, vol. 11, pp. 997–1013, 2002. doi:10.1109/TIP.2002.801121

[71] B. K. Gunturk, Y. Altunbasak, and R. M. Mersereau, "Multiframe resolution-enhancement methods for compressed video," *IEEE Signal Process. Lett.*, vol. 9, pp. 170–174, 2002. doi:10.1109/LSP.2002.800503

[72] B. K. Gunturk, Y. Antunbasak, and R. Mersereau, "Bayesian resolution-enhancement framework for transform-coded video," in *IEEE Int. Conf. Image Process.*, Thessaloniki, Greece, vol. 2, pp. 41–44, 2001.

[73] B. K. Gunturk, A. U. Batur, Y. Altunbasak, M. H. Hayes, and R. M. Mersereau, "Eigenface-domain super-resolution for face recognition," *IEEE Trans. Image Process.*, vol. 12, pp. 597–606, 2003. doi:10.1109/TIP.2003.811513

[74] R. C. Hardie, K. J. Barnard, and E. E. Armstrong, "Joint MAP registration and high-resolution image estimation using a sequence of undersampled images," *IEEE Trans. Image Process.*, vol. 6, pp. 1621–1633, 1997. doi:10.1109/83.650116

[75] R. C. Hardie, K. J. Barnard, J. G. Bognar, E. E. Armstrong, and E. A. Watson, "High resolution image reconstruction from a sequence of rotated and translated frames and its application to an infrared imaging system," *Opt. Eng.*, vol. 73, pp. 247–260, 1998.

[76] C. J. Harris and M. Stephens, "A combined corner and edge detector," in *Proc. Alvey Vis. Conf.*, Manchester, UK, pp. 147–151, 1988.

[77] H. He and L. P. Kondi, "An image super-resolution algorithm for different error levels per frame," *IEEE Trans. Image Process.*, vol. 15, pp. 592–603, 2006.

[78] F. Humblot and A. Mohammad-Djafari, "Super-resolution using hidden Markov model and Bayesian detection estimation framework," *EURASIP J. Appl. Signal Process.*, vol. ID 36971, p. 16, 2006.

[79] M. Irani and S. Peleg, "Improving resolution by image registration," *CVGIP: Graphical Models Image Process.*, vol. 53, pp. 231–239, 1991.

[80] M. Irani and S. Peleg, "Motion analysis for image enhancement: Resolution, occlusion, and transparency," *J. Vis. Commun. Image Represent.*, vol. 4, pp. 324–335, 1993. doi:10.1006/jvci.1993.1030

[81] ISO/IEC International Standard 14496-10:2005, *Information Technology — Coding of Audio-Visual Objects – Part 10: Advanced Video Coding*, 2005.

[82] ISO/IEC JTC1/SC29 International Standard 13818-2, *Information Technology — Generic Coding of Moving Pictures and Associated Audio Information: Video*, 1995.

[83] ISO/IEC JTC1/SC29 International Standard 14496-2, *Information Technology — Generic Coding of Audio-Visual Objects: Visual*, 1999.

[84] M. I. Jordan, Z. Ghahramani, T. S. Jaakola, and L. K. Saul, "An introduction to variational methods for graphical models," in *Learning in Graphical Models*. Cambridge, MA: MIT Press, 1998. pp. 105–162.

[85] M. G. Kang and S. Chaudhuri, Eds., "Super-resolution image reconstruction," *IEEE Signal Process. Mag.*, vol. 20, pp. 1–113, 2003. doi:10.1109/MSP.2003.1203206

[86] M. G. Kang and A. K. Katsaggelos, "General choice of the regularization functional in regularized image restoration," *IEEE Trans. Image Process.*, vol. 4, pp. 594–602, 1995. doi:10.1109/83.382494

[87] R. E. Kass and A. E. Raftery, "Bayes factors," *J. Am. Stat. Assoc.*, vol. 90, pp. 773–795, 1995.

[88] S. M. Kay, *Fundamentals of Statistical Signal Processing: Estimation Theory*. Englewood Cliffs, NJ: Prentice-Hall, 1993.

[89] D. Keren, S. Peleg, and R. Brada, "Image sequence enhancement using sub-pixel displacement," in *Proc. IEEE Conf. Comput. Vis. Pattern Recognit.*, Ann Arbor, Michigan, pp. 742–746, 1988.

[90] H. Kim, J.-H. Jang, and K.-S. Hong, "Edge-enhancing super-resolution using anisotropic diffusion," in *Proc. IEEE Conf. Image Process.*, Thessaloniki, Greece, vol. 3, pp. 130–133, 2001.

[91] S. P. Kim, N.-K. Bose, and H. M. Valenzuela, "Recursive reconstruction of high resolution image from noisy undersampled multiframes," *IEEE Trans. Acoustics, Speech Signal Process.*, vol. 38, pp. 1013–1027, 1990.

[92] K. P. Kording, C. Kayser, and P. Konig, "On the choice of a sparse prior," *Rev. Neurosci.*, vol. 14, pp. 53–62, 2003.

[93] S. Kullback. *Information Theory and Statistics*. New York: Dover Publications, 1959.

[94] S. Lertrattanapanich and N. K. Bose, "High resolution image formation from low resolution frames using Delaunay triangulation," *IEEE Trans. Image Process.*, vol. 11, pp. 1427–1441, 2002.

[95] W. Li, J.-R. Ohm, M. van der Schaar, H. Jiang, and S. Li, *Verification Model 18.0 of MPEG-4 Visual*. ISO/IEC JTC1/SC29/WG11 N3908, 2001.

[96] A. C. Likas and N. P. Galatsanos, "A variational approach for Bayesian blind image deconvolution," *IEEE Trans. Signal Process.*, vol. 52, pp. 2222–2233, 2004.

[97] D. Lim, "Achieving accurate image registration as the basis for super-resolution," Ph.D. thesis, The University of Western Australia, 2003.

[98] Z. Lin and H.-Y. Shum, "Fundamental limits of reconstruction-based superresolution algorithms under local translation," *IEEE Trans. Pattern Anal. Machine Intell.*, vol. 26, pp. 83–97, 2004.

[99]   Z. Lin and H.-Y. Shum, "Response to the comments on 'Fundamental limits of reconstruction-based superresolution algorithms under local translation," *IEEE Trans. Pattern Anal. Machine Intell.*, vol. 28, p. 847, 2006.

[100]   A. López, R. Molina, A. K. Katsaggelos, A. Rodríguez, J. M. López, and J. M. Llamas, "Parameter estimation in Bayesian reconstruction of SPECT images: An aide in nuclear medicine diagnosis," *Int. J. Imaging Syst. Technol.*, vol. 14, pp. 21–27, 2004.

[101]   B. D. Lucas and T. Kanade, "An iterative image registration technique with an application to stereo vision," in *Proc. Image Understanding Workshop*, Washington, D.C., pp. 121–130, 1981.

[102]   D. G. Luenberger, *Linear and Nonlinear Programming*. Reading, MA: Addison-Wesley, 1984.

[103]   D. J. C. MacKay, "Probable networks and plausible predictions — A review of practical Bayesian methods for supervised neural networks," *Network: Comput. Neural Syst.*, vol. 6, pp. 469–505, 1995.

[104]   S. Mallat, *A Wavelet Tour of Signal Processing*. Academic Press, Cambridge, UK, 1998.

[105]   K. V. Mardia, J. T. Kent, and J. M. Bibby, *Multivariate Analysis*. Academic Press, London, UK, 1979.

[106]   G. Marrero-Callicó, "Real-time and low-cost super-resolution algorithms onto hybrid video encoders" Ph.D. thesis, University of Las Palmas de Gran Canaria, Spain, 2003.

[107]   R. Marsh, T. R. Young, T. Johnson, and D. Smith, "Enhancement of small telescope images using super-resolution techniques," *Publications Astronom. Soc. Pacific*, vol. 116, pp. 477–481, 2004.

[108]   J. Mateos, A. K. Katsaggelos, and R. Molina, "Resolution enhancement of compressed low resolution video," in *IEEE Int. Conf. Acoustics, Speech, Signal Process.*, Istanbul, Turkey, vol. 4, pp. 1919–1922, 2000.

[109]   J. Mateos, A. K. Katsaggelos, and R. Molina, "Simultaneous motion estimation and resolution enhancement of compressed low resolution video," in *IEEE Int. Conf. Image Process*, Vancouver, BC, Canada, vol. 2, pp. 653–656, 2000.

[110]   J. Mateos, R. Molina, and A. K. Katsaggelos, "Bayesian high resolution image reconstruction with incomplete multisensor low resolution systems," in *2003 IEEE Int. Conf. Acoustic, Speech Signal Process. (ICASSP2003)*, Hong Kong, vol. III, pp. 705–708, 2003.

[111]   R. Molina, "On the hierarchical Bayesian approach to image restoration. Applications to Astronomical images," *IEEE Trans. Pattern Anal. Machine Intell.*, vol. 16, pp. 1122–1128, 1994.

[112]   R. Molina, A. K. Katsaggelos, L. D. Alvarez, and J. Mateos, "Towards a new video compression scheme using super-resolution," in *Proc. SPIE Conf. Vis. Commun. Image Process.*, San Jose, California, USA, vol. 6077, pp. 1–13, 2006.

[113] R. Molina, A. K. Katsaggelos, and J. Mateos, "Bayesian and regularization methods for hyperparameter estimation in image restoration," *IEEE Trans. Image Process.*, vol. 8, pp. 231–246, 1999.

[114] R. Molina, J. Mateos, and A. K. Katsaggelos, "Blind deconvolution using a variational approach to parameter, image, and blur estimation," *IEEE Trans. Image Process.*, November 2006.

[115] R. Molina, J. Mateos, and A. K. Katsaggelos, "Super resolution reconstruction of multispectral images," in *Virtual Observatory: Plate Content Digitization, Archive Mining and Image Sequence Processing*. Heron Press, Sofia, Bulgary, 2006, pp. 211–220.

[116] R. Molina, J. Núñez, F. J. Cortijo, and J. Mateos, "Image restoration in Astronomy. A Bayesian perspective," *IEEE Signal Process. Magazine*, vol. 18, pp. 11–29, 2001.

[117] R. Molina, M. Vega, J. Abad, and A. K. Katsaggelos, "Parameter estimation in Bayesian high-resolution image reconstruction with multisensors," *IEEE Trans. Image Process.*, vol. 12, pp. 1642–1654, 2003.

[118] R. Molina, M. Vega, J. Mateos, and A. K. Katsaggelos, "Hierarchical Bayesian super resolution reconstruction of multispectral images," in *2006 Eur. Signal Process. Conf. (EUSIPCO 2006)*, Florence (Italy), September 2006.

[119] F. Murtagh, A. E. Raftery, and J.-L. Starck, "Bayesian inference for multiband image segmentation via model-based cluster trees," *Image Vis. Comput.*, vol. 23, pp. 587–596, 2005.

[120] R. M. Neal, "Probabilistic inference using Markov chain Monte Carlo methods," Technical Report CRG-TR-93-1, Department of Computer Science, University of Toronto, 1993. Available online at http://www.cs.toronto.edu/~radford/res-mcmc.html.

[121] N. Negroponte, *Being Digital*. Knopf, New York, USA, 1995.

[122] O. Nestares and R. Navarro, "Probabilistic estimation of optical flow in multiple band-pass directional channels," *Image Vis. Comput.*, vol. 19, pp. 339–351, 2001.

[123] M. Ng, T. Chan, M. G. Kang, and P. Milanfar, "Super-resolution imaging: Analysis, algorithms, and applications," *EURASIP J. App. Signal Process.*, vol. ID 90531, p. 2, 2006.

[124] M. K. Ng, R. H. Chan, and T. F. Chan, "Cosine transform preconditioners for high resolution image reconstruction," *Linear Algebra Appl.*, vol. 316, pp. 89–104, 2000.

[125] M. K. Ng and A. M. Yip, "A fast MAP algorithm for high-resolution image reconstruction with multisensors," *Multidimensional Systems Signal Process.*, vol. 12, pp. 143–164, 2001.

[126] N. Nguyen, "Numerical Algorithms for superresolution," Ph.D. thesis, Stanford University, 2001.

[127] N. Nguyen and P. Milanfar, "A wavelet-based interpolation-restoration method for superresolution," *Circuits, Syst. Signal Process.*, vol. 19, pp. 321–338, 2000. doi:10.1007/BF01200891

[128] N. Nguyen, P. Milanfar, and G. Golub, "Blind superresolution with generalized cross-validation using Gauss-type quadrature rules," in *33rd Asilomar Conf. Signals, Syst. Comput.*, Monterey, CA, vol. 2, pp. 1257–1261, 1999.

[129] N. Nguyen, P. Milanfar, and G. Golub, "A computationally efficient superresolution image reconstruction algorithm," *IEEE Trans. Image Process.*, vol. 10, pp. 573–583, 2001.

[130] N. Nguyen, P. Milanfar, and G. H. Golub, "Efficient generalized cross-validation with applications to parametric image restoration and resolution enhancement," *IEEE Trans. Image Process.*, vol. 10, pp. 1299–1308, 2001.

[131] A. Papoulis, "Generalized sampling expansion," *IEEE Trans. Circuits Syst. Video Technol.*, vol. 24, pp. 652–654, 1977.

[132] S. C. Park, M. G. Kang, C. A. Segall, and A. K. Katsaggelos, "High-resolution image reconstruction of low-resolution DCT-based compressed images," in *IEEE Int. Conf. Acoustics, Speech, Signal Process.*, Orlando, Florida, vol. 2, pp. 1665–1668, 2002.

[133] S. C. Park, M. G. Kang, C. A. Segall, and A. K. Katsaggelos, "Spatially adaptive high-resolution image reconstruction of low-resolution DCT-based compressed images," in *Proc. IEEE Int. Conf. Image Process.*, Ochester, New York, USA, vol. 2, pp. 861–864, 2002.

[134] S. C. Park, M. G. Kang, C. A. Segall, and A. K. Katsaggelos, "Spatially adaptive high-resolution image reconstruction of low-resolution DCT-based compressed images," *IEEE Trans. Image Process.*, vol. 13, pp. 573–585, 2004.

[135] S. C. Park, M. K. Park, and M. G. Kang, "Super-resolution image reconstruction: A technical overview," *IEEE Signal Process. Mag.*, vol. 20, pp. 21–36, 2003. doi:10.1109/MSP.2003.1203207

[136] A. J. Patti and Y. Altunbasak, "Super-resolution image estimation for transform coded video with application to MPEG," in *Proc. IEEE Int. Conf. Image Process.*, Kobe, Japan, vol. 3, pp. 179–183, 1999.

[137] A. J. Patti and Y. Altunbasak, "Artifact reduction for set theoretic super-resolution with edge adaptive constraints and higher-order interpolants," *IEEE Trans. Image Process.*, vol. 10, pp. 179–186, 2001. doi:10.1109/83.892456

[138] A. J. Patti, M. I. Sezan, and A. M. Tekalp, "Robust methods for high-quality stills from interlaced video in the presence of dominant motion," *IEEE Trans. Circuits Syst. Video Technol.*, vol. 7, pp. 328–342, 1997. doi:10.1109/76.564111

[139] A. J. Patti, M. I. Sezan, and A. M. Tekalp, "Superresolution video reconstruction with arbitrary sampling lattices and nonzero aperture time," *IEEE Trans. Image Process.*, vol. 6, pp. 1064–1076, 1997. doi:10.1109/83.605404

[140] J. Pearl, *Probabilistic Reasoning in Intelligent Systems: Networks of Plausible Inference.* Morgan Kaufman Publisher, San Francisco, CA, USA, 1988.

[141] R. R. Peeters, P. Kornprobst, M. Nikolova, S. Sunaert, T. Vieville, G. Malandain, R. Deriche, O. Faugeras, M. Ng, and P. Van Hecke, "The use of super-resolution techniques to reduce slice thickness in functional MRI," *Int. J. Imaging Syst. and Technol.*, vol. 14, pp. 131–138, 2004.

[142] P. J. Phillips, H. Moon, S. A. Rizvi, and P. J. Rauss, "The FERET evaluation methodology for face recognition algorithms," *IEEE Trans. Pattern Anal. Machine Intell.*, vol. 22 pp. 1090–1104, 2000. doi:10.1109/34.879790

[143] P. J. Phillips, H. Wechsler, J. Huang, and P. J. Rauss, "The FERET database and evaluation procedure for face recognition algorithms," *Image Vis. Comput. J.*, vol. 16, pp. 295–306, 1998.

[144] L. C. Pickup, S. J. Roberts, and A. Zisserman, "A sampled texture prior for image super-resolution," in *Advances in Neural Information Processing Systems*, MIT Press, Cambridge, MA, 2003, pp. 1587–1594.

[145] J. Pollack, "Subpixel rendering for high-res mobile displays," *Electronic Design*, 01.13.2005 www.elecdesign.com.

[146] J. Portilla, V. Strela, M. Wainwright, and E. P. Simoncelli, "Image denoising using scale mixtures of Gaussians in the wavelet domain," *IEEE Trans. Image Process.*, vol. 12, pp. 1338–1351, 2003. doi:10.1109/TIP.2003.818640

[147] W. H. Press, S. A. Teukolsky, W. T. Vetterling, and B. P. Flannery, *Numerical Recipes in C*, 2nd edn. Cambridge University Press, New York, USA, 1992.

[148] H. Raiffa and R. Schlaifer, *Applied Statistical Decision Theory.* Boston: Division of Research, Graduate School of Business, Administration, Harvard University, 1961.

[149] D. Rajan and S. Chaudhuri, "Generation of super-resolution images from blurred observations using an MRF model," *J. Math. Imaging Vis.*, vol. 16, pp. 5–15, 2002. doi:10.1023/A:1013961817285

[150] D. Rajan and S. Chaudhuri, "Simultaneous estimation of super-resolved scene and depth map from low resolution defocused observations." *IEEE Trans. Pattern Anal. Machine Intell.*, vol. 25 pp. 1102–1117, 2003. doi:10.1109/TPAMI.2003.1227986

[151] J. Reichel, H. Schwarz, and M. Wien. *Scalable Video Coding—Working Draft 3*. JVT-P201, Poznañ, PL, July, pp. 24–29, 2005.

[152] B. D. Ripley, *Spatial Statistics*. John Wiley, New York, USA, 1981.

[153] R. A. Roberts and C. T. Mullis, *Digital Signal Processing*. Addison-Wesley, 1987.

[154] M. A. Robertson and R. L. Stevenson, "DCT quantization noise in compressed images," in *IEEE Int. Conf. Image Process.*, Thessaloniki, Greece, vol. 1, pp. 185–188, 2001.

[155] D. Robinson and P. Milanfar, "Fundamental performance limits in image registration," *IEEE Trans. Image Process.*, vol. 13, pp. 1185–1199, 2004. doi:10.1109/TIP.2004.832923

[156] D. Robinson and P. Milanfar, "Statistical performance analysis of super-resolution," *IEEE Trans. Image Process.*, vol. 15, pp. 1413–1428, 2006.

[157] V. K. Rohatgi and A. K. Md. E. Saleh, *An Introduction to Probability and Statistics*, 2nd edn. New York: John Wiley & Sons, 2001.

[158] J.J.K. Ó Ruanaidh and W. J. Fitzgerald, *Numerical Bayesian Methods Applied to Signal Processing*. Springer Series in Statistics and Computing, 1st edn. New York: Springer, 1996.

[159] K. Sauer and J. Allebach, "Iterative reconstruction of band-limited images from non-uniformly spaced samples," *IEEE Trans. Circuits Syst.*, vol. 34, pp. 1497–1505, 1987. doi:10.1109/TCS.1987.1086088

[160] R. R. Schultz, L. Meng, and R. L. Stevenson, "Subpixel motion estimation for super-resolution image sequence enhancement," *J. Vis. Commun. Image Represent.*, vol. 9, pp. 38–50, 1998. doi:10.1006/jvci.1997.0370

[161] R. R. Schultz and R. L. Stevenson, "Extraction of high resolution frames from video sequences," *IEEE Trans. Image Process.*, vol. 5, pp. 996–1011, 1996. doi:10.1109/83.503915

[162] G. Schwarz, "Estimating the dimension of a model," *Ann. Stat.*, pp. 461–464, 1978.

[163] C. A. Segall, "Framework for the post-processing, super-resolution and deblurring of compressed video," Ph.D. thesis, Northwestern University, Evanston, IL., 2002.

[164] C. A. Segall, M. Elad, P. Milanfar, R. Webb, and C. Fogg, "Improved high-definition video by encoding at an intermediate resolution," in *Proc. SPIE Conf. Vis. Commun. Image Process.*, San Jose, California, USA, vol. 5308, pp. 1007–1018, 2004.

[165] C. A. Segall, A. K. Katsaggelos, R. Molina, and J. Mateos, "Super-resolution from compressed video," in *Super-Resolution Imaging*, Chap. 9. S. Chaudhuri, Ed. Kluwer Academic Publishers, Boston/Dordrecht/London, 2001, pp. 211–242.

[166] C. A. Segall, R. Molina, and A. K. Katsaggelos, "High-resolution images from low-resolution compressed video", *IEEE Signal Process. Mag.*, vol. 20, pp. 37–48, 2003. doi:10.1109/MSP.2003.1203208

[167] C. A. Segall, R. Molina, A. K. Katsaggelos, and J. Mateos, "Bayesian high-resolution reconstruction of low-resolution compressed video," in *Proc. IEEE Int. Conf. Image Process.*, Thessaloniki, Greece, vol. 2, pp. 25–28, 2001.

[168] C. A. Segall, R. Molina, A. K. Katsaggelos, and J. Mateos, "Reconstruction of high-resolution image frames from a sequence of low-resolution and compressed observations," in *Proc. IEEE Int. Conf. Acoustics, Speech, Signal Process.*, Orlando, Florida vol. 2, pp. 1701–1704, 2002.

[169] C. A. Segall, R. Molina, A. K. Katsaggelos, and J. Mateos, "Bayesian resolution enhancement of compressed video," *IEEE Trans. Image Process.*, vol. 13, pp. 898–911, 2004. doi:10.1109/TIP.2004.827230

[170] E. P. Simoncelli, "Bayesian multi-scale differential optical flow," in *Handbook of Computer Vision and Applications*. Academic Press, San Diego, CA, 1999.

[171] E. P. Simoncelli, E. H. Adelson, and D. J. Heeger, "Probability distributions of optical flow," in *Proc. IEEE Computer Soc. Conf. Comput. Vis. Pattern Recognit.*, Maui, Hawaii, pp. 310–315, 1991.

[172] F. Sroubek and J. Flusser, "Resolution enhancement via probabilistic deconvolution of multiple degraded images," *Pattern Recognit. Lett.*, vol. 27, pp. 287–293, 2006. doi:10.1016/j.patrec.2005.08.010

[173] H. Stark and P. Oskoui, "High resolution image recovery from image-plane arrays, using convex projections," *J. Opt. Soc. Am. A*, vol. 6, pp. 1715–1726, 1989.

[174] A. J. Storkey, "Dynamic structure super-resolution," in *Advances in Neural Information Processing Systems 15 (NIPS2002)*. MIT Press, Cambridge, MA, pp. 1295–1302, 2002.

[175] C. Strecha, R. Fransens, and L. V. Gool, "A probablilistic approach to large displacement optical flow and occlusion detection," in *Eur. Conf. Comput. Vis.*, Copenhagen, Denmark, vol. 1, pp. 599–613, 2002.

[176] E. B. Sudderth, A. T. Ihler, W. T. Freeman, and A. S. Willsky, "Nonparametric belief propagation," in *IEEE Conf. Comput. Vis. Pattern Recognit.*, Madison, Wisconsin, vol. 1, pp. 605–612, 2003.

[177] G. Sullivan and S. Sun, *Ad Hoc Report on Spatial Scalability Filters*. JVT-P201, Poznañ, PL, July 24–29, 2005.

[178] A. Tamtaoui and C. Labit, "Constrained disparity and motion estimators for 3dtv image sequence coding," *Signal Process. Image Commun.*, vol. 4, pp. 45–54, 1991. doi:10.1016/0923-5965(91)90059-B

[179] A. M. Tekalp, *Digital Video Processing*. Signal Processing Series. Prentice Hall, Upper Saddle River, NJ, 1995.

[180] A. M. Tekalp, M. K. Ozkan, and M. I. Sezan, "High-resolution image reconstruction from lower-resolution image sequences and space varying image restoration," in *Proc.*

IEEE Int. Conf. Acoustics, Speech Signal Process., San Francisco, USA, vol. 3, pp. 169–172, 1992.

[181] M. Tipping and C. Bishop, "Bayesian image super-resolution," In *Advances in Neural Information Processing Systems 15*, S. Thrun, S. Becker, and K. Obermayer, Ed. Cambridge, MA: MIT Press, 2003, pp. 1279–1286.

[182] M. E. Tipping, "Sparse Bayesian learning and the relevance vector machine," *J. Machine Learning Res.*, vol. 1 pp. 211–244, 2001. doi:10.1162/15324430152748236

[183] B. C. Tom, N. P. Galatsanos, and A. K. Katsaggelos, "Reconstruction of a high resolution image from multiple low resolution images," in *Super-Resolution Imaging*, Chap. 4, S. Chaudhuri, Ed. Kluwer Academic Publishers, Boston/Dordrecht/London, 2001, pp. 73–105.

[184] B. C. Tom and A. K. Katsaggelos, "Reconstruction of a high resolution image from multiple-degraded and misregistered low-resolution images," in *Proc. SPIE Conf. Vis. Commun. Image Process.*, Chicago (IL), USA, vol. 2308, pp. 971–981, 1994.

[185] B. C. Tom and A. K. Katsaggelos, "Reconstruction of a high-resolution image by simultaneous registration, restoration, and interpolation of low-resolution images," in *Proc. IEEE Int. Conf. Image Process.*, Washington, DC, USA, vol. 2, pp. 539–542, 1995.

[186] B. C. Tom and A. K. Katsaggelos, "An iterative algorithm for improving the resolution of video sequences," in *Proc. SPIE Conf. Visual Communications Image Process.*, Orlando, Florida, pp. 1430–1438, 1996.

[187] B. C. Tom and A. K. Katsaggelos, "Resolution enhancement of monochrome and color video using motion compensation," *IEEE Trans. Image Process.*, vol. 10, pp. 278–287, 2001. doi:10.1109/83.902292

[188] H. L. Van Trees, *Detection, Estimation, and Modulation Theory. Part 1*. New York: Wiley, 1968.

[189] C.-J. Tsai, N. P. Galatsanos, T. Stathaki, and A. K. Katsaggelos, "Total least-squares disparity-assisted stereo optical flow estimation," in *Proc. IEEE Image Multidimensional Digital Signal Process. Workshop*, Alpbach, Austria, July 12–16, 1998.

[190] R. Y. Tsai and T. S. Huang, "Multiframe image restoration and registration," *Adv. Comput. Vis. Image Process.*, vol. 1, pp. 317–339, 1984.

[191] H. Ur and D. Gross, "Improved resolution from subpixel shifted pictures," *Comput. Vision, Graphics, Image Process.*, vol. 54, pp. 181–186, 1992.

[192] P. Vandewalle, S. Süsstrunk, and M. Vetterli, "A frequency domain approach to registration of aliased images with application to super-resolution," *EURASIP J. Appl. Signal Process.*, vol. ID 71459, p. 14, 2006.

[193] M. Vega, J. Mateos, R. Molina, and A. K. Katsaggelos, "Bayesian parameter estimation in image reconstruction from subsampled blurred observations," in *Proc. IEEE Int. Conf. Image Process.*, Barcelona, Spain, pp. 969–973, 2003.

[194] M. Vega, R. Molina, and A. K. Katsaggelos, "A Bayesian superresolution approach to demosaicing of blurred images," *EURASIP J. Appl. Signal Process.*, vol. ID 25072, p. 12, 2006.

[195] L. Wang and J. Feng, "Comments on 'Fundamental limits of reconstruction-based superresolution algorithms under local translation,'" *IEEE Trans. Pattern Anal. Machine Intell.*, vol. 28, p. 846, 2006. doi:10.1109/TPAMI.2006.91

[196] Q. Wang, X. Tang, and H. Shum, "Patch based blind image super resolution," in *Proc. IEEE Int. Conf. Comput. Vis. (ICCV)*, Beijing, China, vol. 1, pp. 709–716, 2005.

[197] Z. Wang and F. Qi, "On ambiguities in super-resolution modeling", *IEEE Signal Process. Lett.*, vol. 11, pp. 678–681, 2004.

[198] Z. Wang and F. Qi, "Analysis of mutiframe super-resolution reconstruction for image anti-aliasing and deblurring," *Image Vis. Comput.*, vol. 23, pp. 393–404, 2005.

[199] Y.-W. Wen, M. K. Ng, and W.-K. Ching, "High-resolution image reconstruction from rotated and translated low-resolution images with multisensors," *Int. J. Imaging Syst. Technol.*, vol. 14, pp. 75–83, 2004. doi:10.1002/ima.20010

[200] R. Willett, I. H. Jermyn, R. Nowak, and J. Zerubia, "Wavelet-based superresolution in Astronomy," in *Proc. Astron. Data Anal. Software Syst.*, Strasbourg, France, October 2003.

[201] N. A. Woods, N. P. Galatsanos, and A. K. Katsaggelos, "Stochastic methods for joint registration, restoration, and interpolation of multiple undersampled images," *IEEE Trans. Image Process.*, vol. 15, pp. 201–213, 2006.

[202] I. Zakharov, D. Dovnar, and Y. Lebedinsky, "Super-resolution image restoration from several blurred images formed in various conditions," in *2003 Int. Conf. Image Process.*, Barcelona, Spain, vol. 3, pp. 315–318, 2003.

[203] W. Zhao and H. S. Sawhney, "Is super-resolution with optical flow feasible?" in *ECCV '02: Proc. 7th Eur. Conf. Comput. Vis.-Part I, LNCS 2350*, Copenhagen, Denmark, pp. 599–613, 2002.

Index

Author Biography

Aggelos K. Katsaggelos received the Diploma degree in electrical and mechanical engineering from the Aristotelian University of Thessaloniki, Greece, in 1979 and the M.S. and Ph.D. degrees both in electrical engineering from the Georgia Institute of Technology, in 1981 and 1985, respectively. In 1985 he joined the Department of Electrical Engineering and Computer Science at Northwestern University, where he is currently professor. He was the holder of the Ameritech Chair of Information Technology (1997–2003). He is also the Director of the Motorola Center for Seamless Communications and a member of the Academic Affiliate Staff, Department of Medicine, at Evanston Hospital.

Dr. Katsaggelos has served the IEEE in many capacities (i.e., current member of the Publication Board of the IEEE Proceedings, editor-in-chief of the IEEE Signal Processing Magazine 1997–2002, member of the Board of Governors of the IEEE Signal Processing Society 1999-2001, and member of the Steering Committees of the IEEE Transactions on Image Processing 1992–1997). He is the editor of Digital Image Restoration (Springer-Verlag 1991), co-author of Rate-Distortion Based Video Compression (Kluwer 1997), co-editor of Recovery Techniques for Image and Video Compression and Transmission, (Kluwer 1998), co-author of Super-resolution for Images and Video (Claypool, 2007) and Joint Source-Channel Video Transmission (Claypool, 2007). He is the co-inventor of twelve international patents, a Fellow of the IEEE, and the recipient of the IEEE Third Millennium Medal (2000), the IEEE Signal Processing Society Meritorious Service Award (2001), an IEEE Signal Processing Society Best Paper Award (2001), and an IEEE International Conference on Multimedia and Expo Paper Award (2006). He is a Distinguished Lecturer of the IEEE Signal Processing Society (2007–08).

Rafael Molina was born in 1957. He received the degree in mathematics (statistics) in 1979 and the Ph.D. degree in optimal design in linear models in 1983. He became Professor of computer science and artificial intelligence at the University of Granada, Granada, Spain, in 2000. His areas of research interest are image restoration (applications to astronomy and medicine), parameter estimation in image restoration, super resolution of images and video, and blind deconvolution. He is currently the Head of the computer science and Artificial Intelligence Department at the University of Granada.

Javier Mateos was born in Granada, Spain, in 1968. He received the degree in computer science in 1991 and the Ph.D. degree in computer science in 1998, both from the University of Granada. He was an Assistant Professor with the Department of Computer Science and Artificial Intelligence, University of Granada, from 1992 to 2001, and then he became a permanent Associate Professor. He is conducting research on image and video processing, including image restoration, image, and video recovery and super-resolution from (compressed) stills and video sequences.

Printed in the United States
by Baker & Taylor Publisher Services